β-Carbolines

β-Carbolines: A Privileged Scaffold for Modern Drug Discovery provides a summary of methods for the synthesis of various natural products engineered with the diverse β-carboline scaffold and their biological evaluation.

β-carboline and its derivatives have generated considerable interest in recent years for their versatile properties in chemistry and pharmacology. This is due to their ability to act as agonists to benzodiazepine receptors, hydroxy serotonin receptors, and intercalate into DNA and to inhibit CDK, topisomerase, and monoamine oxidase enzymes. Their pharmacological properties include anxiolytic, hypnotic, anticonvulsant, antiviral, antiparasitic, and antimicrobial attributes. This book provides a summary of methods for the synthesis of various natural products engineered with the diverse β-carboline scaffold and their biological evaluation.

Features:

- Focuses on the different application and therapies of β-carbolines.
- Provides a summary of methods for the synthesis of various natural products engineered with the diverse β-carbolines.
- Discusses methods of biological evaluation of β-carbolines.
- Explores different chemical transformations occurring with synthesis using β-carbolines.
- Covers both heterocyclic synthesis and multicomponent synthesis related to β-carboline.

This text is key reading for researchers in medicinal chemistry and drug discovery looking to understand the broad potential of β-carbolines.

β-Carbolines

A Privileged Scaffold for Modern Drug Discovery

Edited by
Keshav Lalit Ameta

CRC Press
Taylor & Francis Group
Boca Raton London New York

CRC Press is an imprint of the
Taylor & Francis Group, an **informa** business

Cover illustration: Shutterstock

First edition published 2025
by CRC Press
2385 Executive Center Drive, Suite 320, Boca Raton, FL 33431

and by CRC Press
4 Park Square, Milton Park, Abingdon, Oxon, OX14 4RN

CRC Press is an imprint of Taylor & Francis Group, LLC

ISBN: 978-1-138-48238-8 (hbk)
ISBN: 978-1-032-41075-3 (pbk)
ISBN: 978-1-351-05803-2 (ebk)

DOI: 10.1201/9781351058032

Typeset in Times
by Newgen Publishing UK

Dedication

To my parents, who gave me the gift of dreams and the ability to realize them.

Contents

Foreword

In the ever-expanding realm of chemistry of bioactive heterocycles and its applications, the exploration of β-carboline skeleton, representing the basic structural unit of many bioactive heterocycles, stands as a captivating journey into the intricate tapestry of molecular diversity. As we embark on this intellectual odyssey through the pages of *β-Carboline: A Privileged Scaffold for Modern Drug Discovery,* we find ourselves immersed in a world where bioactive heterocycles create symphonies of unique biological applications. This edited volume is a testament to the collaborative efforts of brilliant researchers and academicians who have dedicated their time and expertise to unravelling the secrets hidden within β-Carboline moieties.

The chapters within this book are not mere compilations of data or just lists of facts; rather, they are thought-provoking journeys of researchers who understand the complexities of synthesis, characterization, and biological study of differently substituted β-carbolines. From foundational ideas to state-of-the-art developments, the authors have expertly constructed a timeline that pushes the limits of human comprehension and goes beyond accepted wisdom.

I extend my gratitude to the contributors who have shared their intellectual jewels, shaping this book into a treasure trove of knowledge. May the readers find inspiration within these pages, sparking new ideas and forging ahead on their own quests for understanding.

I cordially welcome you to explore the realm of "β-carbolines" and go beyond conventional knowledge on an engrossing story of molecular discovery, where the indole-pyridine fused framework becomes the main characters. This is all in the spirit of scientific research.

Wishing you a happy reading experience and scientific pursuits as abundant and fulfilling as the substances found in these pages.

Prof. (Dr.) Rama Shankar Dubey
Vice-Chancellor,
Central University of Gujarat,
Gandhinagar, Bharat-382030.

Preface

Heterocycles offer an extensive variety of structural variations that facilitate the development of varied chemical libraries. The hunt for new medications with enhanced efficacy, fewer side effects, and unique methods of action benefits from this diversity. Heterocycles are found in many therapeutic products. To optimize their interactions with target biomolecules, increase bioavailability, and improve pharmacological characteristics, medicinal chemists often add heterocyclic moieties to drug candidates. Biologically active heterocyclic compounds have been a primary area of interest for organic chemists for many years due to their numerous medicinal properties. In the realm of medicinal chemistry, nitrogen, sulfur, and oxygen-containing five- and six-membered heterocyclic molecules have held great importance.

β-carbolines are a large group of natural and synthetic indole alkaloids having common tricyclic pyrido[3,4-b]indole ring motif with varying degrees of aromaticity, some of which are widely distributed in nature including various plants, marine creatures, insects, foodstuffs, mammalians as well as human tissues and body fluids. This group has occupied an exceptional position in heterocyclic chemistry, and its derivatives have fascinated considerable interests in recent years for their versatile properties in chemistry and pharmacology. β-carboline is a modern artist of current drug discovery architectures. However, in the last 50 years, with the introduction of different biological evaluation, the importance of β-carboline scaffold for drug discovery has been recognized and considerable efforts from both academic and industrial researchers have been focused especially on the design and development of natural product inspire divers β-carboline scaffold.

Thus, the purpose of the present book is to provide a succinct summary of methods for the synthesis of various natural products to inspire diverse β-carboline scaffolds and their biological evaluation. I trust that both undergraduate and graduate students, researchers and faculties will find in these pages an interesting guide and an updated state of the art on the enormous β-carboline scaffold.

Editor
Prof. Keshav Lalit Ameta

About the Editor

Keshav Lalit Ameta is working as a Professor at the Department of Applied Chemistry, School of Applied material Sciences, Central University of Gujarat, Gandhinagar, India. Prior to the current position, he has served at Sardar Patel University, Vallabh Vidyanagar, Gujarat and Mody University of Science and Technology, Lakshmangarh, Rajasthan. Prof. Ameta has published over 100 research articles in the field of organic chemistry and catalysis in the journals of international repute. He received his master and doctorate degree in Organic Chemistry from M. L. Sukhadia University, Udaipur, Rajasthan, India. He has vast experience of teaching both graduate and postgraduate level students. His research area involves synthesis, characterization and biological evaluation of different sized bioactive heterocyclic systems. In addition, to this he has keen interests in heterogeneous catalyzed organic synthesis and photocatalysis.

Contributors

Dinesh K. Agarwal
Department of Pharmacy
Pacific Institute of Pharmacy
Pacific University
Udaipur, Rajasthan, India, 313001.

Shikha Agarwal
Department of Chemistry
Synthetic Organic Chemistry Laboratory, MLSU
Udaipur, 313001, India.

Vnira R. Akhmetova
Laboratory of Heteroatomic Compounds Institute of Petrochemistry and Catalysis
Russian Academy of Science
141 Prospekt Octyabrya, Ufa 450075, Russian Federation

Nail S. Akhmadiev
Laboratory of Molecular Design and Biological Screening of Candidate Substances for the Pharmaceutical Industry
Institute of Petrochemistry and Catalysis, Russian Academy of Science, 141 Prospekt Octyabrya, Ufa 450075, Russian Federation

Chetna Ameta
Department of Chemistry
University College of Science,
M. L. Sukhadia University
Udaipur (Raj.)

Surendra Kumar Bagaria
Department of Chemistry
Govt. Science College
Sikar, Rajasthan 332001, India.

Saroj R. Bembalkar
Department of Chemistry
B. U. Patil Arts and Science College
Sultanpur 431101 MS, India.

Devidass S. Bhagat
Department of Forensic Chemistry
Government Institute of Forensic Science
Aurangabad 431004 MS, India.

Keyur Bhatt
Department of Chemistry
Faculty of Science, Ganpat University, Kherva-384012
Mehsana, Gujarat, India

Ravikumar M. Borade
Department of Forensic Chemistry
Government Institute of Forensic Science
Aurangabad 431004 MS, India.

Dharmendra
Department of Chemistry
University College of Science, M. L. Sukhadia University
Udaipur (Raj.)

Surabhi Dhadda
Department of Chemistry
Faculty of Basic and Applied Sciences
Vivekananda Global University, Jagatpura, Jaipur, Rajasthan 303012, India.

Santosh B. Gaikwad
Department of Chemistry
L P G Arts and Science College, Shirpur (Jain)
Washim-444 504 (Maharashtra), India.

Jagannath S. Godse
Department of Chemistry
Deogiri College
Aurangabad – 431 005
(Maharashtra), India.

Rajita Ingle
Department of Chemistry
Deogiri College, Station Road
Aurangabad 431 005, MS, India.

Dinesh Kumar Jangid
Department of Chemistry
University of Rajasthan, Jaipur
India-302004.

Nidhi Jangir
Department of Chemistry
University of Rajasthan, Jaipur
India-302004.

Kuldeep V Joshi
Department of Applied Chemistry
School of Applied material
Sciences, Central University
of Gujarat,
Gandhinagar-382030,
Gujarat, India

Swati B. Kale
Department of Applied Science
Government Polytechnic
Aurangabad 431004 MS, India.

Dhanraj P. Kamble
Department of Chemistry
S. B. E. S. College of Science
Aurangabad 431001 MS, India.

Monika Kumawat
Department of Chemistry
University College of Science
M. L. Sukhadia University,
Udaipur (Raj.)

Mukesh Kumar
Department of Chemistry
Sahu Jain Degree College
Najibabad, Bijnor-246763, UP, India.

Danil V. Leont'ev
Laboratory of Heteroatomic Compounds
Institute of Petrochemistry and
Catalysis
Russian Academy of Science
141 Prospekt Octyabrya, Ufa 450075,
Russian Federation

Dattatraya Pansare
Department of Chemistry
Deogiri College, Station Road
Aurangabad 431 005, MS, India.

Jaymin Parikh
Department of Chemistry
Faculty of Science, Ganpat University
Kherva-384012, Mehsana,
Gujarat, India

Rajendra P. Pawar
Department of Chemistry
Shiv Chhatrapati College
Aurangabad, 431 001, MS, India.

Aniket Sarkate
Department of Chemical Technology
Dr. B.A.M. University Aurangabad
414004, MS, India.

Shivangi Sharma
Department of Applied Chemistry
Amity School of Applied Sciences,
Amity University Madhya Pradesh
Gwalior, Madhya Pradesh-474 005,
India

Mubarak Shaikh
Department of Chemistry
Radhabai Kale Mahila Mahavidyalaya
Ahmednagar, 414001, MS, India.

Rohini Shelke
Department of Chemistry
Radhabai Kale Mahila Mahavidyalaya
Ahmednagar, 414001, MS, India.

Shivani Soni
Synthetic Organic Chemistry Laboratory
Department of Chemistry,
UCOS, MLSU
Udaipur, India 313001

Shivendra Singh
Department of Applied Chemistry
Amity School of Applied Sciences,
Amity University Madhya Pradesh
Gwalior, Madhya Pradesh-474
005, India

Pankaj Teli
Synthetic Organic Chemistry Laboratory
Department of Chemistry,
UCOS, MLSU
Udaipur, India 313001

Shankar Thopate
Department of Chemistry
Radhabai Kale Mahila Mahavidyalaya
Ahmednagar, 414001, MS, India.

Sunil U. Tekale
Department of Chemistry
Deogiri College
Aurangabad – 431 005
(Maharashtra), India.

Sunita Teli
Synthetic Organic Chemistry Laboratory
Department of Chemistry,
UCOS, MLSU
Udaipur, India 313001

Sanjay B. Ubale
Department of Chemistry
Deogiri College
Aurangabad – 431 005
(Maharashtra), India.

1 β-Carbolines as Anti-Cancer Agents

Synthesis and Biological Study

Ravikumar M. Borade, Devidas S. Bhagat, Swati B. Kale, Dhanraj P. Kamble, Saroj R. Bembalkar, and Rajendra P. Pawar

1.1 INTRODUCTION

A broad class of indole alkaloids, including natural and produced, referred to as "carboline derivatives," have a common tricyclic pyrido-[3, 4-b] indoles ring unit in their structural component. The β-carboline alkaloids were first discovered in the seeds of *Peganumharmala*, which can be found in Northwest China, North Africa, and the Middle East. This plant has been used for hundreds of years in traditional medicine to treat conditions such as malaria and esophageal cancer (Wesson et al. 1996), which suggests that these compounds have become important scientific development triggers for the synthesis of numerous anticancer agents (Pfau and Skog 2004; Cao et al. 2004). β-Carbolines are used as an emmenagogue and an abortifacient in various parts of South America (Grella et al. 1998).

β-carbolines is a group of indole alkaloids that can be either natural or synthetic. It is most commonly defined by nitrogen-containing six-membered rings. Fully aromatic carbolines (Cs) are unsaturated carbolines, whereas dihydrocarbolines (DHCs) are partially saturated alkaloids and tetrahydrocarbolines (THCs) are fully saturated alkaloids (Cain et al. 1982; Gooyit et al. 2015; Jiang et al. 2002; Laine et al. 2014; Stohler et al. 1995). The pyrido ring or indole ring of tricyclic compounds are substituted in a variety of reported derivatives of carbolines (Cao et al. 2007). Several carboline saturated or unsaturated tricyclic ring alkaloids systems from a number of terrestrial plants have been used to extract and characterize their main bioactive components (Wang et al. 2017). Several other types of carboline alkaloids have also been isolated and identified from marine invertebrates, such as soft corals (Lignopsis), bryozoans (Cribellina, Catenicella), tunicates (Ritterella, Pseudodistoma, Eudistoma, Didemnum, Lissoclinum), and hydroids (Aglaophenia) (Aiello et al. 1987; Blunt et al. 2009; Cheng et al. 2012; Suarez-Jimenez et al. 2012). Carbolines are classified as α-, β-, γ-, or δ-carbolines based on the position of the nitrogen in the ring structure. Similar to natural or synthetic cases, monomers display a linkage primarily

DOI: 10.1201/9781351058032-1

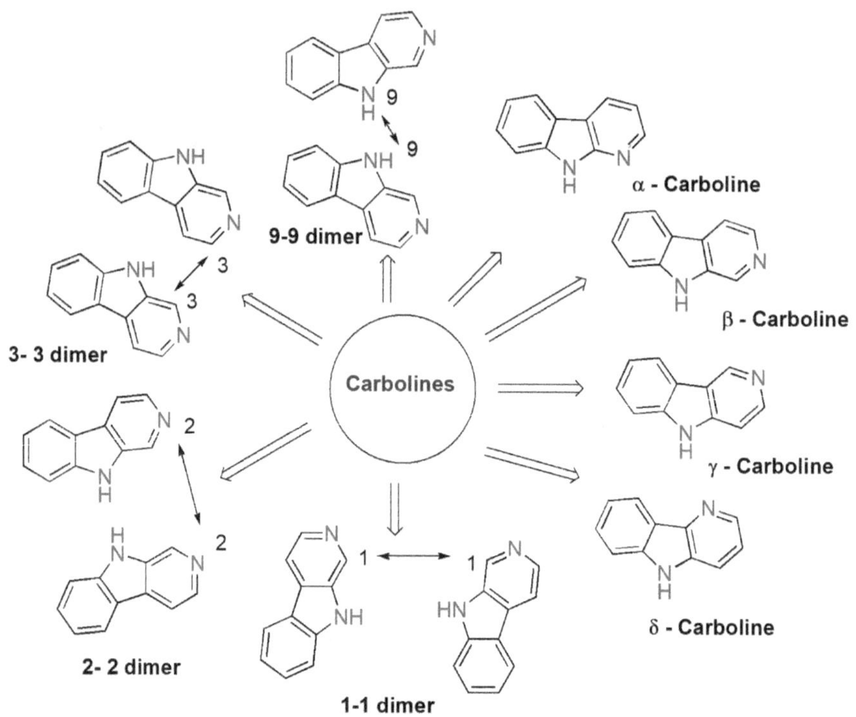

FIGURE 1.1 Structural diversity of carbolines.

through the 1-1, 2-2, 3-3, or 9-9 positions depicted in Figure 1.1, and monomeric and dimeric configurations seem to have a single β-carboline core (Dai et al. 2018; Devi et al. 2018).

A β-carboline alkaloid was first obtained in 1841 and since then several β-carbolines have been isolated from numerous sources generally comprising plants from families like *Rutaceae* (Smirnova O B, 2011), *Simaroubaceae* (Ohishi et al. 2015), *Amaranthaceae* (Porta et al. 2017), *Caryophyllaceae* (Chen et al. 2010), *Rubiaceae* (deOliveira Figueiredo et al. 2014), and *Zygophyllaceae* (Wang et al. 2014); marine creatures such ashydroids (Skropeta and Wei 2014), bryozoans (Till and Prinsep 2009), soft corals (Gao et al. 2013), and sponges (Tanaka et al. 2014); along with various microorganisms (Savi et al. 2015), insects (Marques et al. 2005), food products (Kim et al. 2016), alcoholic beverages (Tsuchiya 2016), tobacco smoke (Herraiz and Chaparro 2005), and human tissue and body fluids (Lamounier et al. 2015). β-carbolines also posses wide-ranging biological activities including sedative (Smith et al. 2013), anxiolytic (Petersen et al. 1982), hypnotic (Kaijima et al. 1984), anticonvulsant (Pogosyan et al. 2007), antitumor (Zheng et al. 2014), antiviral (Nazari Formagio et al. 2009), antiparasitic (Gooyit et al. 2015), and antimicrobial properties (Lee et al. 2015; Montorsi et al. 2014; Park and Nam 2015; Rundfeldt and LÖscher 2014). Recent investigations have shown β-carbolinedimers to be significantly more active than their corresponding monomers (Lu et al. 2017).

1.2 ANTICANCER ACTIVITY OF β-CARBOLINES

Cancer is a leading cause of mortality and morbidity. Globally, roughly 20 million individuals are diagnosed with cancer each year, and 9.5 million people die from the disease (Www.Who.Int/News-Room/Fact-Sheets/Detail/Cancer, n.d.). Depending on the type and severity of the cancer, various treatment methods, including chemotherapy, radiation, and occasionally surgery, are used to control it in the present. Even though this discipline has seen a lot of research, there are still a lot of topics that need to be studied. The main issue that has hindered this research from expanding on its early breakthroughs is "drug resistance."

The majority of research on drug resistance in cancer focuses on cellular resistance, which is primarily explained by the genetic makeup of the cancerous cells. Toxic chemotherapy may cause mutations in sensitive tumors, and adaptive responses such as increased expression of the therapeutic target and activation of alternative compensatory signaling pathways can also result in mutations. The main method for determining mechanisms of medication resistance is to expose the residual cancer cells to fatal agents, and then use various cellular and molecular biology tools to detect the altered genes. This led to the discovery of three primary mechanisms of drug resistance: first, an increase in the energy-dependent efflux of hydrophobic drugs out of the cells; secondly, there is a reduction in the absorption of hydrophilic anticancer medications such as cisplatin, nucleoside analogs, and folate antagonists; thirdly, there are alterations to cellular processes such as alterations in the cell cycle, a decline in apoptosis, increased expression of DNA damage repair, and modified metabolism. Resistance causes cancer to go into remission, which increases the cost on society and negatively impacts the patient's mental health. (Housman et al. 2014; Szakacs et al. 2006).

Combination therapy is now used in cancer therapeutics because it avoids the issue of resistance and is more efficient than a single agent. Research should next focus on the removal of cancer progenitor cells because they are frequently medication resistant and may cause remission in cancer patients (Housman et al. 2014). Several studies on the use of β-carboline derivatives as potential cancer treatments are currently being published. The present chapter describes the synthesis and biological assessment of numerous β-carboline-based compounds, classifying them according to their intended actions.

1.3 SYNTHESIS OF β-CARBOLINE DERIVATIVES AS ANTICANCER AGENTS

1.3.1 DNA Targeting Agents

1.3.1.1 Intercalating Agents

In prokaryotic and eukaryotic cells, β-carbolines have been found to have biological and pharmacological effects that are supported by their ability to intercalate DNA, which alters the enzymatic mechanisms involved in DNA repair or DNA replication. Sulfur was used in the dehydrogenation of compound **1** to produce completely unsaturated methyl-β-carboline-3-carboxylates. After reduction with $LiAlH_4$ in dry

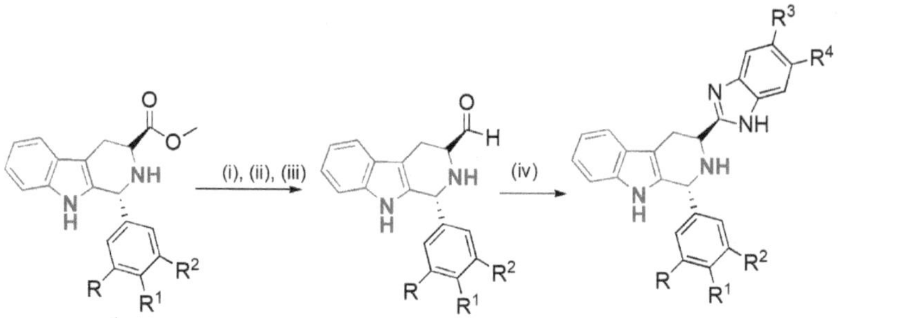

1 2 3

Reaction Conditions: (i) Sulfer, Xylene, Reflux, 10 -12 hr (ii) LAH, Dry THF, 0 °C to rt, 4hr (iii) DMP, DCH, rt, 2 hr (iv) OPD,EtOH,$La(NO_3).6H_2O$, 60 °C, 30 - 40 min

SCHEME 1.1 Synthesis of β-carboline-benzimidazole conjugates.

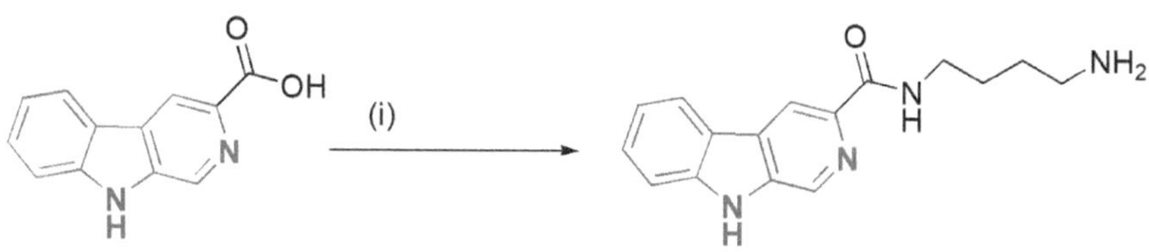

4 5

Reaction Conditions: (i) $H_2N(CH_2)_4NH_2$, $SOCl_2$, butane

SCHEME 1.2 Synthesis of 3-substituted β-carbolines.

tetrahydrofuran (THF), a variety of β-carboline-benzimidazole conjugate compounds containing substituted benzimidazole moieties and aryl rings at the C3 and C1 positions were produced via Dess Martin periodinate-mediated oxidation. The required substituted benzimidazole linked β-carboline conjugates product **3** is produced by condensations of the synthesized chemical **2** with various o-phenylenediamines (Devi et al. 2018; Digwal et al. 2016; Kamal et al. 2014), as shown in Scheme 1.1.

In tests for cytotoxic action against a panel of human cancer cell lines, all the synthetically produced derivatives of compound **3** showed IC_{50} values ranging from 0.36 to 6.3 M. The pBR322 plasmid DNA was seen to be cleaved by these chemicals when they were subjected to UV radiation, according to a biophysical experimental research and molecular modeling (Funayama et al. 1996). The conjugates of β-carboline and benzimidazole have subsequently demonstrated potential anticancer action (Kamal et al. 2014; Sobhani et al.2002).

From β-carboline-3-carboxylic acid **4**, Xiao et al. reported produce molecules of β-carboline that are 3-substituted. To treat β-carboline-3-carboxylic acid 4, thionyl chloride was used to produce the appropriate acid chloride, which was then combined with the necessary component to generate acid chloride (Scheme 1.2). An alkylamine side chain added to the β-carbolines C-3 position made the molecule more flexible

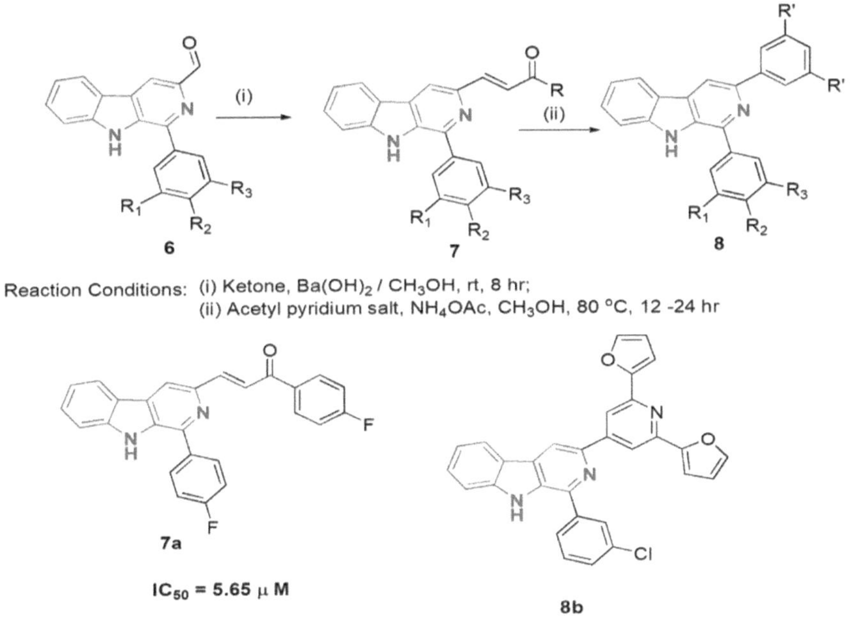

SCHEME 1.3 Synthesis of C3-linked β-carboline-pyridine derivatives.

and increased the intercalating activity **5** of the produced compounds. Using biochemical experimental techniques, examinations of thermal denaturation, binding mode, and binding strength were combined with molecular modeling to investigate the compound **5** affinity for binding DNA. According to experimental results, compound **5** had the highest intrinsic binding constant (K = 4.503 X 10^4 M^{-1}), the lowest binding energy for docking, and the highest rate of inhibition for HL-60 and BGC cells. It also demonstrated the strongest CT-DNA stability (Xiao et al. 2001).

Through a modified Khronke reaction, Nagula et al. reported the synthesis of C3-linked β-carboline pyridine derivatives. Using barium hydroxides, compound **6** is condensed with acetophenones to produce (E)-1-(3,4,5-substituted phenyl)-3-(1-(3,4,5-substituted phenyl)9H-pyrido[[3],[4-b]]indol-3-yl]propen-1-ones **7**, which on addition reacted with acetyl pyridinium salts in anhydrous methanol and ammonium acetate to give derivatives of C-3-linked β-carbolines **8** (Scheme 1.3). Testing the DNA binding affinities of the prepared compounds **7** and **8** revealed that compound **7a** greatly improved the affinity as compared to doxorubicin. Molecular modeling studies demonstrated the DNA intercalation of these compounds through their interaction with planar β-carboline rings and DNA base pairs (Nagula et al. 2018; Srivastava et al. 1999).

1-amino β-carboline motifs were tested for their ability to inhibit the growth of K-562, HOP-92, HT29, M14, UACC-62, OVCAR-3, and MDA-MB-231/ATCC. Compound **10** (GI_{50} value ranging from 0.38 to 5.2 μM) was obtained by direct amination of 1-chloro-β-carboline **9** by heating the chloride with an excess of the amine for several hours at 170 °C, either without a solvent or with a few drops of

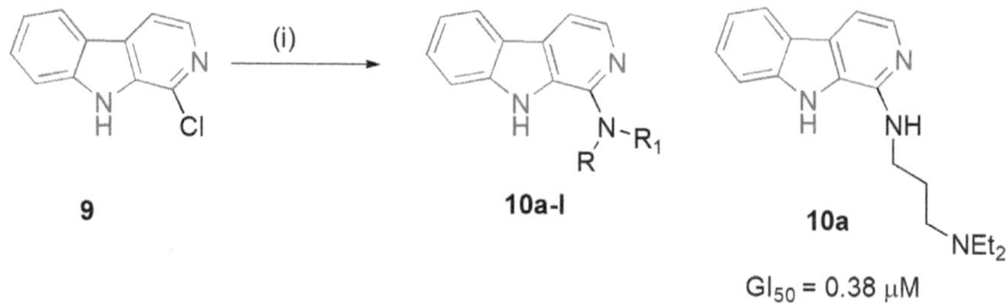

Reaction Conditions: (i) Amines, 170 °C

SCHEME 1.4 Synthesis of 1-amino- β-carbolines.

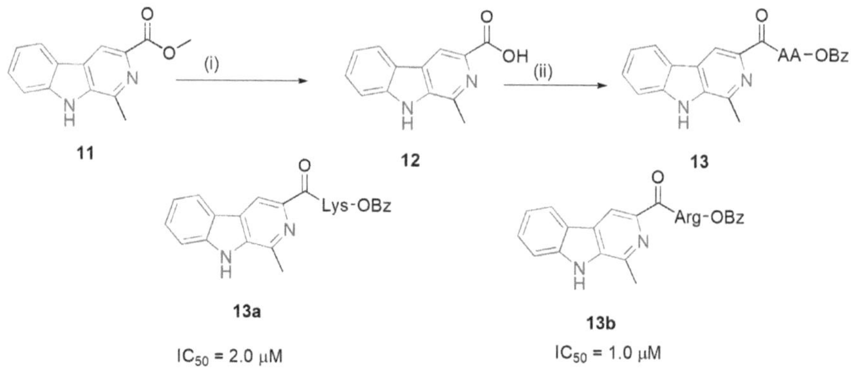

Reaction conditions: (i) aq. NaOH; (II) DCC/DMAP, HCl:AA-OBz

SCHEME 1.5 Synthesis of β-carboline amino acid ester conjugates.

N-methyl pyrrolidin-2-one (Scheme 1.4). Compound **10a** exhibited DNA intercalating characteristics because it possessed a planar aromatic structure that could easily stack in base pairs (Boursereau and Coldham 2006).

The cytotoxic effect and synthesis of β-carboline amino acid ester conjugates were studied by Zhao *et al.* Compound **12** (Scheme 1.5) was synthesized by hydrolysis of methyl 1-methyl-β-carboline-3-carboxylate **11** in base medium, followed by coupling with a suitable amine in the presence of (dimethylamino)pyridine (DMAP) and dicyclohexylcarbodiimide (DCC). Compound **12** showed that cytotoxic activities are significantly enhanced when cationic amino acids, such as arginine and lysine, are added to produce the new series of compounds **13**, **13a,** and **13b,** respectively; this is in contrast to uncharged amino acids. When all of the synthesized compounds **13** were tested in vitro using the human cancer cell lines HeLa, MCF-7, and HepG2, it was discovered that compounds **13a** and **13b** interacted with DNA and had the lowest IC_{50} values (Zhao et al. 2006), in the range of 1 to 7 μM, when compared to other compounds.

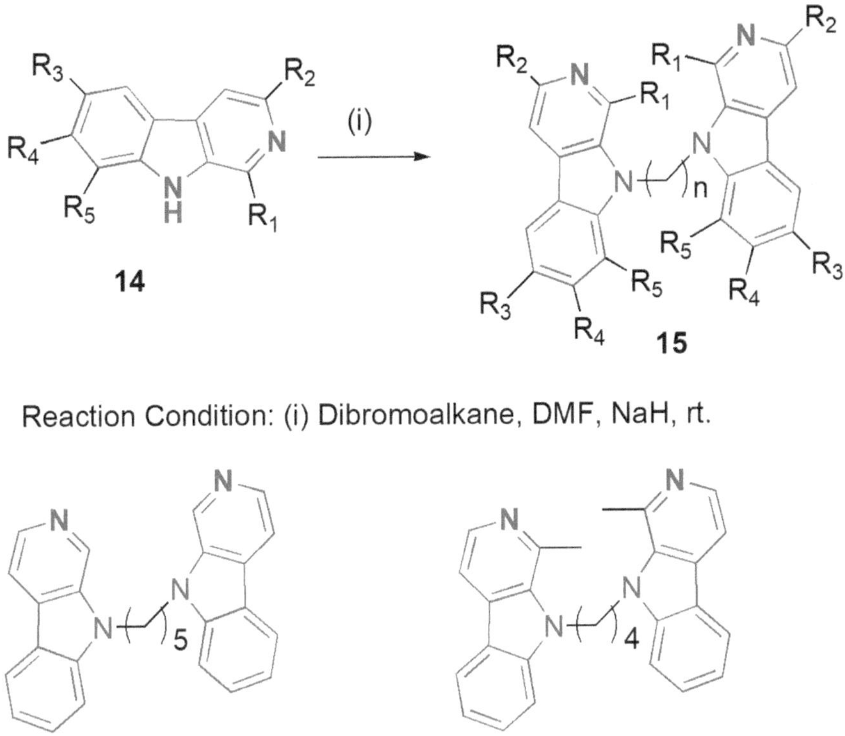

15a
IC_{50} = 10.2 μM

15b
IC_{50} = 4.8 μM

SCHEME 1.6 Synthesis of bivalent β-carbolines.

The synthesis of many bivalent β-carboline derivatives with spacer lengths ranging from three to ten carbon units between two indole ring nitrogens was reported by Shi et al., who also assessed the derivatives' cytotoxic potential against a range of cancer cell lines. Symmetric β-carbolines **15** were synthesized by reacting monovalent β-carbolines **14** with the appropriate dibromo alkane in anhydrous DMF (Scheme 1.6). All the prepared bivalent β-carbolines were evaluated for their antitumor potential using different cancer cell lines. Structure-Activity Relationship (SAR) studies reveal that the C1 position is essential to the activity; appropriate substitution boosted the efficacy against the tumor. A linker with a chain length of four to six methylene units had an effect on the antitumor activity. With IC_{50} values less than 20 μM, these drugs demonstrated excellent efficacy against the BGC, A-375, KB, and SKOV-3 cell lines. A few of the **15** symmetrical β-carbolines with outstanding activity were also selected for in vivo testing in mice harboring the malignancies B16 melanoma, Lewis' lung cancer sarcoma 180, H22 liver cancer, and CT-26 colon cancer.

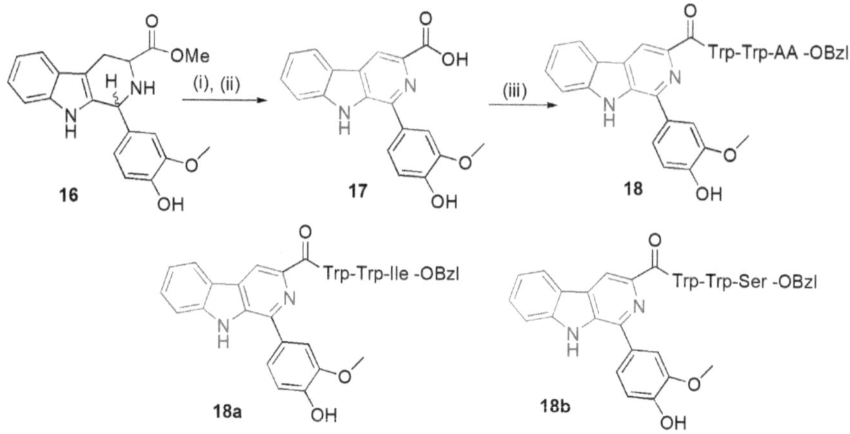

Reaction conditions: (i) SeO_2, AcOH ; (ii) 6M NaOH; (iii) H_2N-Trp-Trp-AA-OBzl, NMM, DCC, HOBt

SCHEME 1.7 Synthesis of β-carboline-3-carboxyl-Trp-Trp-AA-OBzl compounds.

For pharmaceutical action, the outcomes were compared to the reference standard using cyclophosphamide (CTX). At dosages of 30 and 40 mg/kg, certain compounds showed tumor inhibition rates of 50.8% and 56.2% against mice with Lewis lung cancer, CT-26 colon cancer, and H22 liver cancer, respectively, against these mice (Shi et al. 2013).

Chen et al. reported synthesis of tripeptide benzyl ester by condensation of a number of amino acids, and then assessing the biological properties of the unique β-carboline linked tripeptide. Compound **18** (Scheme 1.7) is prepared by 1-(4-hydroxy-3-methoxyphenyl)-tetrahydro-β-carboline-3-carboxylate **16** interacts with selenium oxide in the presence of acetic acid, subsequently 1-(4-hydroxy-3-methoxyphenyl)-β-carboline- 3-carboxylic acid **17** is produced by hydrolyzing the reaction with NaOH. Employing the MTT (3-[4,5-dimethylthiazol-2-yl]-2,5 diphenyl tetrazolium bromide) assay on the cancer cell lines HT-29, A-549, K-562, and HL-60, all **18** prepared compounds were assessed; the outcomes showed moderate activity. While certain compounds showed the highest efficacy in the in vivo study when tested on S-180 mice, displaying tumor inhibition of 63.4% and 62.9%, respectively, which was comparable to the positive control, adriyamycin (61.1%), an anticancer drug. These compounds operate by DNA intercalation. β-carboline and Trp-Trp, the two structural units, both have planar polycyclic aromatic pharmacophore that can stack at the DNA pairs' intercalation sites. The process in which the β-carboline derivatives synthesized in this study intercalated with the DNA of calf thymus (CT) was examined. A variety of well-known, simple, yet accurate procedures were used in spectroscopy, including relative viscosity tests with and without CT DNA and other data indicating intercalative binding of the β-carboline derivatives to DNA (Chen et al. 2014).

1.3.1.2 DNA Groove-binding Agents

Numerous DNA-interactive C3-tethered 1, 2, and 3-triazolo-β-carboline motifs were synthesized via the click reaction, and their cytotoxicity was examined in vitro. Compound **19** reacted with sodium azide in DMF via Cu(I)-catalyzed azide-alkyne cycloaddition (CuAAC), and then it reacted with different alkynes to form C-3 linked 1,2,3-triazolo-β-carboline hybrids motifs **20** (Scheme 1.8). Each of the **20** compounds was evaluated for its in vitro cytotoxicity against the gastric cancer cell line HGC-27, PC-3, HeLa, MCF-7, and HT-29. Only a few of the compounds exhibited IC_{50} values less than 10 μM. Compound **20a** was shown to be the most potent against HT-29, with an IC50 value of 3.67 μM. Molecular docking studies revealed that these drugs were actually bound to a tiny groove in DNA. Experiments such as circular dichroism, fluorescence titration, UV-visible titration, and viscometrical titration corroborated the previous results and demonstrated that compound **20** exhibits a strong electrostatic bond with DNA (Shankaraiah et al. 2016).

By IC_{50} value of 0.13 μM, compound **19e** from this new series was discovered to have potential cytotoxic property against the triple-negative breast tumor cell line (MDA-MB-231). Notably, the DCFDA assay revealed that **19e** promoted the production of ROS. Moreover, identifying **19e** moiety JC-1 enables to observe the breakdown of the mitochondrial membrane potential. Furthermore, **19e** was found to suppress colony pattern and cell migration in a dose-reliant manner in clonogenic and wound healing experiments. The typical intercalation of **19e** with CT-DNA with a binding constant of $1x10^5$ M^{-1} was then identified by molecular modeling and DNA binding affinity experiments using relative viscosity, circular dichroism, and UV-visible spectroscopy.

The cytotoxic activity of 3-(1H-benzo[d]imidazole-2-yl)-β-carboline metal complex was tested by Jin et al. using cisplatin as the standard reference. The metal β-carboline complexes **22** to **25** shown in Scheme 1.9 were produced as a result of the reaction between the β-carboline imidazole amalgam **21** and a number of metal salts. When compared to the usual cisplatin treatment, the anticancer evaluation report against several cancer cell lines was unsatisfactory (Jin et al. 2014).

The structural unit of naturally occurring eudistomin compounds contains β-carboline and exhibits a numerous biological attributes. The prepared eudistomin K derivative showed excellent efficacy when tested against numerous malignant cell lines, including LI-210, Molt-4F, MT-4, and P-388 leukemic cells. Successful overall synthesis of Eudistomin U was noted and its cytotoxicity examined. A Suzuki

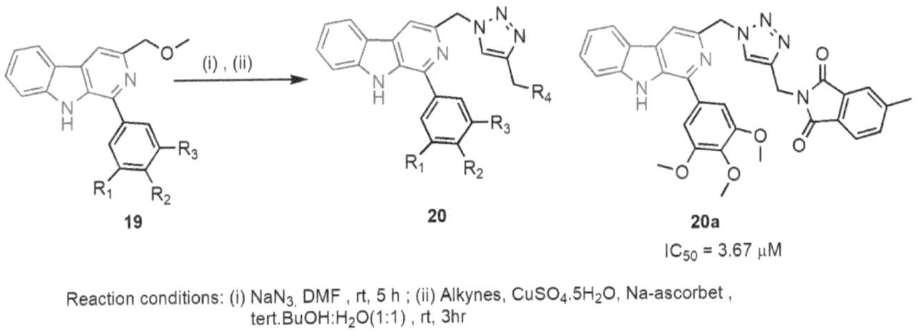

SCHEME 1.8 Synthesis of C3-tethered triazolo-β-carboline derivatives.

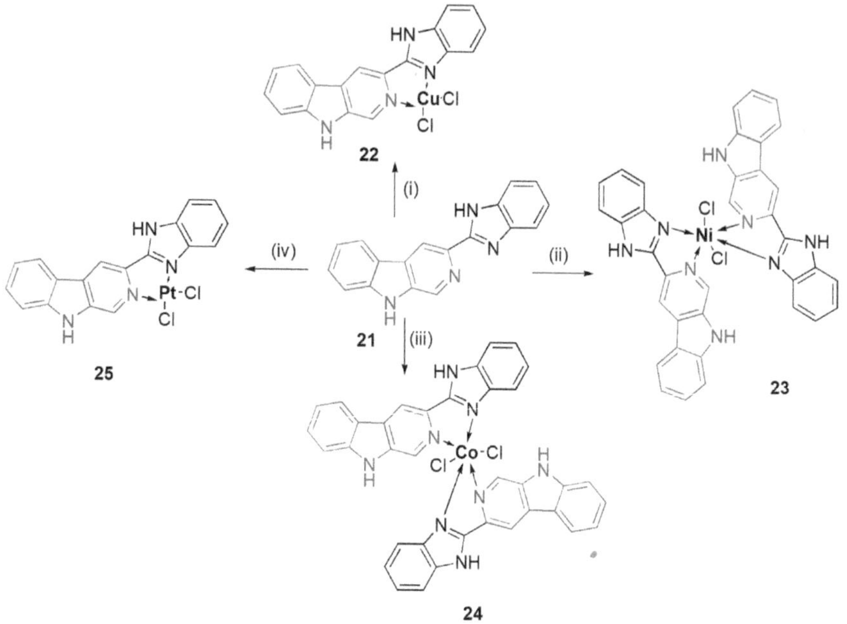

Reaction conditions: (i) $NiCl_2.6H_2O$, $THF:H_2O(4:1)$, rt, 4hr; (ii) $CuCl_2.2H_2O$, $DMF:H_2O(1:1)$, rt, 4hr; (iii) $CoCl_2.6H_2O$, $THF:H_2O(4:1)$, rt, 2hr; (iv) $CisPt(DMSO)_2Cl_2$, MeOH, rt, 20 min

SCHEME 1.9 Synthesis of β-carboline complexes with 3-(1H-benzo[d]imidazole-2yl)-metal.

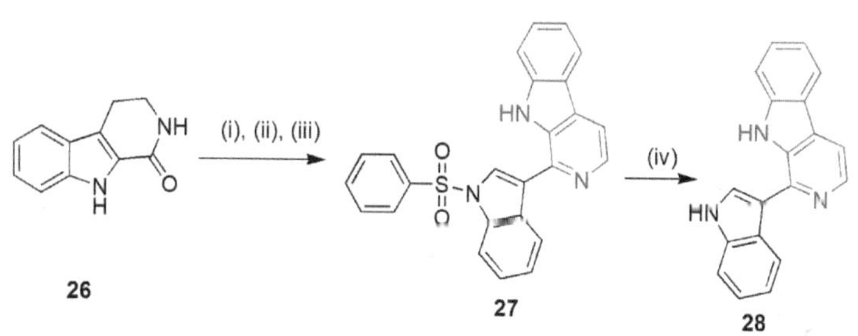

Reaction conditions:(i) DDQ, dioxane; (ii) Tf2O, pyridine, (iii) Indole boronic acid; (iv) KOH, MeOH, THF

SCHEME 1.10 Synthesis of eudistomin U.

cross-coupling reaction among β-carboline and indoleboronic acid followed by DDQ-mediated oxidation of lactum **26** produced compound **27**, which was then transformed into the desired product **28** in the presence of a base (Scheme 1.10). The prepared compound has IC_{50} value of 15.6 M against the C19 cancer cell line when its potency was evaluated using the MTT cell viability test (Rinehart et al. 1984; Schupp et al. 2003).

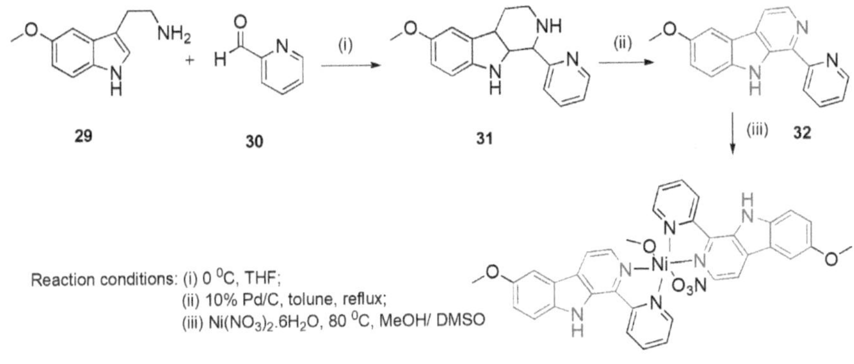

SCHEME 1.11 Synthesis of novel nickel (II) complex of 6-methoxy-1-pyridine-β-carboline.

Jing-Mei Yang et al. fabricated a novel nickel (II) complex of 6-methoxy-1-pyridine-β-carboline **33,** which is depicted in Scheme 1.11. The intermediate **31** was prepared from commercially available 5-methoxytryptamine **29** and pyridine-2-carbaldehyde **30** by using Pictet–Spengler cyclization (Yang et al. 2018). 6-Methoxy-1-pyridine-β-carboline **32** was found from intermediate **31** by dehydrogenation, which was catalyzed by Pd/C. Finally nickel (II) complex of 6-methoxy-1-pyridine-β-carboline **33** was made by hydrothermal synthesis in 12:1 methanol/DMSO using $Ni((II)NO_3)_2•6H_2O$ and 6-methoxy-1-pyridine-β-carboline 32.

The produced complex is subjected to cytotoxicity evaluations against six cancer cell lines (MGC-803, Hep G2, T24, OS-RC-2, NCI-H460, and SK-OV-3) as well as the human normal liver cell line HL-7702. Compared to ligand **32** and cisplatin, the IC_{50} values of MGC-803, Hep G2, T24, OS-RC-2, NCI-H460, and SK-OV-3 were generally lower, falling within the micromolar range (3.77–15.10 μM).

Ramya Tokala *et al.* reported production of novel β-carboline-thiazolidinedione hybrids motifs and evaluated for in vitro cytotoxicity possibilities against certain human cancer cell lines, specifically, PC-3, A549, MG-63, HCT-15, MDA-MB-231, A431, and PANC-1 along with a normal human cell line (L-132) (Tokala et al. 2018). β-carboline-linked 2,4-thiazolidinedione hybrids motifs **39** were prepared by a parallel synthesis principal used in the Knoevenagel condensation reaction. The condensation reaction between 1-aryl-9H-pyrido[3,4-b]indole-3-carbaldehydes **36** and various substituted 2,4-thiazolidinediones was performed (Scheme 1.12). Readily available L-tryptophan **34** esterified using thionyl chloride in methanol gives L-tryptophan ester **35**, which subjected to Pictet–Spengler reaction with different aldehydes by with catalytic quantity of TFA produces diastereomers **36**. The compound **36** aromatized by means of $KMnO_4$ to afford **37** followed by reduction of ester group with $LiAlH_4$ offers the consequent alcohols **38**. Activated MnO_2 then converted these alcohols to their related aldehydes **39**.

Through the Pictete–Spengler condensation of L-tryptophan with formaldehyde, the free ligand Norharman was generated quickly and then refluxed in ethanol to produce **Ru1** through **Ru4** (Chen et al. 2013). These findings showed that the 2-N

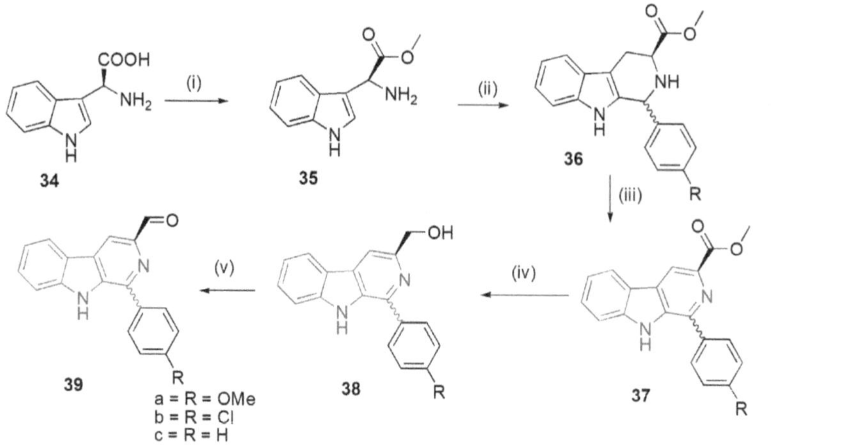

Reaction conditions: (i) $SOCl_2$,MeOH, rt, 4 h, 90%; (ii) R-CHO, 5% TFA / CH_2Cl_2, rt, 6 h, 70 - 80%; (iii) $KMnO_4$, THF, rt, 8 h, 70 - 80%; (iv) $LiAlH_4$, THF, 0 °C - rt, 75 - 85%; (v) MnO_2, CH_2Cl_2, rt, 6 h, 75 - 85%

SCHEME 1.12 Synthesis of novel β-carboline-thiazolidinedione hybrids motifs.

atoms on Norharman are coupled with ruthenium. The DFT approach was used to optimize the complexes **Ru1** through **Ru4** geometries depicted in Figure 1.2.

According to Siyu Zhu et al., new β-carboline-fused imidazolium motifs **45** were synthesized, based on the fascaplysin structure, and tested for cytotoxicity (Zhu et al. 2022). The 3-(4,5-dimethylthiazol-2-yl)-2,5-diphenyltetrazolium bromide (MTT) assay was used to assess the cytotoxicity of **45** in lung carcinoma (A549), gastric carcinoma (BGC-823), murine colon carcinoma (CT-26), liver carcinoma (Bel-7402), and breast carcinoma (MCF-7) cell lines. The findings demonstrated that the majority of the target compounds exhibited outstanding activity against one or more cancer cell lines. The compounds' strong inhibitory effect on VEGFR2 was confirmed by docking studies because it bound to the important amino acids in the active site (Ile892, Leu889, and Asp1046) and interacted hydrophobically with the hydrophobic pocket of the receptor. Scheme 1.13 depicts the target novel β-carboline-fused imidazolium derivatives synthesis processes. L-tryptophan was used as the starting material for synthesis of harmane using Pictet–Spengler cyclization, which was then followed by oxidation and decarboxylation in one step with the help of active manganese dioxide (MnO_2). The key intermediates, β-carboline-1-carboxaldehydes **43**, were produced by first synthesizing the N9-alkylation products **41** from the harmane **40**. The methyl group at position 1 of the compounds **41** was then further oxidized to its corresponding carboxaldehyde using selenium dioxide (SeO_2) in anhydrous dioxane. Finally, the required β-carboline N-fused imidazolium **44**, **45** was synthesized by reacting **43** with the appropriate primary amine, formaldehyde, and acetyl chloride while utilizing anhydrous alcohol as a solvent.

As reported by Dighe et al., the Mannich reaction between methyl 1-(dimethoxymethyl)-9H-pyrido[3,4-b]indole-3-carboxylate **46**, formaldehyde, and primary amines synthesized a number of new N-heterocyclic carbenes based on

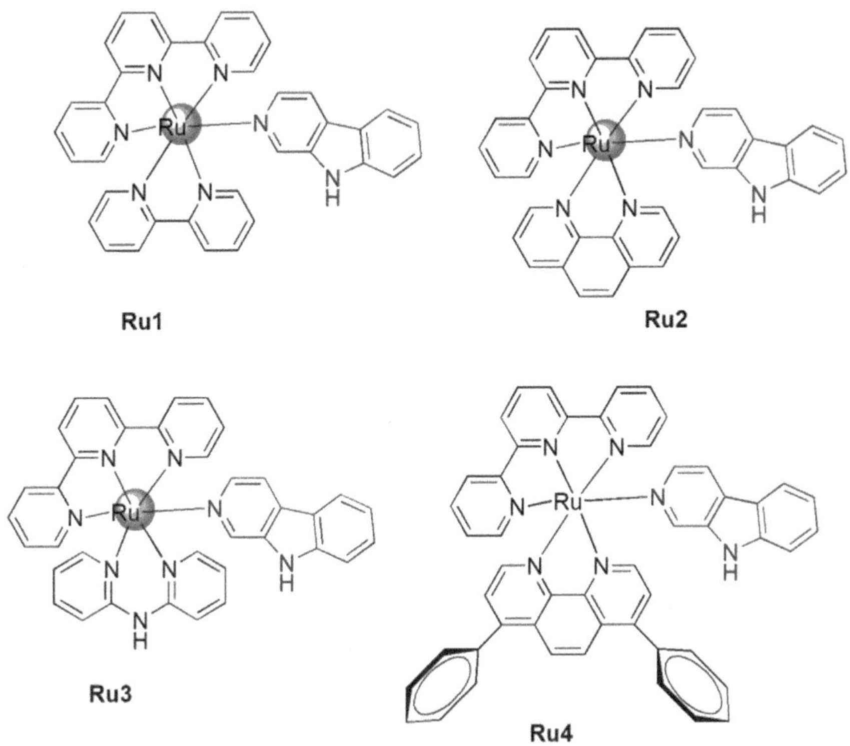

FIGURE 1.2 Geometries of complexes Ru1, Ru2, Ru3 and Ru4.

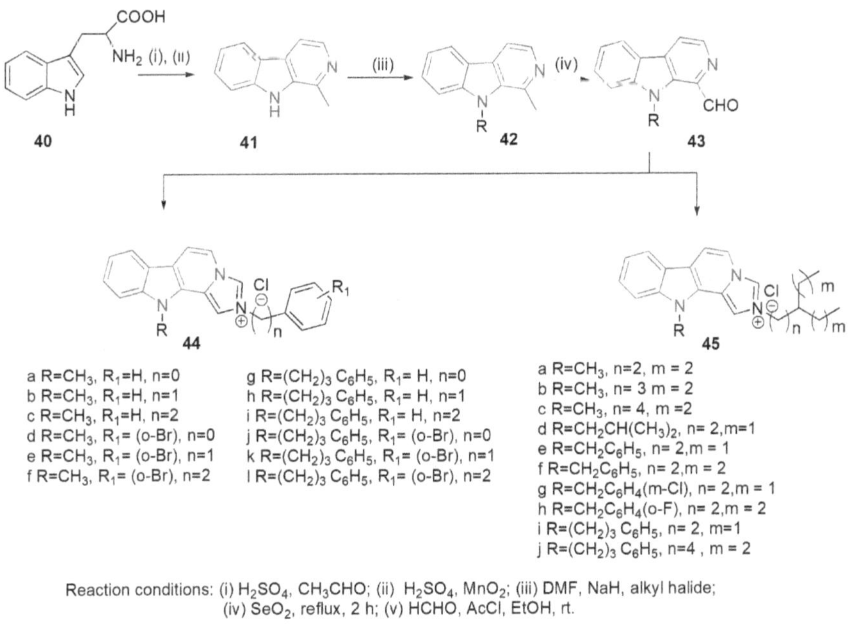

SCHEME 1.13 Synthesis of β-carboline-fused imidazolium motifs.

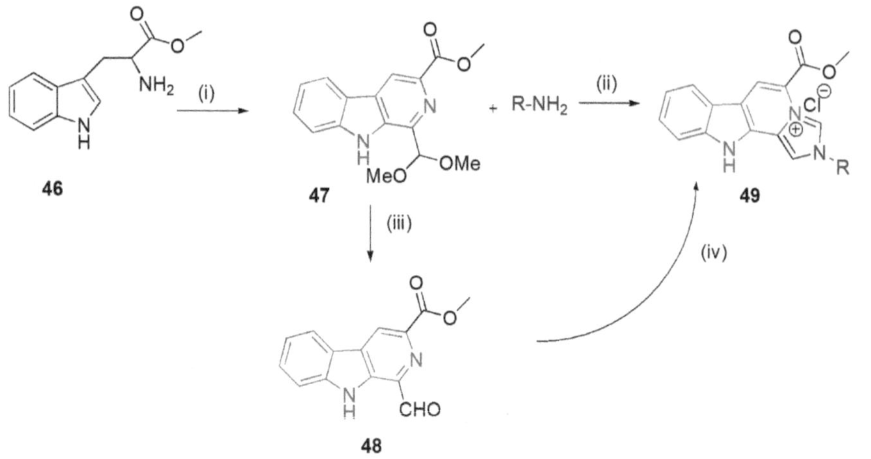

SCHEME 1.14 Synthesis of N-heterocyclic carbenes based on β-carbolines.

β-carbolines **49** as shown in Scheme 1.14 (Dighe et al. 2015). Using human breast cancer and lung cancer cell lines, the antiproliferative effectiveness of each synthesized β-carboline motif was evaluated. After 24 hours of treatment, three motifs of **49** displayed IC_{50} values of less than 10 μM against human breast cancer MDA-MB-231 cells. However, owing to their great affinity for HSA, which was examined spectroscopically with compound 3h, these analogs had no in vivo effect in animal models. The starting compound methy-1-formyl-9H-carboxylate **46** react with formaldehyde and aniline in ethanolic HCl solvent in situ by replacing HCl with AcCl, this reaction condition gave a product **49** in 92% yield.

By introducing various alkylamino(methyl) groups into the 1,3,4-oxadiazole unity of **50**, Savariz et al. reported the synthesis and anticancer activity assessment of many new Mannich bases (Savariz et al. 2010) **51** (Scheme 1.15). The aim of the current report is to synthesize six series of β-carboline derivatives employing the Mannich bases formation reaction and to investigate the effects of various 3-alkylamino (methyl) substituents on the oxadiazole ring with the expectation that performing so will improve their antitumor activity. More particularly, the production and anticancer action in vitro of 3-alkylamino(methyl)-2-thioxo-1,3,4-oxadiazol-5-yl groups at C-3 on β-carboline derivatives **51**. The antibacterial properties of Mannich bases **51** and **50** were also assessed.

1.3.1.3 Topoisomerase Inhibitors

Sathish et al. outlined the synthesis of podophyllotoxin-linked β-carboline motifs for potential applications as DNA topoisomerase II inhibitors and anticancer medicines (Sathish et al. 2018). Compounds **53** are produced when β-carboline acids **52** react with 4-amino-podophyllotoxin in the presence of the coupling mixture of EDCI and HOBt as shown in Scheme 1.16. Using the MTT approach, the in vitro cytotoxicity

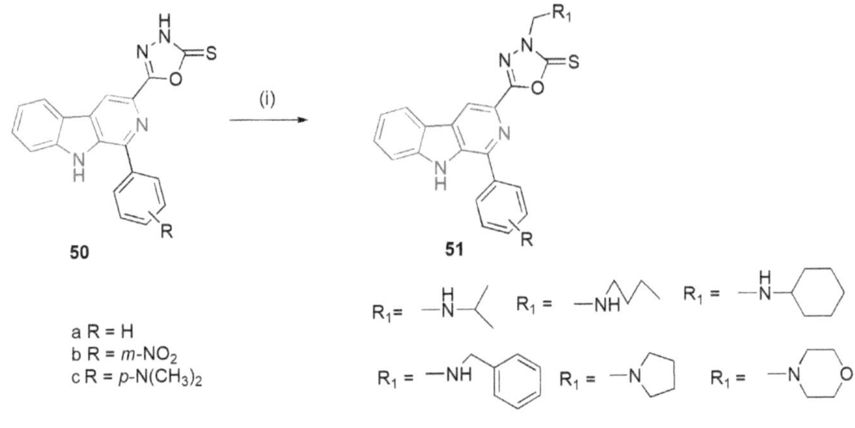

Reaction conditions: (i) Amine, HCHO(37%),EtOH, 60 °C, 24 h

SCHEME 1.15 Synthesis Mannich base based β-carboline derivatives.

of all synthesized podophyllotoxin linked β-carboline motifs against several human cancer cell lines, including A-549, DU-145, MDA MB-231, HT-29, and HeLa, was assessed. Most of the 30 motifs exhibited excellent cytotoxicity and selectivity against DU-145 cells as compared to doxorubicin, etoposide, and podophyllotoxin, with an IC_{50} value ranging from <10 μM to <5 μM.

Additionally, Kamal et al. prepared a dithiacarbamate with β-carboline motifs that have the ability to block DNA topoisomerase II and induce apoptosis (Kamal et al. 2015). The compound **54** has been transformed into the expected compound **55**, depicted in Scheme 1.17, through the addition of carbon disulfide, alkyl halides such as methyl iodide, allyl bromide, and benzyl bromide, as well as TEA in pyridine. The cytotoxic effects of all the produced compounds on a variety of cancer cell lines, namely A-549, MCF-7, DU-145, and HeLa, were studied. These synthesized compounds **55** were then tested for their ability to induce apoptosis using Annexin V-FITC, Hoechst staining assays, and DNA-binding studies. It was found that these compounds bind to DNA more strongly and inhibit topoisomerase II, which causes cell death.

The impact of Harman and norharmanon on the functions of DNA topoisomerases was studied by Funayama et al. It was found to inhibit DNA topoisomerase I activity, with corresponding ED_{50} values of 23.8 and 34.4 mg/ml (Funayama et al. 1996). However, topoisomerase II-mediated stability of the complex that can be broken as well as topoisomerase II inhibitory action were not seen with these drugs. Scheme 1.18 depicted the synthesis route of amino acid functionalized β-carboline motifs.

1.3.1.4 Combilexins

One of these compounds, called combilexin, has the ability to attach to DNA in two different ways: by binding to the DNA minor groove and by intercalating DNA. A set of substituted phenyl and chalcone/(N-acetyl)-pyrazole moiety containing

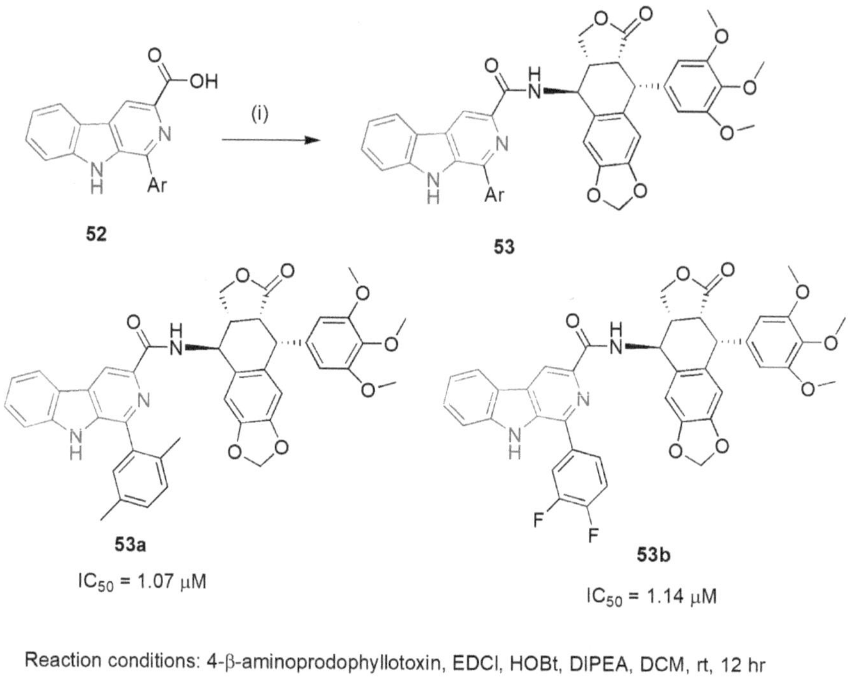

SCHEME 1.16 Synthesis of podophyllotoxin linked β-carbolines.

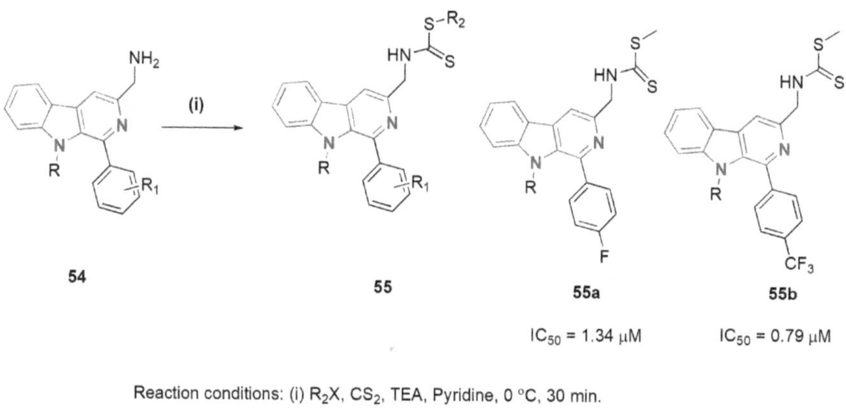

SCHEME 1.17 Synthesis of dithiacarbamate containing β-carbolines.

β-carboline hybrid motifs **62** were synthesized by Kamal et al. Using the 3-(4,5-dimethylthiazol-2-yl)-2,5-diphenyl tetrazolium bromide (MTT) assay (Scheme 1.19), all the prepared hybrid products **62** and **63** were evaluated against several cancer cell lines, including A-549, DU-145, MCF-7, HeLa, ACHN, and HEK-293 celllines. Doxorubicin and harmine served as positive controls. When some compounds were

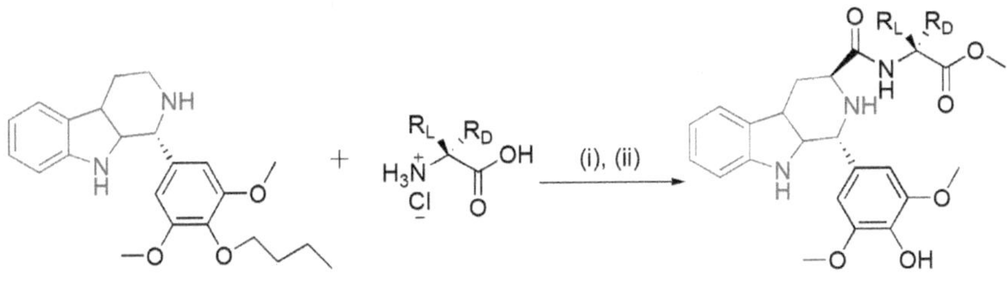

56 57 58

Reaction conditions: (i) ByPop, TEA, DCM, reflux; (ii) TBAF/THF

SCHEME 1.18 Synthesis of amino acid functionalized β-carbolines.

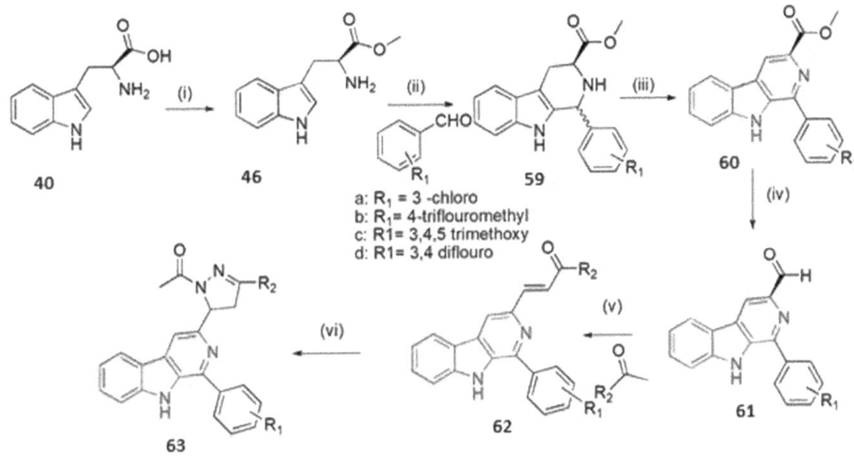

40 46 59 60 61 62 63

Reaction conditions: (i) $SOCl_2$, MeOH, 0°C -RT, 4h, yield 90%;(ii) PTSA, toluene, reflux, 12 h; (iii) sulfer, xylene, reflux, overnight, then 4°C, 3 h, 70- 80% ; (iv) a) $LiAlH_4$, THF, 0 °C -RT, 4h; b) DMP, CH_2Cl_2, 0°C -RT, 3h, 80 - 85% ; (v) Ba(OH)2, MeOH,RT, 64 - 85%; (vi) $NH_2NH_2.H_2O$, acetic acid, reflux, 51 - 85%

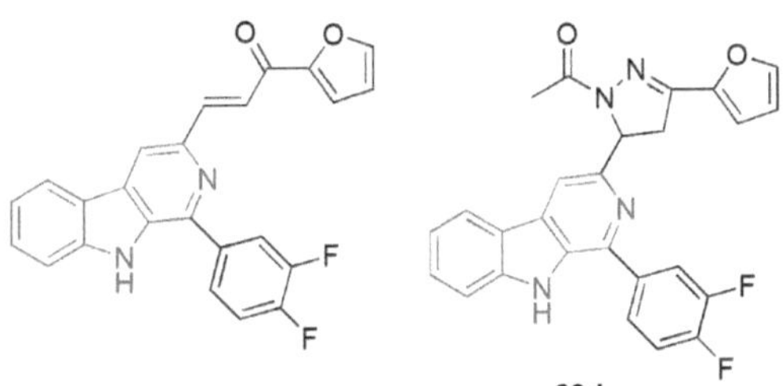

62d 63d

SCHEME 1.19 Synthesis of Combilexin.

exposed to UV light, pBR322 plasmid DNA was observed to be cleaved. DNA fragmentation testing, Hoechst staining, and Annexin V-FITC analysis all supported the malignant cells apoptosis. A few compounds had topoisomerase I inhibitory action in addition to combilexin characteristics. From molecular docking and binding, it was found that topoisomerase I inhibition was mostly caused by the development of a ternary complex of DNA and topo I as well as a bimodal contact with DNA (Kamal et al. 2014; Luo et al. 2009).

As DNA binding and topoisomerase inhibitors, β-carboline-bis-indole motifs were synthesized by via Amberlite IR-120H as a reusable heterogeneous catalyst in reflux conditions (Kovvuri et al. 2018). Dess-Martin periodinate was employed to mediate the oxidation of **64** to produce methyl-β-carboline-3-carboxaldehyde **65**, which was then treated with various substituted indoles in a water-methanol solvent to produce **66** (Scheme 1.20).

Through the use of the MTT assay, all of the prepared β-carboline-bis-indole motifs were tested for their ability to inhibit the proliferation of a number of human cancer cell lines, including A-549, DU-145, HeLa, and MCF-7. These compounds demonstrated moderate-to-good anticancer activity, with doxorubic serving as the positive control and IC_{50} values in the scale of 1.80 to 44.20 μM. Following the 4-methyl (**66a**) substitution, the compound with the 4-fluoro (**66b**) substitution on the phenyl ring at C1 position displayed tremendous antiproliferative action against DU-145 cells. The anticancer activity was significantly affected by the quantity of methoxy groups on the phenyl moiety.

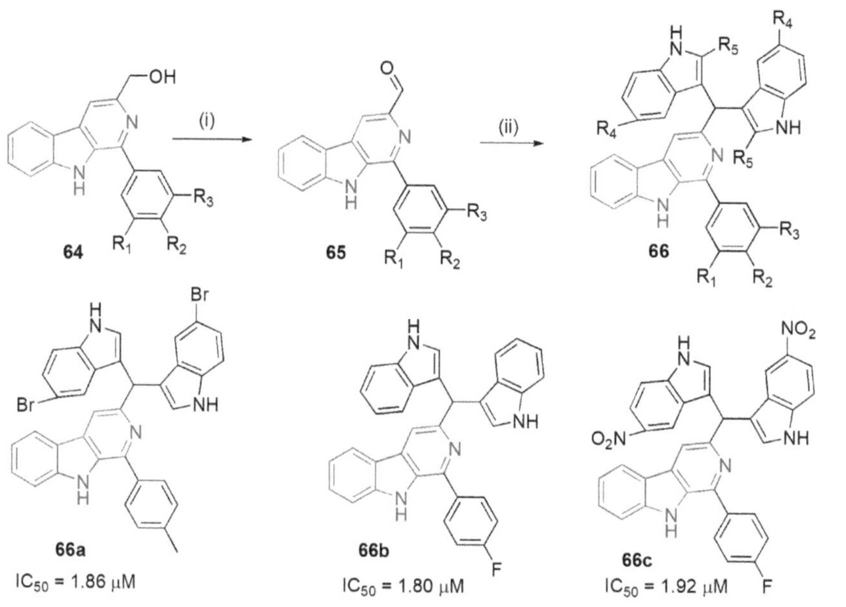

Reaction conditions: DMP, DCM, 27 °C, 2hr; (ii) Indoles, H_2O-EtOH, cat. Amberlite IR-120, 80 °C, 6hr

SCHEME 1.20 Synthesis of β-carboline-bisindole derivatives.

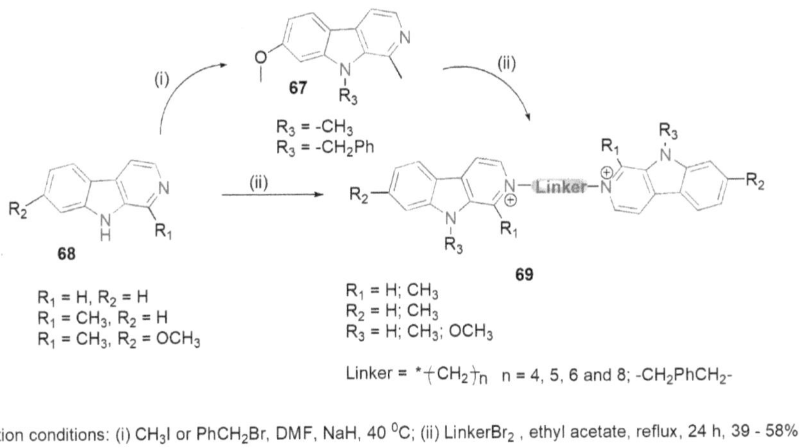

SCHEME 1.21 Synthesis of bivalent β-carboline derivatives.

When the 4-methyl group **66a** on the phenyl ring was replaced with an added electron-rich methoxy group, the activity against DU-145 was reduced by 1.4 times. These compounds interact with DNA topoisomerase I to produce a ternary enzyme-DNA-ligand complex structure, which inhibits it, according to molecular docking and binding studies.

Du H et al. reported the synthesis of novel bivalent β-carboline derivatives and investigated their impacts on the cell cycle, apoptosis induction, and antiproliferative action on a group of cancer cell lines (Du et al. 2016). The investigation showed that bivalent β-carbolines are more effective at inhibiting cell proliferation than monomer. Additionally, the antiproliferative effects of bivalent β-carbolines were enhanced when a methyl or benzyl group was added to position-N9. Harmine **67** was isolated from *Peganumharmala L.*, and from harmine, N9-substituted analogues **68** and **69** were prepared. Then, based on monomeric β-carbolines chain varying from n as 4 to 8, the bivalent β-carbolines **69** were synthesized by reacting with the appropriate dibromoalkane in ethyl acetate (Scheme 1.21). These compounds have various groups as spacers between the nitrogens of the pyridine ring.

1.3.1.5 Scission

Al-Allaf et al. documented the fabrication of trans-palladium complex **70** and harmine **67**, along with a cytotoxic evaluation against P388, L1210, and K562. In order to produce the palladium complex **70**, $PdCl_2$ was added to DMSO and heated to 90°C for two hours in order to completely dissolve. Following the filtration of the reaction mixture, the filtrate was mixed with the HCl solution and allowed to cool to ambient temperature in order to carry out the whole precipitation process. Furthermore, the multifaceted trans-$[Pd(DMSO)_2Cl_2]$ **70** was dissolved in gently heated DMSO, and harmine **67** was added slowly while being violently agitated. Once the mixture cooled, drops of a cold HCl solution were applied until the entire required component **71** (Scheme 1.22) precipitated.

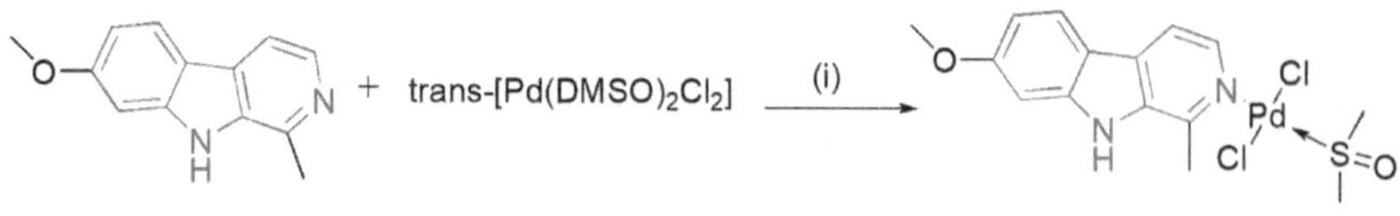

67 **70** **71**

IC_{50} = 0.36 μM

Reaction conditions: (i) DMSO, 0.05M HCl

SCHEME 1.22 Synthesis of harmine trans-palladium complex.

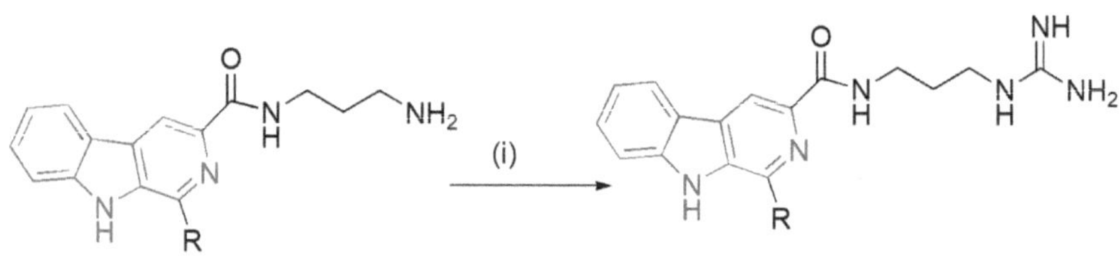

72 **73**

R = H, CH_3

Reaction conditions: (i) *S*-methyl urea, 2N NaOH

SCHEME 1.23 Synthesis of β-carbolines with terminal guanidinium.

Compound **71** had a significant cytotoxic impact (IC_{50} value ranging in between 0.364–0.38 μM) in comparison to platinum-containing drugs such as carboplatin and cisplatin (Al-Allaf and Rashan 1998). A number of platinum (II) and palladium (II) complexes with different β-carbolines, such as harmaline, harmalol, harmine, and harman, were additionally prepared and described for their antiproliferative effects against three tumor cells, with IC_{50} values ranging from 0.2 to 2.0 mg/mL (Al-Allaf et al. 1990). According to Yu et al., the reaction of **72** with S-Methyl urea in sodium hydroxide base yielded β-carboline motifs with terminal guanidinium changeover **73** (Scheme 1.23). On tat-TAR, compound **73** displayed inhibitory effects. The capillary electrophoresis assay demonstrated that these drugs also have the ability to block tat-TAR interactions in addition to their TAR-binding characteristics (Yu et al. 2004).

β-carboline-carbohydrate hybrid motifs synthesis and anticancer efficacy were reported by Toshima et al. Trimethylsilyl trifluoromethane sulfonate was used in the reaction between compound **74** and carbohydrate derivatives to yield compound **75** (Scheme 1.24). It was found that when exposed to long-wavelength UV radiation, a β-carboline-carbohydrate cross molecule containing 2, 6-dideoxy sugars cleaves DNA at the guanine site. Since many naturally occurring DNA binding anticancer drugs have different forms of 2,6-dideoxy sugars, many 2,6-dideoxy sugars were employed as the source of carbohydrates for the design of such hybrid motifs. Furthermore, these compounds **52** displayed increased DNA cleaving ability due to the hydrophobicity of sugar molecule that has been found to be favorable for binding to DNA grooves (Toshima et al. 2002).

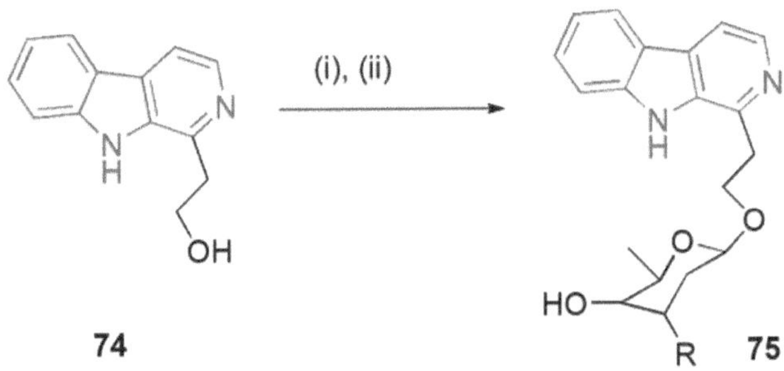

74 75

R = NMe, $OCONH_2$, OH

Reaction conditions: Carbohydrates, TMSOTf, THF, DCM; (ii) NaOMe, NaOH

SCHEME 1.24 Synthesis of β-carboline-carbohydrate hybrid.

1.3.2 Kinases as Target

By phosphorylating amino acids with ATP, protein kinases control the biological activity of proteins. This process converts proteins from their inactive state to their active state by generating conformational changes. Protein kinase inhibitors are now acting as prospective targets for the treatment of cancer because kinases play a significant role in modulating a variety of cell processes, including apoptosis, proliferation, DNA damage and repair, cell motility, etc. Therefore, occasionally they function as cogenesis. Novel forms of 7, 8-dichloro-1-oxo-β-carboline motifs have been produced and their ability to inhibit kinases tested. The main reason for the inherent affinity of kinase inhibitors for kinases is their ATP-mimetic binding to the kinase hinge section. However, the substituted 7, 8-dichloro-1-oxo-β-carboline motifs have unique structural properties of the alkaloid bauerine C, whose kinase inhibitory effect is independent of the standard ATP-mimetic hinge interactions. The presence of halogen bonds with kinase backbone residues and an unexpected, inverted binding mode were both revealed by the cocrystal. Compound **76** was cyclized with sodiumhydride (NaH), and after reacting with N-Boc-4-piperidone using methyl iodide, compound **77** was produced. Compound **78** was then produced by performing out the deprotection reaction in the presence of TFA (Scheme 1.25). The substances exhibit exceptional selectivity against a variety of human protein kinases as inhibiting specific kinases like PIM1. These motifs were suggested to be useful conceptual frameworks for the synthesis of some kinase inhibitors, such as the PIM family, CLKs, DAPK3 (ZIPK), BMP2K (BIKE), etc., by biochemical and structural studies (Huber et al. 2012).

1.3.2.1 Haspin Kinase Inhibitors

Haspase kinase inhibitors engage in recreation in the treatment of cancer and are among the best probes for illuminating the cellular functions of proteins. It was communicated to Cuny et al. who evaluated β-carboline as a potential haspase kinase inhibitor (Cuny et al. 2012). In stronger base NaOH, compound **79** was reacted with

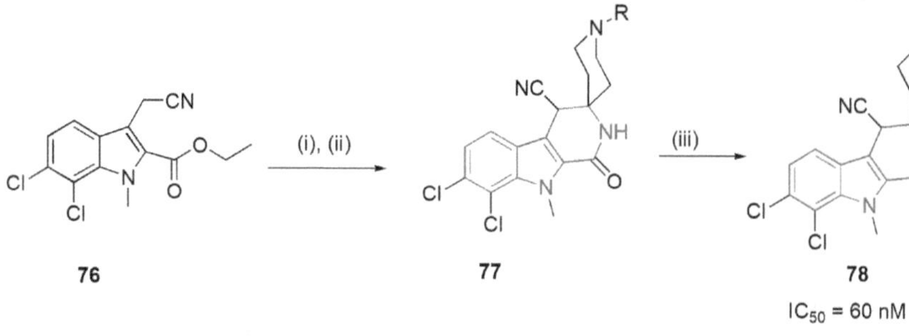

Reaction conditions : (i) NaH, DMF; (ii) CH_3I, N-Boc-4-piperidone; (iii) TFA, DCM

SCHEME 1.25 Synthesis of 7,8-dichloro-1-oxo-β-carbolines.

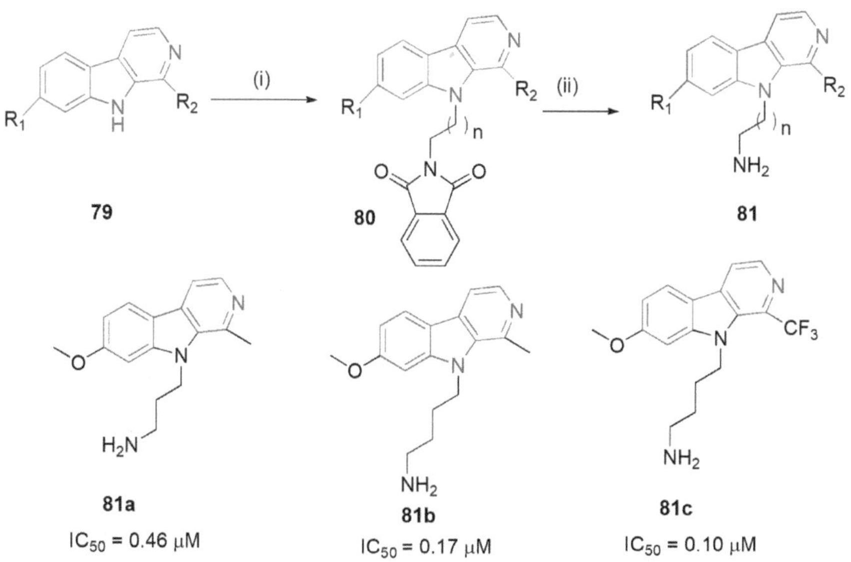

Reaction conditions ; (i) $Br(CH_2)_nNPhth$, NaOH, DMF, rt; (ii) $NH_2NH_2.H_2O$, EtOH, 65 °C

SCHEME 1.26 Synthesis of β-carboline derivatives.

phthalimide-protected alkylamines to form compound **80**, which was then deprotected with hydrazine hydrate to yield compound **81** (Scheme 1.26). High-throughput screening (HTS) was used to determine the synthetic β-carboline motifs ability to inhibit haspase-kinase at various concentrations. A compound with an IC_{50} value of 0.10 μM was discovered to be the most active of those that had potential activity. It was discovered that the substance with amine at position N-9 significantly lowered the power of the enzyme when tested against human DYRK2.

As B-Raf inhibitors, Xin et al. synthesized a number of 1-carboxamide- and 6-sulfonamide-substituted β-carboline motifs that play a crucial role in the MAPK signal-transduction conduit that controls cell proliferation, differentiation, and survival (Xin et al. 2012). Starting with the indole derivative **82**, compound **83** was prepared through cyclization reaction, and on esterification yielded compound **84**, which on dehydrogenated with the use of the potassium permanganate to form aromatic ester **85**. Ester **85** on direct amidation by methanolic ammonium hydroxide yields compound **86**, which on a reaction with chlorosulfonic acid at 0 ^{0}C temperature gives compound **87**. Compound **87** treated with different amines gives target motifs **88** (Scheme 1.27). The IC_{50} values for all the produced compounds ranged from 1.62 to 91.55 μM, indicating extremely strong inhibitory effects. One-carboxamide-6-(N-(3-(dimethylamino) propyl)-sulfamoyl)-β-carboline **88a**, which contains two dimethyl aminopropyl groups and has an IC_{50} value of 1.62 μM, was found to be the most powerful compound. With an IC_{50} value of 2.81 μM, it was shown that substituting diethyl aminopropyl for both dimethyl aminopropyl groups led to a reduction in activity. With wild-type B-Raf kinase, molecular docking studies were carried out to determine the binding model of these drugs. With the hinge backbone carbonyl oxygen and amide of CYS531, respectively, H-bonding between the NH of the carboxamide and the nitrogen atom of pyridine was seen when the most effective molecule, **88a**, was docked with B-Raf kinase. Additionally, the NH and nitrogen atoms of sulfonamide may form two hydrogen bonds with the amide and carbonyl oxygen of ASP593, and the oxygen atom of sulfonamide can establish two hydrogen bonds with the amine of LYS482. This is identical to the binding models found in the crystal structures of the DFG loop and hinge region in association with the B-Raf inhibitors Sorafenib and PLX4032, respectively. It was then decided that new B-Raf inhibitors may be constructed using the β-carboline ring as the hinge binder. The innovative synthesis protocol of 1-carboxamide and 6-sulfonamide-substituted β-carboline drug **91** is shown in Scheme 1.28.

1.3.2.3 Cyclin-dependent Kinase Inhibitors

Cyclin-dependent kinases (CDKs) are heterodimeric protein kinases that control cell cycle succession through phosphorylation. It has been demonstrated in numerous cases that CDKs are frequently dysregulated in human cancers; as a result, CDK inhibitors are given to stop cell proliferation during cancer treatment. ATP-Mg^{2+} and CDK inhibitors fight for the ATP-binding site of CDKs. When harmine and 1-substituted β-carboline motifs were made and their anticancer efficacy assessed, 5-methylimidazol-4-yl **92** (Figure 1.3) was found to have the most CDK inhibitory activity, inhibiting CDK5 by 94% at a concentration of 50 μM and CDK2 by 89% at the same concentration (Lo and Zhang, 2012).

Iridium complex is a new anticancer drug that improves focus by inducing apoptosis and interacts with DNA or protein kinase. Liang et al. reported the preparation of two cyclometalated iridium (III) complexes with 2-(2-thienyl) pyridine (thp) **93** and β-carbolines alkaloids as functional ligands **94** (Figure 1.4). Two equivalents of β-carboline and the corresponding dimer of cyclometalated Ir(III) were refluxed in a mixture of DCM and MeOH (2:1) before being subjected to an anion exchange reaction using NH_4PF_6. Complexes 69 and 70 were tested against the A-549, A-549cisR, cisplatin-resistant A-549, HepG2, HeLa, and LO2 cancer cells (He et al. 2014)

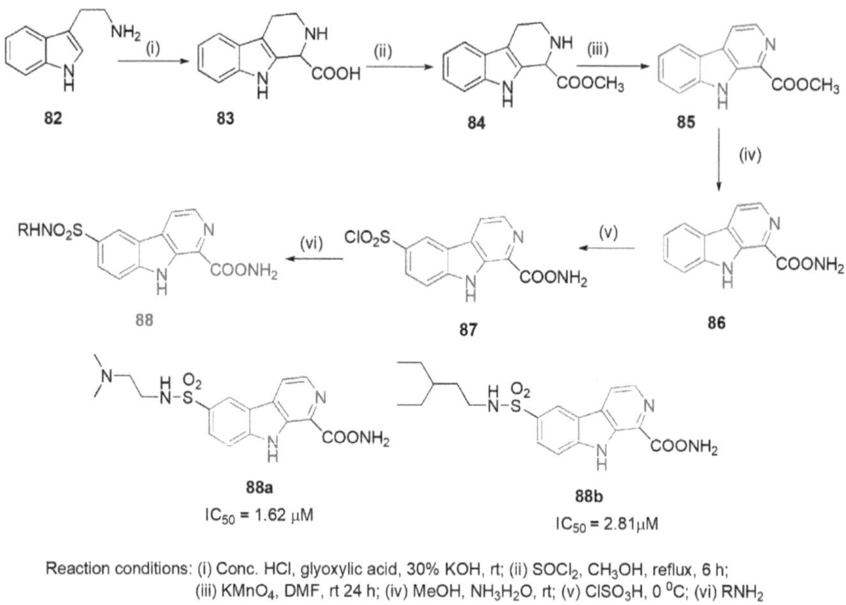

SCHEME 1.27 Synthesis of 1-carboxamide-6-sulfonamide substituted β-carbolines.

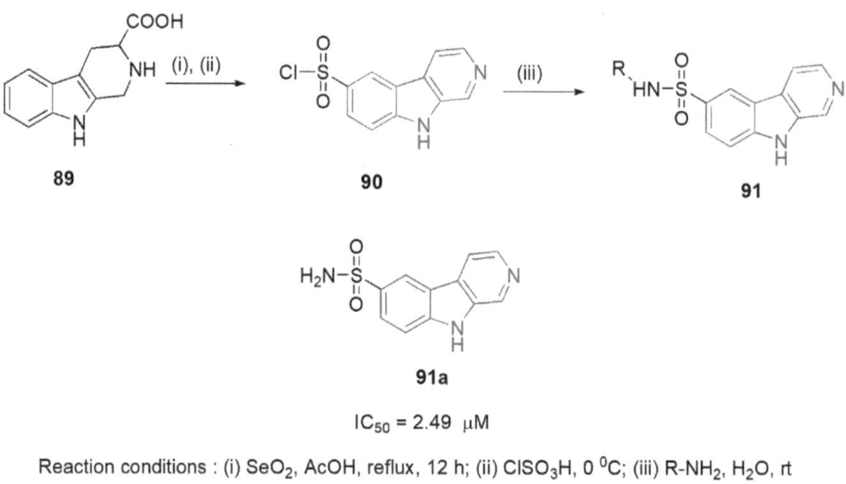

SCHEME 1.28 Synthesis of β-carbolines 91.

1.3.3 Tubulin/Microtubule as a Target

The synthesis of 1-substituted 1, 2, 3, 4-tetrahydro **95** and 3, 4-dihydro β-carboline **96** motifs was carried out by Shen et al. with starting indole derivative **82** (Scheme 1.29). The Pictet–Spengler reaction method was employed to synthesize the desired compounds **95**, which were then subjected to DDQ oxidation to produce compounds

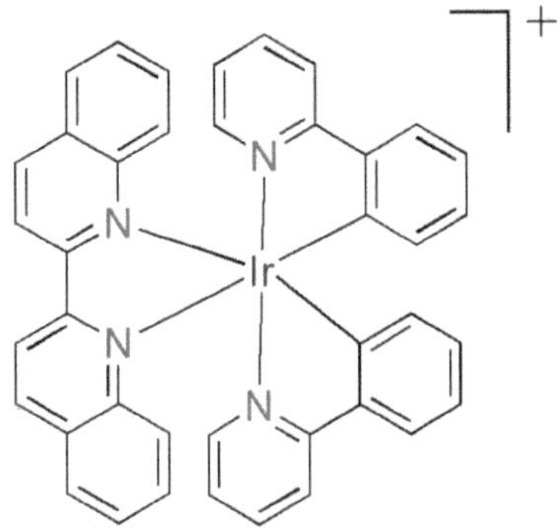

92

FIGURE 1.3 Structure of iridium (III) biquinoline complex.

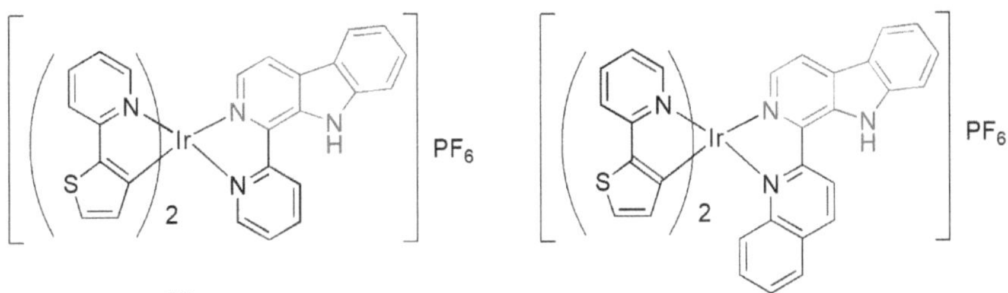

93 **94**

FIGURE 1.4 Structures of the Ir(III) complexes 93 and 94.

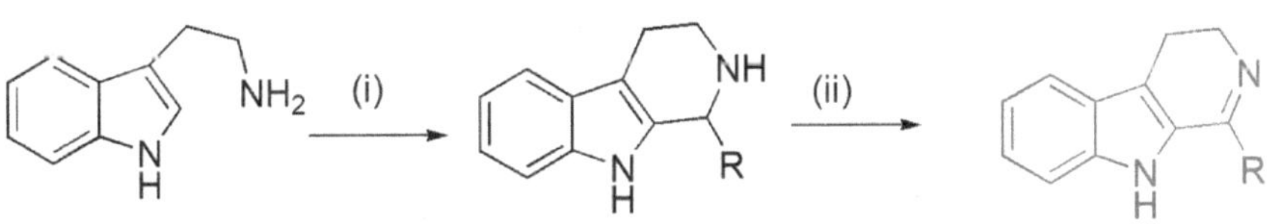

82 **95** **96**

Reaction conditions: (i) RCHO, CF_3COOH ; (ii) DDQ

SCHEME 1.29 Synthesis of β-carbolines.

96. The anticancer efficacy of the produced compounds was tested against the human tumor cell lines KB-16, A-549, and HT-29 as well as the murine P-388 tumor cell line86. Compound **96** demonstrated great activity because it has a planar structure due to this double bond between C-1 and C-2, which is essential for cytotoxicity and may make it easier to intercalate with the DNA molecule (Shen et al. 2005).

1.3.4 Kinesin (Eg5) Inhibitors

Based on the known structure of the Eg5 inhibitor HR22C16, Shankaraiah et al. described the preparation and anticancer assessment of a variety of hybrid tetrahydro-β-carboline-hydantoin motifs. Compound **97** yields hydration-fused tetrahydro-β-carbolines **98** after its treatment with different isocyanates (Shankaraiah et al. 2014). Favorable cis product was produced in the current reaction, and it was easily transformed into trans-hydantoin-fused tetrahydro-carbolines **99** during base-catalyzed epimerization at the C11 position in reflux conditions using acetonitrile and K_2CO_3 (Scheme 1.30). Through the MTT assay, the in vitro cytotoxicity of all the synthesized compounds was assessed against the cell lines A549, ME-180, HeLa, MCF-7, and PC-3. The compounds reported IC_{50} values of fewer than 20 mM against the PC-3, A549, and MCF7 cell lines, exhibiting equipotent action to that of the reference standard etoposide. It effectively binds to the Eg5 motor proteins allosteric site, which is predominantly made up of the residues Glu116, Gly117, Glu118, Trp127, Ala133, Ile136, Pro137, Tyr211, Leu214, and Glu215, according to molecular docking studies. One compound showed high specificity towards prostate cancer cells and its IC_{50} was five times higher in normal prostate epithelial cells (RWPE) than the standard drug. Other compounds were discovered to be extremely active against PC-3 cell lines and were subsequently screened for in vitro cytotoxicity on ordinary RWPE.

1.3.5 HDAC Inhibitors

As HDAC inhibitors, Ling et al. reported the synthesis of β-carboline hydroxamic acid. When compound **100** is treated with $NH_2(CH_2)nCOOMe$, a Curtius rearrangement occurs to produce compound **101**. Compound **101** then reacts with NH_2OK

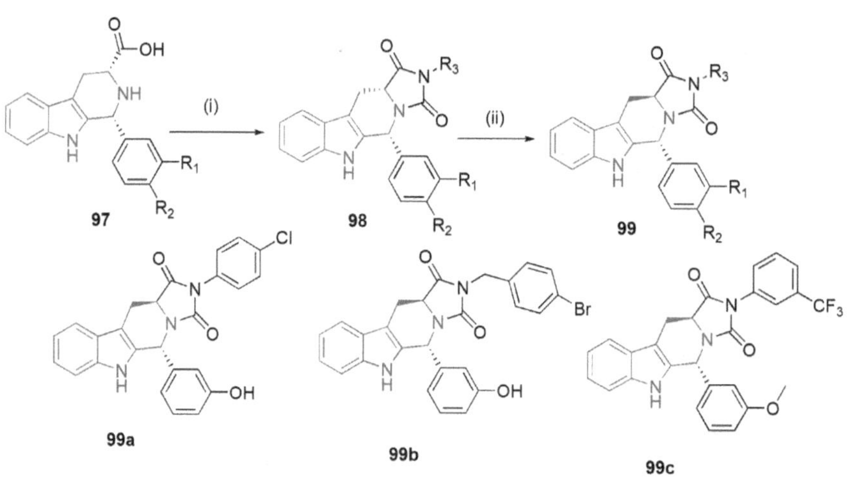

Reaction conditions: R_3-NCO, CH_3CN/ DMSO, reflux, 4hr; (ii) K_2CO_3, CH_3CN, reflux, 2hr

SCHEME 1.30 Synthesis of tetrahydro-β-carboline-hydantoin hybrids.

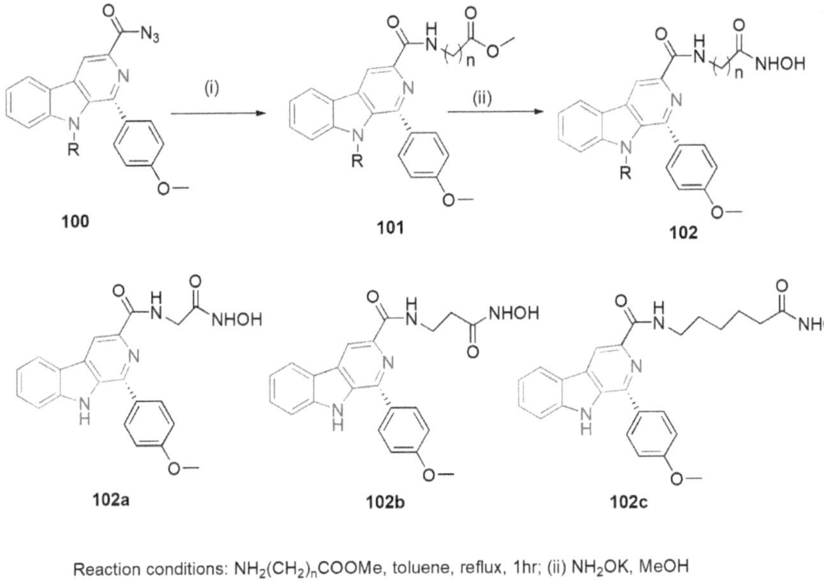

SCHEME 1.31 Synthesis β-carboline/hydroxamic acid hybrids.

in MeOH to yield hydroxamic acid derivatives **102** (Scheme 1.31). The effectiveness of each produced drug was tested against an HDAC-enriched HeLa cell nuclear extract. Typically, HDAC inhibitors have three domains: the zinc binding group, the hydrophobic aromatic substitution on the cap group, and the saturated or unsaturated linker domain. In this instance, the zinc binders cap group was identified by the β-carboline ring and hydroxamic acid fragments. The compounds ability to block an enzyme when it contains one or two methylene units was minimal, whereas compounds with more than two methylene units displayed increased inhibitory efficacy (Ling et al. 2015). A series of β-carboline and N-hydroxycinnamamide hybrid motifs were described by Ling et al. as potential HDAC inhibitors for the treatment of drug-resistant hepatocellular cancer. In the presence of a base, compound **103** reacted with methyl bromoethylcinnamate to yield compound **104**, which was then treated with NH_2OK in methanol (Scheme 1.32). Compound **104a** demonstrated potent and precise antiproliferative effects on HepG2, Bel7402, and drug-resistant Bel7402/5FU cells. In Bel7402/5FU cells, it controlled the appearance of proteins associated with apoptosis and significantly increased autophagic flux activity (Ling et al. 2019).

Ji Liu et al. reported the synthesis of a number of new hydroxamate derivatives **111** based on β-carboline that were used as HDAC inhibitors depicted in Scheme 1.33 (Liu et al. 2017). The majority of these inhibitors were shown to have strong histone deacetylase inhibitory effects as well as superior antiproliferative activity among IC_{50} in the low micromole range. In line with its effective HDAC inhibition, compound **111k** also led to a rise in the acetylation of histone H3 and α-tubulin. Importantly, compound **111** caused hypochromism through an electrostatic contact with CT-DNA, which raises the possibility of causing DNA damage.

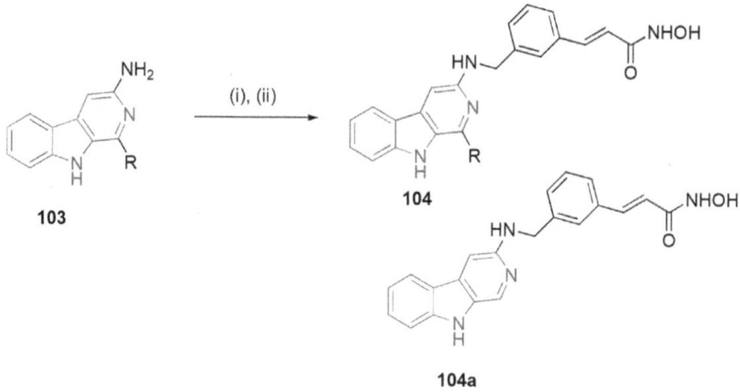

Reaction conditions: (i) Methyl bromomethyl-cinnamate, K_2CO_3, CH_3CN, 50 °C; (ii) NH2OK, MeOH, rt. 12 hr

SCHEME 1.32 Synthesis of β-carboline/N-hydroxycinnamamide hybrids.

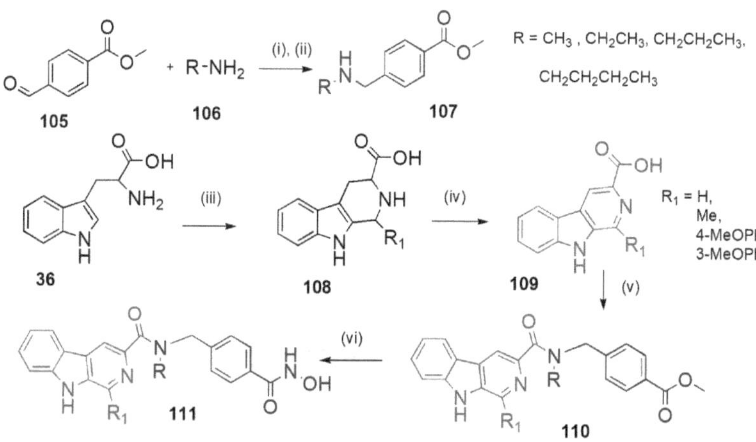

Reaction conditions: (i) AcOH,THF, rt, 1-2h; (ii) $NaBH_4$, THF, rt, 2h, 56-73%; (iii) H^+ or OH^-, RCHO, reflux, 2-4 h, 81-86%; (iv) $KMnO_4$, DMF, rt, 6h, 67-75%; (v) EDCI, DMAP, DCM, rt, 5-8h; (vi) NH_2OK, MeOH, rt, 10-15h, 52-67%

SCHEME 1.33 Synthesis of hydroxamate derivatives.

Finally, **111k** substantially increased the expression of proteins associated with apoptosis in HepG2 cells. Scheme 1.33 shows the synthesis process for compound **111**. In the presence of acetic acid, the initial methyl 4-formylbenzoate **105** interacted with various alkyl amines **106** to produce the imines, which were subsequently reduced by $NaBH_4$ to produce the secondary amines **107**. The preparation of substituted β-carbolines **109** took place in two steps. First, using different substituted aldehydes (formaldehyde, acetaldehyde, or 4-methoxyphenaldehyde), a Pictet–Spengler reaction was used to transform commercially available L-tryptophan **36** into 1-substituted-1,2,3,4-tetrahydro-β-carboline3-carboxylic acid

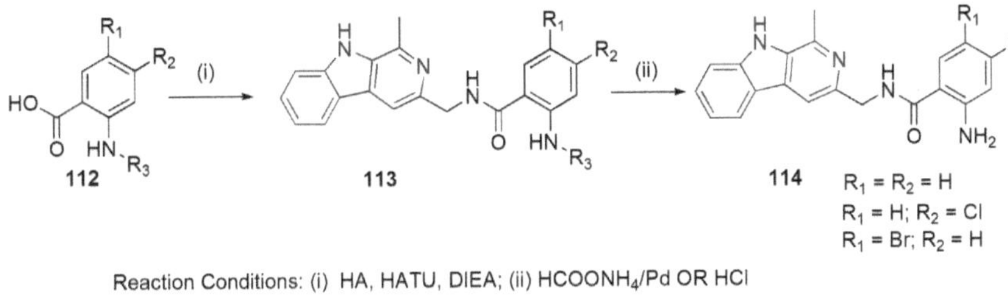

SCHEME 1.34 Synthesis of Amide based on anthranilic acid β-carboline.

108. After intermediates **108** were oxidized by KMnO4 in DMF, compounds **109** were produced. These compounds **109** then interacted with secondary amines **107** in the presence of EDCI and 4-dimethylaminopyridine (DMAP) to generate amides **110**. Finally, the target compounds **111** were produced by treating intermediates **110** with NH_2OK.

According to Maja Beus et al., novel amide-type hybrid compounds based on anthranilic acid β-carboline heterocyclic scaffolds were designed and synthesized. The Pictet–Spengler reaction was used to construct the harmane scafold from tryptophan ester. A related tetrahydro-β-carboline product **113** was produced by heating methyl 3-(1H-indol 3-yl)-2-aminopropanoate **112**, acetaldehyde dimethyl acetal (ADMA), and trifuoroacetic acid (TFA) under microwave irradiation. This product **113** was subsequently aromatized using $KMnO_4$ at room temperature and made compound **114** (Scheme 1.34). The synthesized compounds **114** were used to assess the anticancer efficacy of a number of cancer cell lines. Eight human solid tumor and leukemia cell lines were used for the in vitro experiments (glioblastoma LN-229, pancreatic adenocarcinoma Capan-1, colorectal carcinoma HCT-116, lung carcinoma NCI-H460, acute lymphoblastic leukemia DND-41, acute myeloid leukemia HL-60, chronic myeloid leukemia K-562, and non-Hodgkin lymphoma Z-138). The noncancerous cell model was human lung fibroblasts (HEL 299), which are normal cells. The IC_{50} values were calculated when the cells were exposed to various doses of the substances.

1.3.6 Cell Cycle as Target

Ikeda et al. described the synthesis of 8-amino-3-benzylamino-6-chloro-β-carboline **125**, a precise intermediate molecule of 3-benzylamino-β-carboline derivatives with various types of substituents on the N atom of the 8-amino group (Ikeda et al. 2012). The Pictet–Spengler reaction of L-tryptophan was one of three steps used to build the skeleton of the β-carboline motif **36**. Three-step processes using the Curtius rearrangement were used to synthesize **116**, 3-amino-6-chloro-β-carboline **120** after anchoring a chloro group on the 6-position of ethyl β-carboline-3-carboxylate. Compound **122** was prepared by nitrating the 8-position of compound **121** using sodium nitrate and protecting the 3-amino group of that compound by acetyl group. After the acetyl group was removed, the amino group was benzylated by benzaldehyde reductive amination to produce compound **124** with an 83% yield. Within

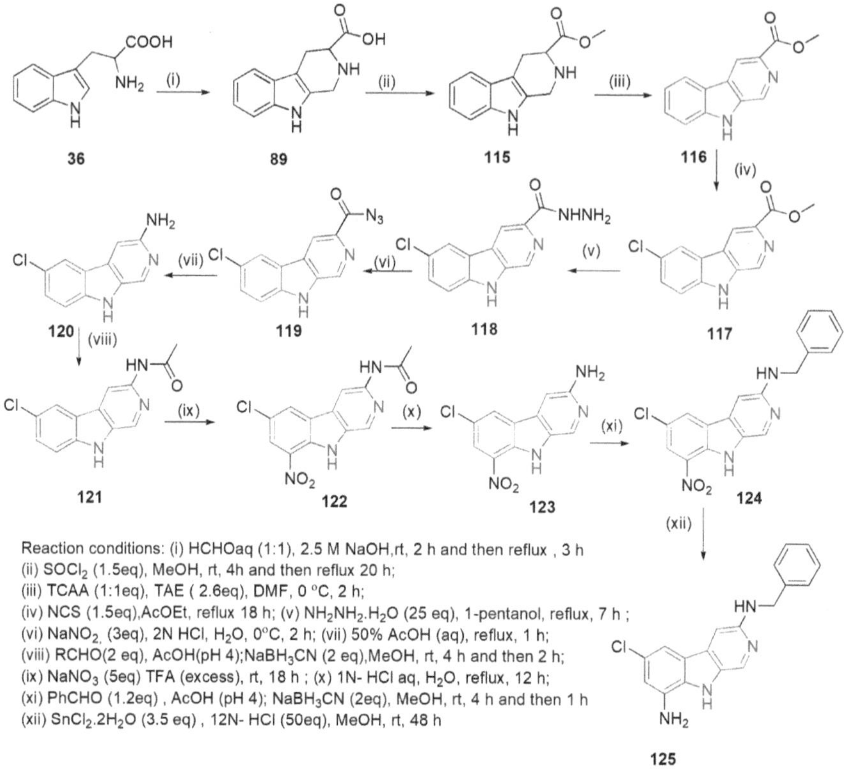

SCHEME 1.35 Synthesis of 6,8-substituted 3-benzylamino-β-carboline derivatives.

12 stages, the complete synthesis of 8-amino-3-benzylamino-6-chloro-carboline **125** was completed, yielding a cumulative 14%. Overall, 72% of the **124** provided yields after the nitro group was reduced (Scheme 1.35).

Scheme 1.36 illustrates the several steps involved in synthesizing 6,8-substituted 3-benzylamino-β-carboline derivatives. Furthermore, sulfonamide derivatives of β-carbolines **126** and **127** are produced when **125** reacts with the proper alkane sulfonyl chloride in the presence of TEA and DMAP.

The anticancer activity of each produced drugs was evaluated against the cancer cell lines HeLa S3, Sarcoma 180, and 293T using the MTT assay. With IC_{50} values of 0.046 and 0.032 μM, respectively, motifs **128** and **129** showed the most activity against the HeLa S-3 cell line among them. The targeted molecule was thought to be holding the cell cycle in the G2/M phase, as indicated by the DNA disintegration prototype, Hoechst 33342 staining, and flow cytometry analysis that revealed the cell death pathway (Figure 1.5). Furthermore, the presence of an alkyl group increases hydrophobicity, which raises cell membrane permeability and anticancer action.

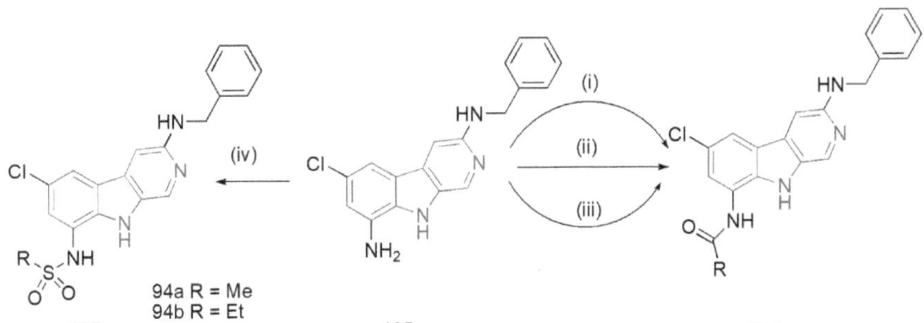

127 125 126

Reaction conditions: (i) $(RCO)_2O$ (1.0 eq), DMAP (1.2 eq), $CHCl_3$;
(ii) RCOOH (1eq), DCC (1.2eq), DMAP (1.2 eq), CH_2Cl_2, rt, 7 - 12 h
(iii) RCOCl, (1eq)/ $CHCl_3$, rt, 0.5 h
(iv) R SO_2Cl (1.2 eq), TEA (10 eq), DMAP (0.2 eq), CH_2Cl_2, rt, 24 h

SCHEME 1.36 Synthesis of 6,8-substituted 3-benzylamino-β-carbolines.

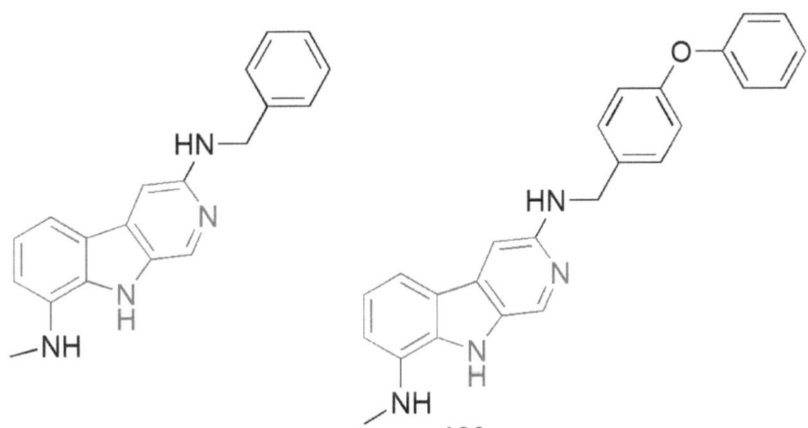

128 129

IC_{50} = 0.046 μM

IC_{50} = 0.032 μM

FIGURE 1.5 Structure of 8-substituted 3-benzylamino-β-carboline derivatives.

1.3.7 Miscellaneous

1.3.7.1 Anticancer β-carbolines Targeting Multiple Biological Objectives

The antitumor activity of a new class of cytotoxic drugs called β-carboline-combretastatin carboxamide conjugates was investigated by Jadala et al. Compound **132** was synthesized by treating compound **130** with various amines **131** by using EDCI and HOBt (Scheme 1.37). Using several tumor cell lines, including HeLa,

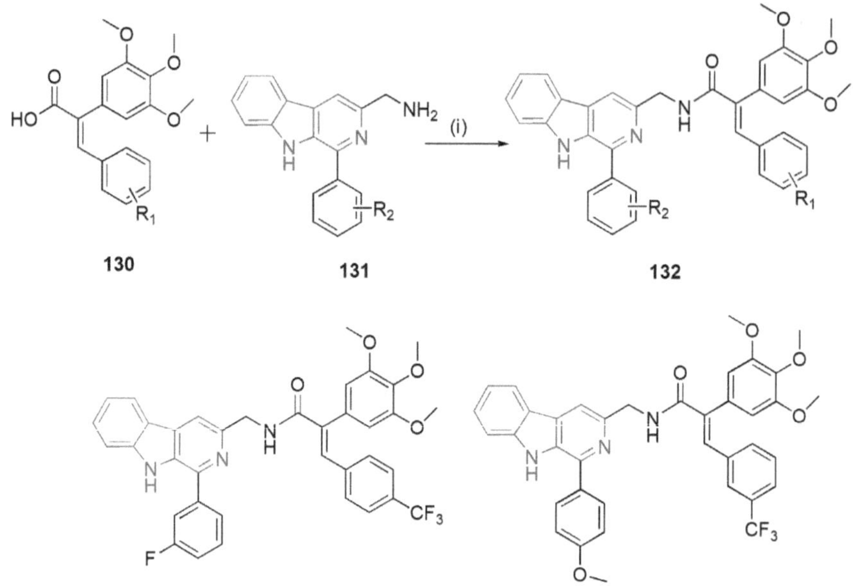

130 131 132

132a IC_{50} = 1.01 μM

132b IC_{50} = 1.50 μM

Reaction conditions : (i) EDCI, OHBt, DMF, rt, 8hr

SCHEME 1.37 Synthesis of β-carboline linked combretastatin conjugates.

DU-145, and A-549, through the SRB assay, all the synthesized motifs **132** were tested for cytotoxic efficacy; Etoposide and harmine served as positive controls. All of the compounds showed reasonable to good activity, with IC_{50} ranges between 1.01 and 50 μM. Certain compounds with IC_{50} values less than 2 μM, meaning IC_{50} values of 1.01 and 1.50 μM against A-549 and DU-145 cells, correspondingly were discovered to be extremely active. To test if these compounds are capable of triggering apoptosis, morphological tests, DCFDA, Annexin V-FITC, and Hoechst staining were performed. Additionally, a flow cytometric investigation showed that compound **132b** caused dose-dependent cell cycle arrest in the G2/M phase (Jadala et al. 2019).

Tetrahydro-β-carboline derivatives were synthesized, according to Zhang et al., and their anticancer and chemopreventive effects, as well as their effects on quinonereductase 1 (QR1) induction, aromatase inhibition, and nitric oxide (NO) production, were evaluated (Zhang et al. 2018). In the synthesis of callophycin A analogs were carried out by esterification of chiral carboxylic acid isomer **133a** (S) or **133b** (R) with corresponding alcohol in the presence of $SOCl_2$ produces **134**, the prepared compound subjected to reaction with different alkyl bromides in acetonitrile at reflux condition in the presence of N,N-diisopropylethylamine (DIPEA), giving the N-alkylated derivatives **135** in 62–82% yields. Finally, in the presence of NaOH in $MeOH/H_2O$ at room temperature, **135** were hydrolyzed to give free carboxylic acids **136** in yields ranging from 81 to 93% (Scheme 1.38).

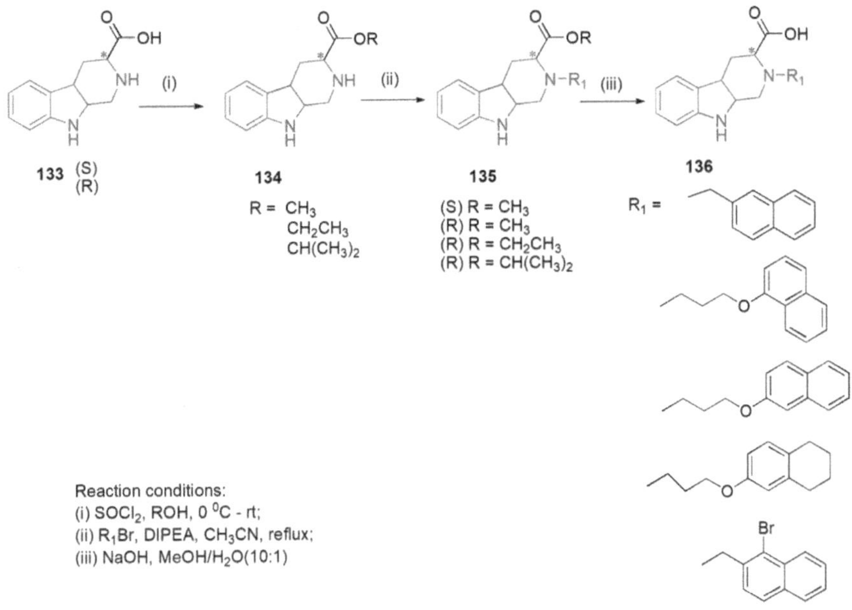

SCHEME 1.38 Synthesis of callophycin A.

The R-isomer of the amide derivative (2-((1-bromonaphthalen-2-yl)methyl)-2,3,4,9-tetrahydro-1H-pyrido[3,4-b]indol-3-yl)(4- methylpiperazin-1-yl)methanone proved to be an excellent potent inhibitor of NO production with a 50% inhibitory concentration, IC_{50}=6.54 μM and had a low cytotoxic effect (IC_{50}=17.98 μM) on RAW 264.7 cells.

According to the NCI procedure, Abdelsalam M. A. et al. reported the synthesis of new β-carboline derivatives and evaluated their anticancer potential (Abdelsalam M A, 2018). The combined topo-I and KSP inhibition was seen in compounds **139b**, **139d**, and **141d**, with compound **139d** being the most effective. Melanoma MDA-MB-435 cells were subjected to active chemicals, which caused Pre-G1 apoptosis and cell cycle arrest at the G2/M phase. Scheme 1.39 shows the synthetic procedure that was followed to prepare the desired β-carboline derivatives. The corresponding 5-substituted amino-1,3,4-oxadiazole derivatives were produced by cyclodesulfurization of the primary intermediate thiosemicarbazides produced by employing freshly prepared yellow mercuric oxide in refluxing dioxane. Additionally, by heating the thiosemicarbazides in a 2.5 M aqueous sodium hydroxide solution, the corresponding sulfanyl-1,2,4-triazole derivatives **139(a-d)** were obtained. Unfortunately, rather than the anticipated thiazolidinone derivatives, the reaction of the thiosemicarbazides **137(a-d)** with ethyl bromoacetate led to the unexpected products β-carboline-triazolylsulfanyl acetates **140(a-d)**. By employing ethyl bromoacetate to S-alkylate the sulfanyltriazoles **139(a-d)**, identical compounds were produced. Utilizing methyl iodide and variably substituted phenacyl bromides, the sulfanyltriazoles **139(a-d)** were S-alkylated to get the desired methylsulfanyltriazole derivatives **141(a-d)**.

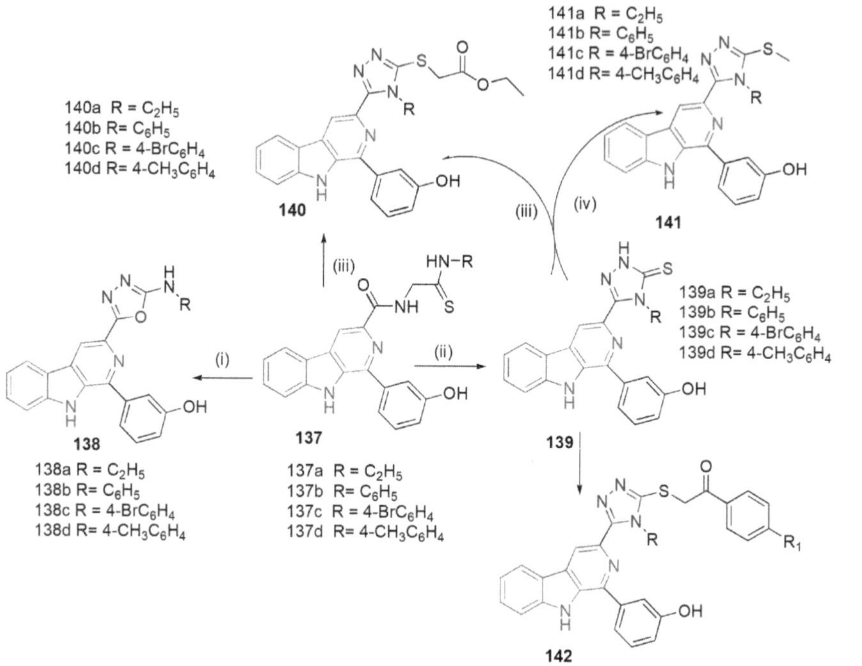

142a R = C_2H_5; R1 = H
142b R = C_2H_5; R1 = Br
142c R= C_6H_5 ; R1 = H
142d R= C_6H_5 ; R1 = Br
142e R = 4-BrC_6H_4 ; R1 = H
142f R = 4-BrC_6H_4 ; R1 = Br
142g R= 4-$CH_3C_6H_4$; R1= H
142h R= 4-$CH_3C_6H_4$; R1= Br

Reaction Conditions : (i) HgO, dioxane, reflux, 5 h, 78 -85%;
(ii) 2.5 M aqueous NaOH; reflux, 4 h, 73 - 84%;
(iii) $BrCH_2COOEt$, NaOAc, EtOH, reflux, 5 h, 78 - 87%;
(iv) CH_3I, Na_2CO_3, EtOH, stiring, rt, 4 h, 60 - 70%;
(v) $R_1C_6H_4COCH_2Br$; K_2CO_3; DMF; stiring, 4 h, 60- 80%

SCHEME 1.39 Synthesis of the target anticancer β-carboline motifs.

1.3.7.2 Compounds with Unexplored Mechanism of Action

Cao et al. synthesized harmine derivatives and studied their cytotoxicity against HeLa, Bel-7402, HepG2, 769-P, 786-O, OS-RC-2, KB, A-549, A-375, HT-29, 5CaBER, Blu-87, and U-251 cells (Cao et al. 2013). The harmine compound 46 was treated with benzyl bromide in ethyl acetate to produce the desired compound **144** (Scheme 1.40). The harmine compound **67** was first treated with HBr in the presence of acetic acid, followed by reactions with alkyl bromide in the presence of sodium hydride and DMF to produce 7-alkoxy-β-carboline motifs **143**. Using cisplatin as the baseline, the cytotoxic activity of each synthesized harmine motif was evaluated. The benzylated compounds **144** showed appealing cytotoxic potencies with IC_{50} lower than 10 μM against all the examined cancer cell lines, while the IC_{50} values of **143** ranged from 10 to 100 μM. Mice with Lewis lung cancer and Sarcoma 180 were used in in vivo anticancer tests with compound **144**, which had recently shown promise in preliminary cytotoxic studies. With LD_{50} values of 12.5, 12.5, 15.0, 5.0, and 6.25 mg/kg, respectively, some compounds exhibited acute cytotoxicity. When compared to harmine I, which inhibited sarcoma 180 and Lewis lung carcinoma, respectively, with

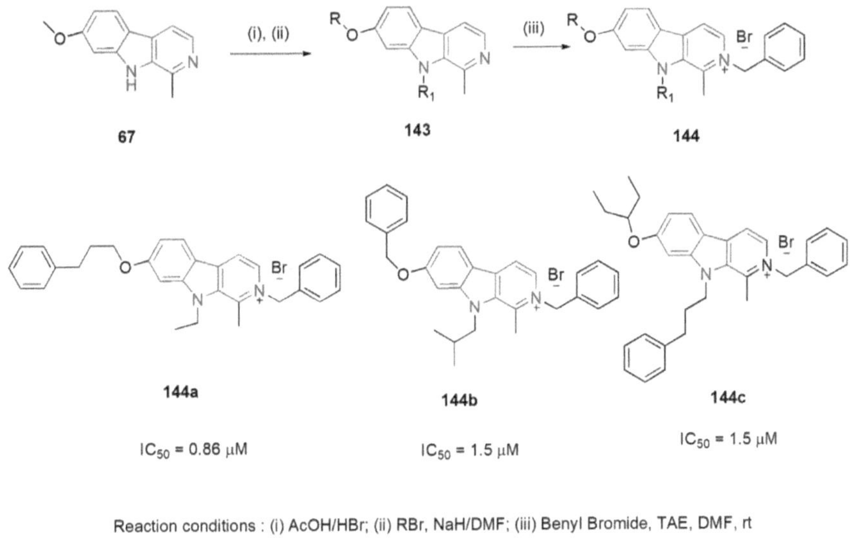

SCHEME 1.40 Synthesis of β-carboline of harminedeivatives.

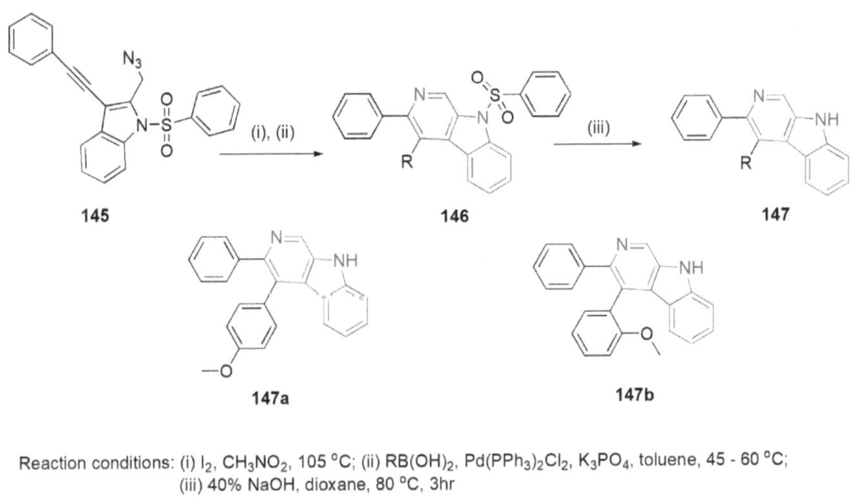

SCHEME 1.41 Synthesis of 3-phenyl-4-substituted-β-carbolines.

inhibition rates of 30.8% and 33.7%, the majority of motifs **143** showed outstanding anticancer activity with over 40% tumor inhibition rate.

The 3-phenyl-4-substituted-1-(phenylsulfonyl)-β-carboline **146** was synthesized by the reaction of compound **145** with iodine under basic conditions, coupled with phenyl boric acid or cyclopropyl boric acid, and then hydrolyzed to produce the final desired products **147** (Scheme 1.41). When the inhibitory activity of compounds **147** was tested against HeLa, MCF-7, and A548, it was discovered that the compounds

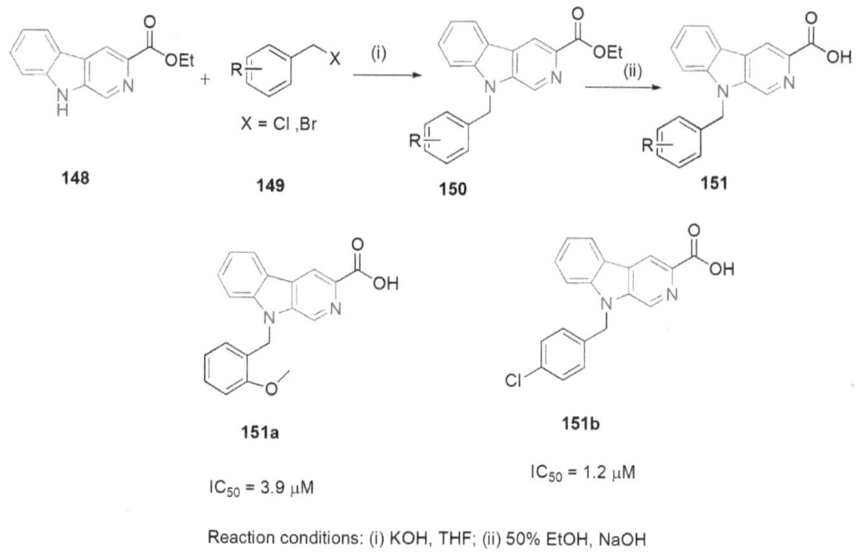

SCHEME 1.42 Synthesis of 3,9-substituted β-carboline derivatives.

had exceptional activity in comparison to the reference harmine, with IC_{50} values of 20.8 μM and 35.3 μM against MCF-7, respectively (Liu et al. 2012).

By reacting compound **148** with benzyl halide **149** in the presence of potassium hydroxide, Chen et al. reported the synthesis of 3,9-substituted-β-carboline motifs **151** (Scheme 1.42). This resulted in ethyl-β-carboline-3-carboxylate derivatives **150**, which upon hydrolysis in alkaline solution gave the corresponding carboxylic acid derivatives **151**. The IC_{50} values for all **151**-produced compounds ranged from 1.2 to >100 μM when tested for anticancer activity against various cancer cell lines. Inhibiting cell development by causing apoptosis, the drug with a 4-chlorophenyl substitution had an IC_{50} value of 1.2 μM against the Hep-3B cell line (Chen et al. 2015).

In their description of the synthesis and cytotoxic evaluation of C-1 aryl substituted β-carbolines motifs against tumor cell lines, Bai et al. Compound **152** was created by reacting the methyl esters of compound **11** with a reducing agent, lithium aluminum hydride (Scheme 1.43). The compounds with the alcoholic group at the C-3 position showed good cytotoxicity, and the majority of them had IC_{50} values that were less than 20 μM. One substance was discovered to be the most effective substance, with IC_{50} values of 0.75, 0.91, 1.00, 1.13, and 2.54 μM, respectively, when tested in the HL-60, SMMC-7721, A-549, MCF-7, and SW-480 cell lines (Bai et al. 2014).

The fabrication of 1, 3, 6-trisubstituted β-carbolines motifs was described by Lungariya et al. In the first step of the synthesis, compound **11** underwent a nitration reaction using a nitrating mixture, followed by a reduction with sodium borohydrate ($NaBH_4$) in the catalytic amounts of Pd/C to give **153**, which was then further acylated with a number of acyl chlorides in anhydrous dichloromethane to produce the preferred compound **154** (Scheme 1.44). The cytotoxic potential of the synthesized motifs against various cancer cell lines was assessed at a concentration of 10 M

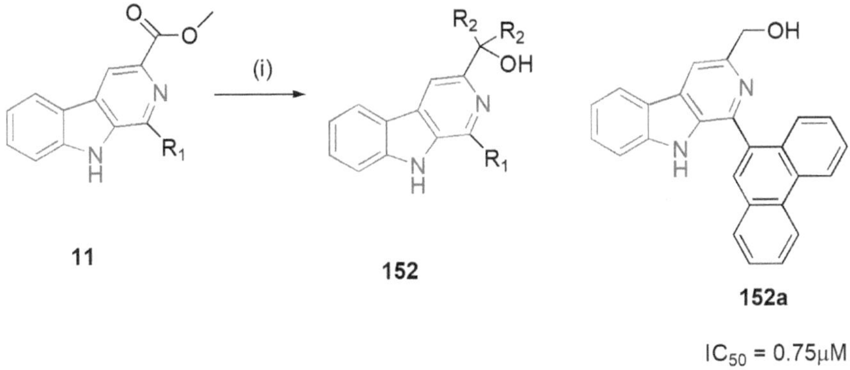

Reaction conditions: (i) $LiAlH_4$, THF, rt

SCHEME 1.43 Synthesis of β-carboline derivatives.

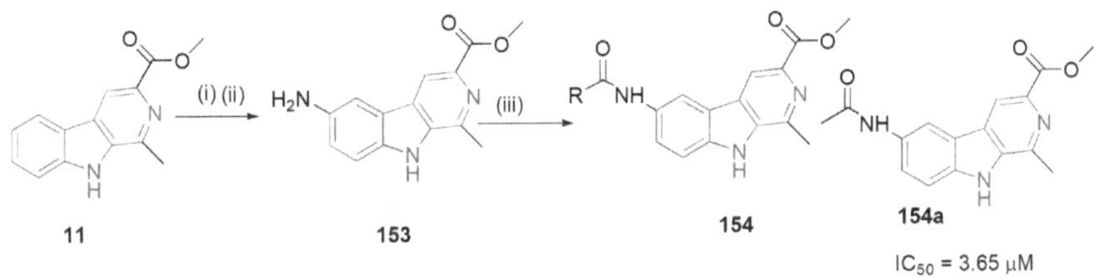

Reaction conditions: (i) Conc. HNO_3, Conc. H_2SO_4, 0 °C, 30min.(ii) Pd/C, $NaBH_4$, MeOH, rt, 3hr; (iii) RCOCl, anhyd.DCM, rt, 6 -10 hr

SCHEME 1.44 Synthesis of 1,3,6-trisubstituted β-carboline derivatives.

using docetaxel as a positive control. With IC_{50} values of 4.72, 3.59, 3.65, and 4.17 μM against the cancer cell lines A- 549, HeLa, Hep G2, and MCF-7, the drug was discovered to be more powerful (Lunagariya et al. 2016).

The most valuable synthon chalcones, which are simple to prepare and have good anticancer treatments, are used in numerous medically significant drug syntheses. Chauhan et al. recently reported on the production and testing of **156** β-carboline-chalcone motifs (Chauhan et al. 2014), wherein one aryl group of the chalcone was substituted with a β-carboline ring, as anticancer agents. Aqueous acetic acid was used to deprotect acetal **155**, and the resulting products, **156**, were produced by a Claisen-Schmidt condensation reaction with substituted 1-phenyle than one (Scheme 1.45).

The effectiveness of each synthesized hybrid motif was tested against a variety of cancer cell lines, including DLD-1, MCF-7, PLC/PRF-5, A-549, SKOV-3, MiaPaca-2, DU-145, and A-172. Alkylation and various substitutions employing groups that donate or remove electrons had an impact on the anticancer efficacy of β-carbolines. According to the SAR research, the molecule having electron-donating groups at the chalcones phenyl ring exhibited strong anticancer activity. As N-alkylation of the hybrid motif results in a decrease in anticancer activity, the

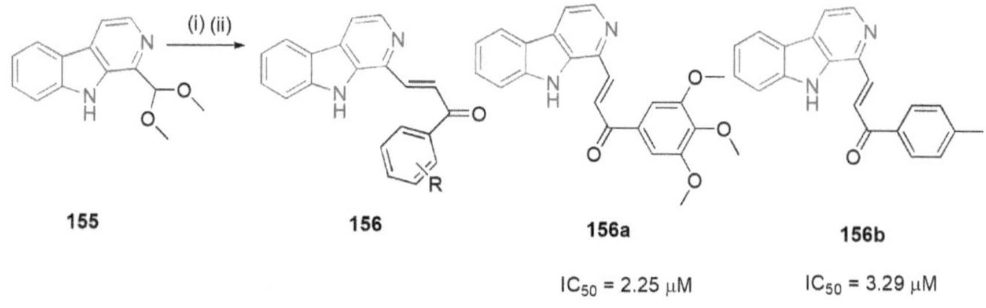

Reaction conditions: (i) AcOH/ H_2O(2:3), 120 °C, 1hr: (ii) 10%NaOH, MeOH, rt, 4hr

SCHEME 1.45 Synthesis of β-carboline based chalcones.

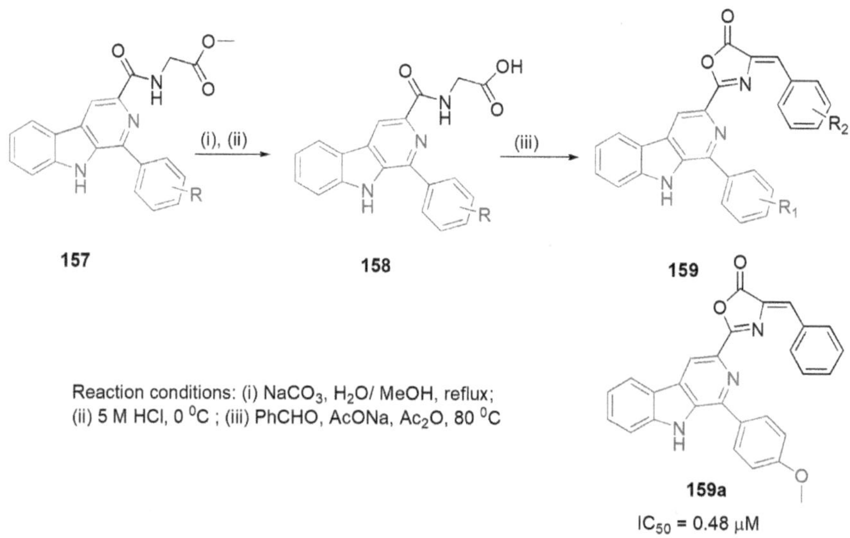

Reaction conditions: (i) $NaCO_3$, H_2O/ MeOH, reflux; (ii) 5 M HCl, 0 °C ; (iii) PhCHO, AcONa, Ac_2O, 80 °C

SCHEME 1.46 Synthesis of β-carboline-4-benzylidene-4H-oxazol-5-one hybrids.

free indole NH group is crucial for anticancer action. All cancer cell lines were possibly responsive to the synthesized compounds in a dose-dependent way. The fabrication of hybrids of β-carboline-4-benzylidene-4H-oxazol-5-one was reported by Savariz et al., who also assessed their anticancer effectiveness (Savariz et al. 2012). Compound **157** was hydrolyzed by sodium carbonate under reflux in methanol/water, and with subsequent acid hydrolysis, compound **158** was produced, which underwent an Erlenmeyer-Plochl reaction with benzaldehyde to get the desired product, β-carboline-3-oxazolone motifs **159** (Scheme 1.46). Using doxorubicin as the reference medication, all prepared series of compound **159** were tested for cytotoxicity against a panel of cancer cell lines. A SAR investigation revealed that the cytotoxic potential of β-carboline-oxazolone hybrids was boosted by electron-donating substituent at the C1 position. The most effective molecule was discovered to have a

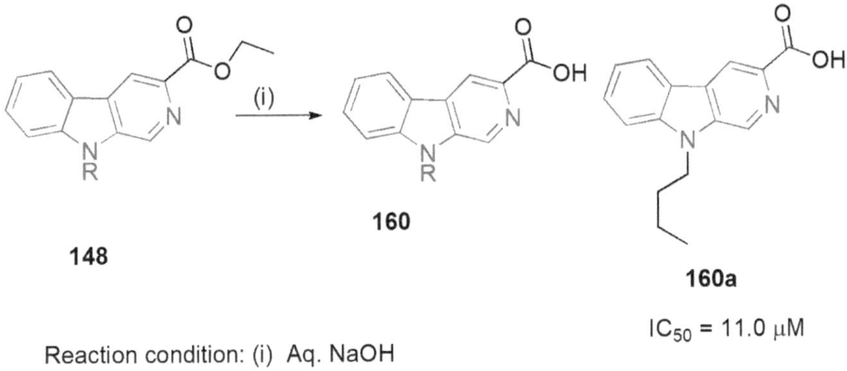

SCHEME 1.47 Synthesis of 9-substituted β-carbolines.

4-methoxyphenyl ring at the C1 position and a phenyl ring on oxazolone, with IC_{50} values against the U251, PC-3, and OVCAR-O3 cancer cell lines of 0.48, 1.50, and 1.07 μM, respectively.

According to a report by Cao et al. on the synthesis of 9-substituted β-carboline motifs **160** through the reaction of **148** with sodium hydroxide (Scheme 1.47), harmines affinity for binding to DNA could be increased by adding a suitable substituent to its N-9 position, which in turn increases Topo I inhibition.When electron-releasing substituents were applied, such compounds started to exhibit improved cleavage efficiency. For instance, a molecule with an n-butyl substituent at the N-9 position demonstrated the strongest anticancer action, with an IC_{50} value of 11.0 μM against Lovo.

Alkyl linked bivalent β-carboline motifs were synthesized, and their biological evaluation was published by Chen et al. Compound **161** produces the Schiff's base when combined with the appropriate sym-diamines, and when this base is reduced with $NaBH_3CN$, the required compound **162** is produced (Scheme 1.48). Using Endostar as the standard reference, the capacity to suppress these drugs was assessed against the human umbilical vein cancer cell line EAHY26. The findings revealed striking antiproliferative actions, and some drugs displayed IC_{50} values between 2.16 and 5.0 μM. The vascular targets of anticancer medication are affected by these bivalent β-carboline motifs (Chen et al. 2016).

Gu et al. also described the synthesis of a number of bivalent β-carboxyline motifs that were tested against a variety of cancer cell lines, including BGC-823, A-375, P-769, and SK-OV-3. In order to create symmetrical bivalent β-carboline motifs **164**, a monovalent β-carboline-3-carboxylic acid **163** was combined with the appropriate dibromoalkane in anhydrous DMF (Gu et al. 2018) (Scheme 1.49). Compound **164** with the highest potential was found to exhibit a 64.2% inhibition rate against Lewis lung carcinoma in mice. At a dose of 26.9 mmol/kg, a compound with a 3-phenylpropyl group at the C-9 position of the β-carboline ring inhibited Lewis lung cancer in mice with a rate of 53.5%.

The preparation of simple dimeric β-carboline motifs and assessment of their cytotoxic potential against various cancer cell lines were both attested by Chatwichien

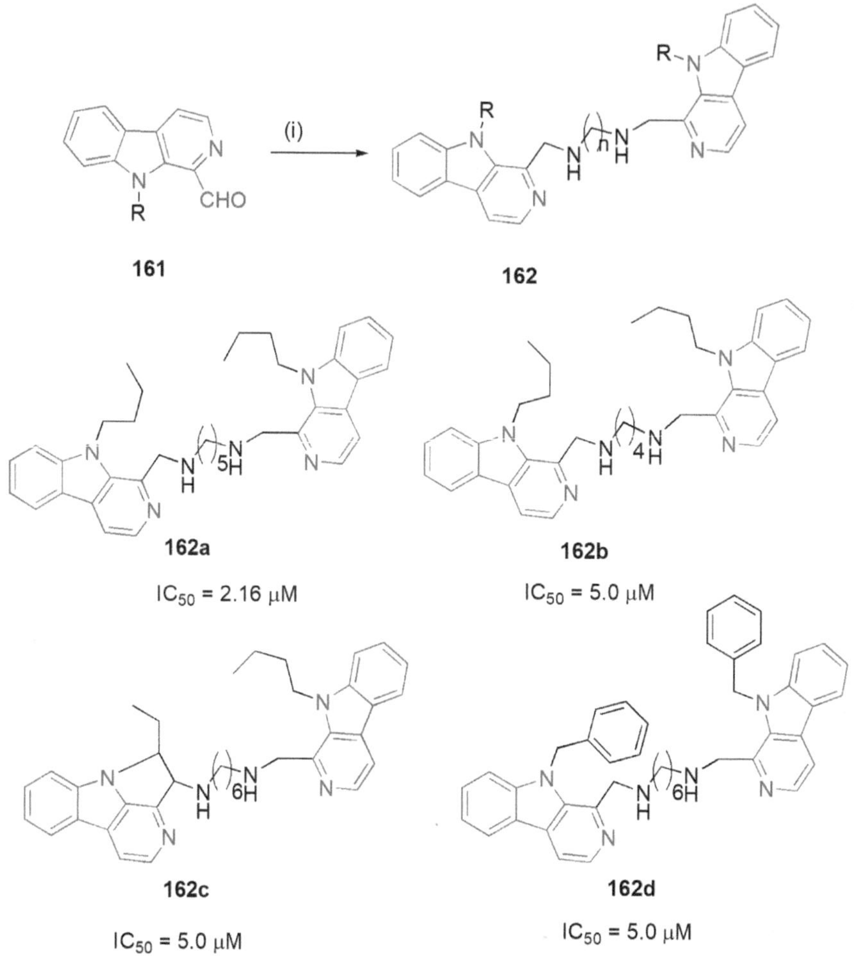

Reaction condition: (i) sym-diamine, $NaBH_3CN$, rt

SCHEME 1.48 Synthesis of alkyl linked bivalent β-carbolines.

et al. Condensation of 1-formyl-β-carboline **165** with linkers was used to create dimers of β-carboline motifs, which were then reduced with sodium cyanoborohydride to produce **166** and **167** (Scheme 1.50). According to studies, dimers are more effective than monomers against the cancer cell lines H1299 and A375, with IC_{50} values of 1.6 and 2.0 μM, respectively (Chatwichien et al. 2015).

Sun et al. described the preparation of piperazine-linked β-carboline, which has superior antitumor activity and an improved solubility profile (Sun et al. 2015). In the synthesis, compound **163** and piperazine were stirred in the presence of anhydrous dichloromethane at 60 °C, and the desired compound **168** was then produced by further reducing anhydrous dichloromethane with $NaBH_3$ at room temperature (Scheme

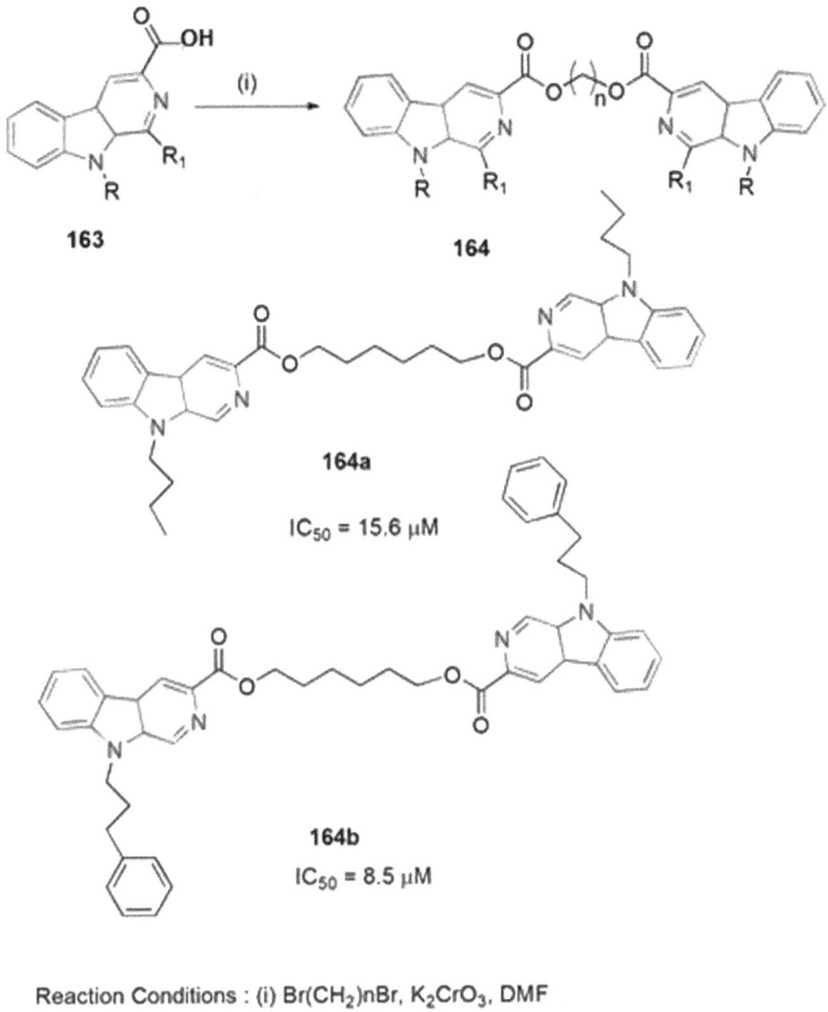

SCHEME 1.49 Synthesis of bivalent β-carbolines.

1.51). Utilizing cisplatin as the standard reference, the anticancer activity of each of the synthesized bivalent β-carboline motifs **168** was assessed against a variety of cancer cell lines. The C1 and C9 positions of β-carbolines were significantly more important in the SAR study's demonstration of their contribution to anticancer action. The cytotoxic potential of β-carboline motifs was improved by adding an alkyl group at the C9 position; the tert-butyl substituent was thought to be the best. With IC_{50} values of 7.62, 8.95, 5.32, 3.02, 8.35, 5.51, 7.62, and 5.5 μM against MCF-7, HepG2, 22RV1, 769-P, A-375, SK-OV-3, BCG-823, and LLC cancer cell lines, respectively, the most potent cytotoxic molecule was examined.

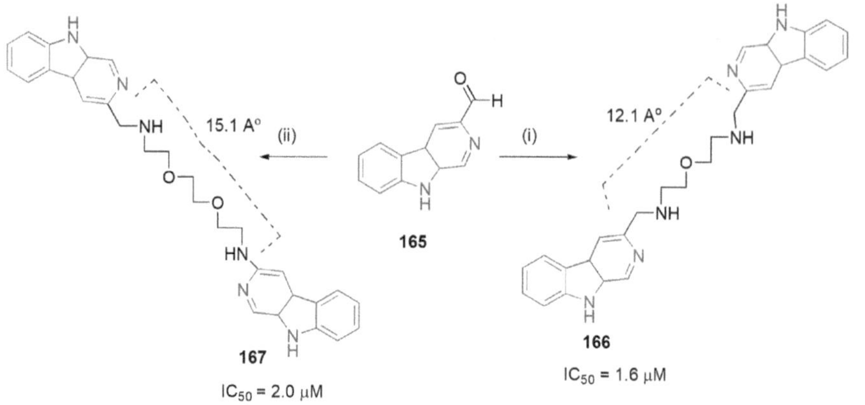

Reaction condition: (i) 2, 2' - oxydiethanamine, MeOH/ DCM, $NaBH_3$, MeOH
(ii) 2,2'-(ethane-1,2 diylbis(oxy)diethanamine, MeOH/ DCM, $NaBH_3$, MeOH

SCHEME 1.50 Synthesis of dimeric β-carbolines.

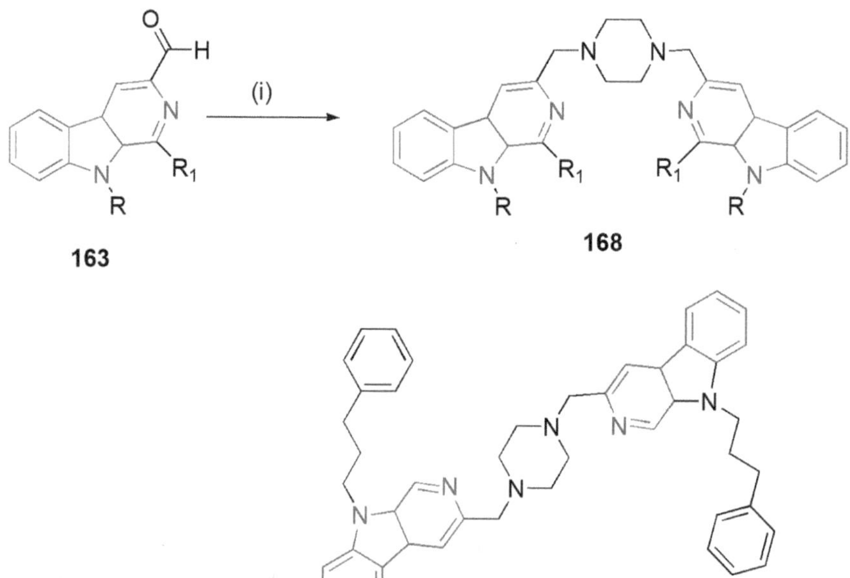

168a

IC_{50} = 3.02 μM

Reaction Conditions : (i) syn-dichloroethane, $NaBH_3CN$, rt.

SCHEME 1.51 Synthesis of piperazine-linked bivalent β-carbolines.

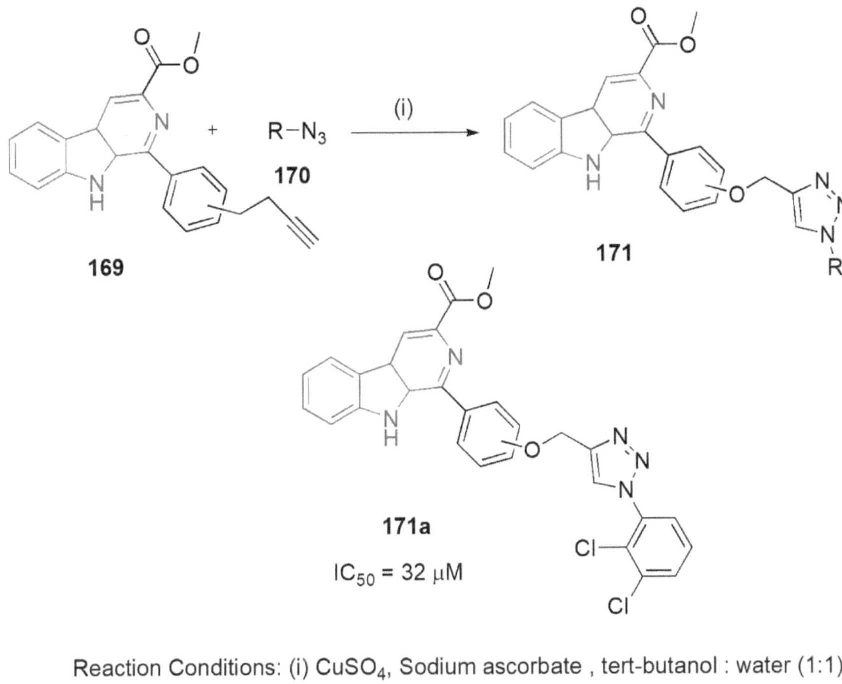

SCHEME 1.52 Synthesis of 1,2,3-trazole-tethered β-carbolines hybrids.

Salehi et al. described a methodical approach to the systematic preparation of H-1, 2, 3-triazole-tethered-β-carboline hybrids and assessed their potential for anticancer action (Salehi et al. 2016). The intended 1H-1,2,3-trizoletethered β-carboline hybrid **171** was produced by treating the o-propargylated β-carbolines **169** with a number of alkyl/aryl azides **170** by a Cu-promoted azidealkyne cycloaddition reaction (Scheme 1.52).

Using paclitaxel as a positive control, all the prepared compounds were tested against cancer cell lines. The most effective molecule out of these **171** was discovered to contain a fragment of 3,4-dichlorophenyl as azide, which had an IC_{50} value of 46 and 32 μM against the HeLa and HepG2 cancer cell lines, respectively.

Drug resistance is brought on by the efflux of medicines from cancer cells by the breast cell resistance protein (ABCG2/BRCP1). Spindler et al. developed ABCG2/BRCP1 as a result to prevent this outflow (Spindler et al. 2016). Using the Pictet–Spengler method, tryptamine or 5-hydroxytryptamine **172** reacts with various aldehydes **173** in the presence of dichloromethane and TFA to yield 1-substituted tertahydro-β-carbolines or 1,6-disubstituted tetrahydro-β-carbolines **174**. These subsequently react with substituted benzoyl chloride 175 using THF and TEA, yielding Tetrahydro-β-carbolines replaced with N-2-acyl-1 or N-2-acyl-1,6-disubstituted with N-2-acyl-1 **176** (Scheme 1.53).

Similar to Ko143 in terms of inhibitory activity, two compounds prefer BRCP1/ABCG2 over ABCB1 over ABCB1. The drugs also reversed SN-38 resistance

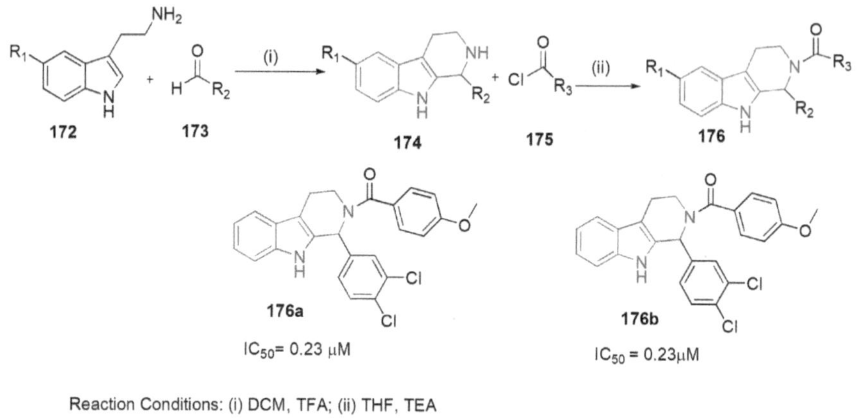

SCHEME 1.53 Synthesis of N-2-acyl-1,6-disubstituted tetrahydro-β-carbolines.

mediated by BRCP1/ABCG2, but the effect of Ko143 was less effective. The sample showed separate membranes with basal ATPase activity from Spodoptera frugiperda ovarian cells (Sf9) infected with a recombinant baculovirus expressing ABCG2/BRCP1.

1.4 CONCLUSIONS AND FUTURE PERSPECTIVES

Cancer has emerged as one of the leading causes mortalities worldwide. Various treatment modalities, such as chemotherapy, radiation therapy, and sometimes surgery, are employed to control the disease, depending on its type and severity. Despite a great deal of unneeded investigation that has been performed in the medical field, there are still numerous areas wherein researchers can make improvements to address this troubling clinical issue. For cancer treatments, combination therapy has become popular since it prevents resistance from growing and outperforms single agents in terms of effectiveness. Eliminating cancer progenitor cells should be the next step in the fight against cancer because they often exhibit drug resistance and have the potential to induce remission. Research needs to be done in this area in order to comprehend the basic mechanisms underlying cancer medication resistance and to identify drugs that can treat tumors without running the risk of developing resistance. In the interim, several novel approaches can be further investigated using currently available drugs. There are now a number of studies being published on the usage of β-carboline derivatives as possible cancer therapies. This chapter outlined the biological evaluation and synthesis of many β-carboline-based compounds, grouping them depending on the expected activities. According to observations collected, the approach employed in the referenced publications to analyze the anticancer effects against different cancer cell lines primarily involved molecular docking analysis. The interactions between anticancer motifs and cancer cells were also examined using contemporary techniques including UV-Vis study. There is potential for more research on cancer treatments using cutting-edge spectroscopic investigations to comprehend how the drugs function on target cells.

REFERENCES

Abdelsalam M. A., O. M. Aboul Wafa, E. S. Badawey, M. S. El-Shoukrofy, M. M. El-Miligy, and N. Gouda. 2018. Design and synthesis of some β-carboline derivatives as multi-target anticancer agents. *Future Med. Chem.* 10(24): 2791–2814.

Aiello A., E. Fattorusso, S. Magno, and L. Mayol. 1987. Brominated β-carbolines from the marine hydroid aglaophenia Plumalinnaeus. *Tetrahedron* 43:5929–5932.

Al-Allaf T. A. K., and L. J. Rashan. 1998. Synthesis and cytotoxic evaluation of the first trans-palladium(II) complex with naturally occurring alkaloid harmine. *Eur. J. Med. Chem.* 33: 817–820.

Al-Allaf T. A. K., M. T. Ayoub, and L. J. Rashan. 1990. Synthesis and characterization of novel biologically active platinum(II) and palladium(II) complexes of some β-carboline alkaloids. *J. Inorg. Biochem.* 38: 47–56.

Bai B., X. Y. Li, L. Liu, Y. Li, and H. J. Zhu. 2014. Syntheses of novel β-carboline derivatives and the activities against five tumor-cell lines. *Bioorg. Med. Chem. Lett.* 24: 96–98.

Blunt J. W., B. R. Copp, W. P. Hu, M. H. G. Munro, P. T. Northcote, and M. R. Prinsep. 2009. Marinenatural products. *Nat. Prod. Rep.* 26:170–244.

Boursereau Y., and I. Coldham. 2006. Synthesis and biological studies of 1-amino β-carbolines. *Bioorg. Med. Chem. Lett.* 14: 5841–5844.

Cain M., R. W. Weber, F. Guzman, J. M. Cook, S. A. Barker, K. C. Rice, J. N. Crawley, S. M. Paul, and P. Skolnick. 1982. β-Carbolines, Synthesis and neurochemical and pharmacological actions on brain benzodiazepine receptors. *J. Med. Chem.* 25: 1081–1091.

Cao R., Q. Chen, X. Hou, H. Chen, H. Guan, Y. Ma, W. Peng, and A. Xu. 2004. Synthesis, acute toxicities, and antitumor effects of novel 9-substituted β-carboline derivatives. *Bioorg. Med. Chem.* 12: 4613–4623.

Cao R., W. Fan, L. Guo, Q. Ma, G. Zhang, J. Li, X. Chen, Z. Ren, and L. Qiu. 2013. Synthesis and structure-activity relationships of harmine derivatives as potential antitumor agents. *Eur. J. Med. Chem.* 60: 135–143.

Cao R., W. Peng, Z. Wang, and A. Xu. 2007. β-Carboline alkaloids: Biochemical and pharmacological functions. *Curr. Med. Chem.* 14: 479–500.

Chatwichien J., S. Basu, M. E.Murphy, M. T. Hamann, and J. D.Winkler. 2015. Design, synthesis and biological evaluation of β-carboline dimers based on the structure of neokauluamine. *Tetrahedron Lett.* 56:3515–3517.

Chauhan S. S., A. K. Singh, S. Meena, M. Lohani, A. Singh, R. K. Arya, S. H. Cheruvu, J. Sarkar, J. R. Gayen, D. Datta, and P. M. Chauhan. 2014. Synthesis of novel β-carboline based chalcones with high cytotoxic activity against breast cancer cells. *Bioorg. Med. Chem. Lett.* 24:2820–2824.

Chen H., P. Gao, M. Zhang, W. Liao, and J. Zhang. 2014. Synthesis and biological evaluation of a novel class of β-carboline derivatives. *New J. Chem.* 38: 4155–4166.

Chen W., G. Zhang, L. Guo, W. Fan, Q. Ma, X. Zhang, R. Du, and R. Cao. 2016. Synthesis and biological evaluation of novel alkyl diamine linked bivalent β-carbolines as angiogenesis inhibitors. *Eur. J. Med. Chem.* 124: 249–261.

Chen Y. F, P. C. Kuo, H. H. Chan, I. J. Kuo, F. W. Lin, C. R. Su, M. L. Yang, D. T. Li, and T. S. Wu. 2010. β-Carboline alkaloids from *Stellariadichotomavar. lanceolata* and their anti-inflammatory activity. *J. Nat. Prod.* 73: 1993–1998.

Chen Y. F., Y. C. Lin, J. P. Chen, H. C. Chan, M. H. Hsu, H. Y. Lin, S. C. Kuo, and L. J. Huang. 2015. Synthesis and biological evaluation of novel 3, 9-substituted β-carboline derivatives as anticancer agents. *Bioorg. Med. Chem. Lett.* 25: 3873–3877.

Chen Y., M. Q. Qin, J. H. Wu, L. Wang, H. Chao, L. N. Ji, and A. L. Xu. 2013. Synthesis, characterization, and anticancer activity of ruthenium(II)- β-carboline complex. *Eur. J. Med. Chem.* 70: 120–129.

Cheng L, C. Wang, H. Liu, F. Wang, L. Zheng, J. Zhao, E.Chu, and X. Lin. 2012. A novel polypeptide extracted from Cionasavignyi induces apoptosis through a mitochondrial-mediated pathway in human colorectal carcinoma cells. *Clin. Colorectal Canc.* 11: 207–214.

Cuny G. D., N. P. Ulyanova, D. Patnaik, J. F. Liu, X. Lin, K. Auerbach, S. S. Ray, J. Xian, M. A. Glicksman, R. L. Stein, and J. M. G. Higgins. 2012. Structure-activity relationship study of β-carboline derivatives as haspin kinase inhibitors. *Bioorg. Med. Chem. Lett.* 22: 2015–2019.

Dai J., W. Dan, U. Schneider, and J. Wang. 2018. β-Carboline alkaloid monomers and dimers: Occurrence, structural diversity, and biological activities. *Eur. J. Med. Chem.* 157: 622–656.

deOliveira Figueiredo P, R. T. Perdomo, F. R. Garcez, M. deFatima Cepa Matos, J. E. deCarvalho, and W. S. Garcez. 2014. Further constituents of *Galianthethalictroides* (Rubiaceae) and inhibition of DNA topoisomerases I and II α by its cytotoxic β-carboline alkaloids. *Bioorg. Med. Chem. Lett.* 24: 1358–1361.

Devi N., S. Kumar, S. K. Pandey, and V. Singh. 2018. 1(3)-Formyl-β-carbolines: Potential aldo-X precursors for the synthesis of β-carboline-based molecular architectures. *Asian J. Org. Chem.* 7: 6–36.

Dighe S. U., S. Khan, I. Soni, P. Jain, S. Shukla, R. Yadav, P. Sen, S. M. Meeran, and S. Batra. 2015. Synthesis of β-carboline-based N-heterocyclic carbenes and their antiproliferative and anti-metastatic activities against human breast cancer cells. *J. Med. Chem.* 58(8): 3485–3499.

Digwal C. S., U. Yadav, A. P. Sakla, P. V. SriRamya, S. Aaghaz, and A. Kamal. 2016. $VOSO_4$ catalyzed highly efficient synthesis of benzimidazoles, benzothiazoles, and quinoxalines. *Tetrahedron Lett.* 57:4012–4016.

Du H., H. Gu, N. Li, and J. Wang. 2016. Synthesis and biological evaluation of bivalent β-carbolines as potential anticancer agents. *Med. Chem. Commun.* 7:636–645. DOI: 10.1039/C5MD00581G.

Funayama Y., K. Nishio, K. Wakabayashi, M. Nagao, K. Shimoi, T. Ohira, S. Hasegawa, and N. Saijo. 1996. Effects of β- and γ-carboline derivatives on DNA topoisomerase activities. *Mutat. Res.* 349: 183–191.

Gao G., X. Yi, R. Huang, F. Yan, B. He, and B. Chen. 2013. Alkaloids from corals. *Chem. Biodivers.* 10: 1435–1447.

Gooyit M., N. Tricoche, S. Javor, S. Lustigman, and K. D. Janda. 2015. Exploiting the polypharmacology of ß-carbolines to disrupt O. volvulus molting. *ACS Med. Chem. Lett.* 6: 339–343.

Grella B., M. Dukat M, R. Young, M. Teitler, K. Herrick-Davis, C. B. Gauthier, and R. A. Glennon. 1998. Investigation of hallucinogenic and related β-carbolines. *Drug Alcohol Depend.* 50: 99–107.

Gu H., N. Li, J. Dai, Y. Xi, S. Wang, and J. Wang. 2018. Synthesis and in vitro antitumor activity of novel bivalent β-carboline-3-carboxylic acid derivatives with DNA as a potential target. *Int. J. Mol. Sci.* 19: 3179-1-16

He L., S. Y. Liao, C. P. Tan, Y. Y. Lu, C. X. Xu, L. N. Ji, and Z. W. Mao. 2014. Cyclometalated iridium(III)-β-carboline complexes as potent autophagy-inducing agents. *Chem. Commun. (J. Chem. Soc. Sect. D).* 50:5611–5614.

Herraiz T., and C. Chaparro. 2005. Human monoamine oxidase is inhibited by tobacco smoke: β-carboline alkaloids act as potent and reversible inhibitors. *Biochem. Biophys. Res. Commun.* 326: 378–386.

Housman G., S. Byler, S. Heerboth, K. Lapinska, M. Longacre, N. Snyder, and S. Sarkar. 2014. Drug resistance in cancer: An overview. *Cancers (Basel)* 6: 1769–1792. www.who.int/news-room/fact-sheets/detail/cancer

Huber K., L. Brault, O. Fedorov, C. Gasser, P. Filippakopoulos, A. N. Bullock, D. Fabbro, J. Trappe, J. Schwaller, S. Knapp, and F. Bracher. 2012. 7,8-Dichloro-1-oxo-β-carbolines as a versatile scaffold for the development of potent and selective kinase inhibitors with unusual binding modes. *J. Med. Chem.* 55: 403–413.

Ikeda R., T. Kimura, T. Tsutsumi, S. Tamura, N. Sakai, and T. Konakahara. 2012. Structure-activity relationship in the antitumor activity of 6-, 8- or 6,8- substituted 3-benzylamino-β-carboline derivatives. *Bioorg. Med. Chem. Lett.* 22: 3506–3515.

Jadala C., M. Sathish, T. S. Reddy, V. G. Reddy, R. Tokala, S. K. Bhargava, N. Shankaraiah, N. Nagesh, and A. Kamal. 2019. Synthesis and in vitro cytotoxicity evaluation of β-carboline-combretastatincarboxamides as apoptosis inducing agents: DNA intercalation and topoisomerase-II inhibition. *Bioorg. Med. Chem.* 27: 3285–3298.

Jiang W. Q., C. Charlet-Fagnere, J. Sapi, J. Y. Laronze, P. Renard, B. Pfeiffer, and S. Leonce. 2002. Cytotoxic bis-3,4-dihydro-β-carbolines and bis-β-carbolines. *J. Enzym. Inhib. Med. Chem.* 17: 369–374.

Jin Q. M., Y. Lu, J. L. Jin, H. Guo, G. W. Lin, Y. Wang, and T. Lu. 2014. Synthesis, characterization, DNA binding ability and cytotoxicity of the novel platinum(II), copper(II), cobalt(II) and nickel(II) complexes with 3-(1H-benzo[d]imidazol2-yl)-β-carboline. *Inorg. Chim. Acta.* 421: 91–99.

Kaijima M., L. DaCosta-Rochette, R H. Dodd, J. Rossier, and R. Naquet. 1984. Hypnotic action of ethyl β-carboline-3-carboxylate, a benzodiazepine receptor antagonist, in cats. *Electroencephalogr. Clin. Neurophysiol.* 58: 277–281.

Kamal A., M. P. NarasimhaRao, P. Swapna, V. Srinivasulu, C. Bagul, A. B. Shaik, K. Mullagiri, J. Kovvuri, V. S. Reddy, K. Vidyasagar, and N. Nagesh. 2014. Synthesis of β-carbolinee-benzimidazole conjugates using lanthanum nitrate as a catalyst and their biological evaluation. *Org. Biomol. Chem.*12: 2370–2387.

Kamal A., M. Sathish, V. L. Nayak, V. Srinivasulu, B. Kavitha, Y. Tangella, D. Thummuri, C. Bagul, N. Shankaraiah, and N. Nagesh. 2015. Design and synthesis of dithiocarbamate linked β-carboline derivatives: DNA topoisomerase II inhibition with DNA binding and apoptosis inducing ability. *Bioorg. Med. Chem.* 23:5511–5526.

Kim D. C., T. Q. Quang, C. S. Yoon, N. T. T. Ngan, S. I. Lim, S. Y. Lee, Y. C. Kim, and H. Oh. 2016. Anti-neuroinflammatory activities of indole alkaloids from kanjang (Korean fermented soy source) in lipopolysaccharide-induced BV2 microglial cells. *Food Chem.* 213: 69–75.

Kovvuri J., B. Nagaraju, V. L. Nayak, R. Akunuri, M. P. N. Rao, A. Ajitha, N. Nagesh, and A. Kamal. 2018. Design, synthesis and biological evaluation of new β-carboline-bisindole compounds as DNA binding, photocleavage agents and topoisomerase I inhibitors. *Eur. J. Med. Chem.* 143: 1563–1577.

Laine A. E., C. Lood, and A. M. P. Koskinen. 2014. Pharmacological importance of optically active tetrahydro-β-carbolines and synthetic approaches to create the C1 stereocenter. *Molecules* 19: 1544–1567.

Lamounier A. P., N. S. Mateus, A. L. M. C. daCunha, A. S. Luna, and R. Q. Aucelio. 2015. Determination of six β-carboline alkaloids in urine and phytotherapic extracts using micellar liquid chromatography with fluorimetric detection. *J. Liq. Chromatogr. Relat. Technol.* 38: 997–1006.

Lee K. I., M. J. Kim, H. Koh, J. I. Lee, S. Namkoong, W. K. Oh, and J. Park. 2015. The antihypertensive drug reserpine induces neuronal cell death through inhibition of autophagic flux. *Biochem. Biophys. Res. Commun.* 462: 402–408.

Ling Y, W. J. Gao, C. Ling, J. Liu, C. Meng, J. Qian, S. Liu, H. Gan, H. Wu, J Tao, H. Dai, and Y. Zhang. 2019. β-Carboline and N-hydroxycinnamamide hybrids as anticancer agents for drug-resistant hepatocellular carcinoma. *Eur. J. Med. Chem.* 168: 515–526.

Ling Y., C. Xu, L. Luo, J. Cao, J. Feng, Y. Xue, Q. Zhu, C. Ju, F. Li, Y. Zhang, Y. Zhang, and X. Ling. 2015. Novel β-carboline/hydroxamic acid hybrids targeting both histone deacetylase and DNA display high anticancer activity via regulation of the p53 signaling pathway. *J. Med. Chem.* 58: 9214–9227.

Liu J., T. Wang, X. Wang, L. Luo, J. Guo, Y. Peng, Q. Xu, J. Miao, Y. Zhang, and Y. Ling. 2017. Development of novel β-carboline-based hydroxamate derivatives as HDAC inhibitors with DNA damage and apoptosis inducing ability. *Med. Chem. Commun.,* 8: 1213–1219.

Liu L., Y. Y. Xu, Z. Q. Yang, J. N. Xiang, and G. Y. Xu. 2012. Synthesis and cytotoxic activity of 3-phenyl-4-substituted-β-carbolines. *Chin. Chem. Lett.* 23: 1230–1232.

Lo K. K. W., and K. Y. Zhang. 2012. Iridium(III) complexes as therapeutic and bioimaging reagents for cellular applications. *RSC Adv.* 2: 12069–12083.

Lu X., X. Pan, Y. Yang, M. Ji, X. Chen, Z. Xiao, and Z. Liu. 2017. Synthesis and cytotoxicity of a novel series of saframycin-ecteinascidin analogs containing tetrahydro β- carboline moieties. *Eur. J. Med. Chem.* 135: 260–269.

Lunagariya N. A., V. M. Gohil, V. Kushwah, S. Neelagiri, S. Jain, S. Singh, and K. K. Bhutani. 2016. Design, synthesis and biological evaluation of 1,3,6- trisubstitutedβ-carboline derivatives for cytotoxic and anti-leishmanial potential. *Bioorg. Med. Chem. Lett.* 26: 789–794.

Luo J., N. L. Solimini, and S. J. Elledge. 2009. Principles of cancer therapy: Oncogene and non-oncogene addiction. *Cell* 136: 823–837.

Marques M. R., M. A. Mendes, C. F. Tormena, B. M. Souza, L. M. Marcondes Cesar, R. Rittner, and M. S. Palma. 2005. Structure determination of a tetrahydro-β-carboline of arthropod origin: A novel alkaloid-toxin subclass from the web of spider Nephilaclavipes. *Chem. Biodivers.* 2:525–534.

Montorsi F., G. Brock, J. U. Stolzenburg, J. Mulhall, I. Moncada, H. R. H. Patel, D. Chevallier, K. Krajka, C. Henneges, R. Dickson, and H. Büttner. 2014. Effects of tadalafil treatment on erectile function recovery following bilateral nervesparing radical prostatectomy: A randomised placebo-controlled study (REACTT). *Eur. Urol.* 65: 587–596.

Nagula S., S. Pankaj, P. Srinivas, N. Shalini, S. Vunnam, K. P. Niggula, and K. Ahmed. 2018. Synthesis of novel C3-linked β-carboline-pyridine derivatives employing KrÖhnkereaction: DNA-binding ability and molecular modeling studies. *Lett. Drug Des. Discov.* 13: 335–342.

NazariFormagio A. S., P. R. Santos, K. Zanoli, T. Ueda-Nakamura, L. T. DüsmanTonin, C. V. Nakamura, and M. H. Sarragiotto. 2009. Synthesis and antiviral activity of β-carboline derivatives bearing a substituted carbohydrazide at C-3 against poliovirus and herpes simplex virus (HSV-1). *Eur. J. Med. Chem.* 44: 4695–4701.

Ohishi K., K. Toume, M. A. Arai, T. Koyano, T. Kowithayakorn, T. Mizoguchi, M. Itoh, and M. Ishibashi. 2015. 9-Hydroxycanthin-6-one, a β-carboline alkaloid from *Eurycomalongifolia*, is the first wnt signal inhibitor through activation of glycogen synthase kinase 3β without depending on casein kinase 1α. *J. Nat. Prod.* 78: 1139–1146.

Park Y. W., and M. S. Nam. 2015. Bioactive peptides in milk and dairy products: A review. *Korean J. Food Sci. Anim. Resour.* 35: 831–840.

Petersen E. N., G. Paschelke, W. Kehr, M. Nielsen, and C. Braestrup. 1982. Does the reversal of the anticonflict effect of phenobarbital by b-CCE and FG 7142 indicate benzodiazepine receptor-mediated anxiogenic properties? *Eur. J. Pharmacol.* 82: 217–221.

Pfau,W., and K. Skog. 2004. Exposure to β-carbolinesnorharman and harman, *J. Chromatogr. B.* 802: 115–126.

Pogosyan S. A., N. P. Grigoryan, and R. G. Paronikyan. 2007. Synthesis and anticonvulsant activity of dihydrochlorides of indoline-30spiro-1-(1, 2, 3, 4-tetrahydro)- β-carboline derivatives. *Pharm. Chem. J.* 41: 527–528.

Porta A, A. M. Petrone, S. Morello, I. Granata, F. Rizzo, D. Memoli, A. Weisz, and B. Maresca. 2017. Design and expression of peptides with antimicrobial activity against *Salmonella typhimurium. Cellular Microbiology,* 19(2): e12645.

Rinehart K. L., J. Kobayashi, G. C. Harbour, R. G. Hughes, S. A. Mizsak, T. A. Scahill, and E. K. Eudistomins. 1984. C, L, Potent antiviral compounds containing a novel oxathiazepine ring from the Caribbean tunicate Eudistomaolivaceum. *J. Am. Chem. Soc.*106:1524–1526.

Rundfeldt C., and W. LÖscher. 2014. The pharmacology of imepitoin: The first partial benzodiazepine receptor agonist developed for the treatment of epilepsy. *CNS Drugs* 28: 29–43.

Salehi P., K. Babanezhad-Harikandei, M. Bararjanian, A. Al-Harrasi, M. A. Esmaeili, and A. Aliahmadi. 2016. Synthesis of novel 1, 2, 3-triazole tethered 1, 3- disubstituted β-carboline derivatives and their cytotoxic and antibacterial activities. *Med. Chem. Res.* 25: 1895–1907.

Sathish M., B. Kavitha, V. L. Nayak, Y. Tangella, A. Ajitha, S. Nekkanti, A. Alarifi, N. Shankaraiah, N. Nagesh, and A. Kamal. 2018. Synthesis of podophyllotoxin linked β-carboline congeners as potential anticancer agents and DNA topoisomerase II inhibitors. *Eur. J. Med. Chem.* 144: 557–571.

Savariz F. C., A. S. N. Formagio, V. A. Barbosa, M. A. Foglio, J. E. deCarvalho, M. C. T. Duarte, B. P. D. Filho, and M. H. Sarragiotto. 2010. Synthesis, Antitumor and Antimicrobial Activity of Novel 1-Substituted Phenyl-3-[3-alkylamino (methyl)-2-thioxo-1,3,4-oxadiazol-5-yl] β-Carboline Derivatives. *J. Braz. Chem. Soc.*, 21(2): 288–298.

Savariz F. C., M. A. Foglio, J. E. DeCarvalho, A. L. T. G. Ruiz, M. C. T. Duarte, M. F. DaRosa, E. Meyer, and M. H. Sarragiotto. 2012. Synthesis and evaluation of new β-carboline-3-(4-benzylidene)-4H-oxazol-5-one derivatives as antitumor agents. *Molecules* 17: 6100–6113.

Savi D. C., K. A. Shaaban, N. Vargas, L. V. Ponomareva, Y. M. Possiede, J. S. Thorson, C. Glienke, and J. Rohr. 2015. Microbispora sp. LGMB259 endophyticactinomycete isolated from Vochysiadivergens (Pantanal, Brazil) producing bcarbolines and indoles with biological activity. *Curr. Microbiol.* 70: 345–354.

Schupp P., T. Poehner, R. Edrada, R. Ebel, A. Berg, V. Wray, and P. Proksch. 2003. W. and X. Eudistomins, Two new β-carbolines from the micronesian tunicate *Eudistomasp. J. Nat. Prod.* 66: 272–275.

Shankaraiah N., S. Nekkanti, K. J. Chudasama, K. R. Senwar, P. Sharma, M. K. Jeengar, V. G. Naidu, V. Srinivasulu, G. Srinivasulu, and A. Kamal. 2014. Design, synthesis and anticancer evaluation of tetrahydro-β-carboline-hydantoin hybrids, *Bioorg. Med. Chem. Lett.*24:5413–5417.

Shankaraiah N., C. Jadala, S. Nekkanti, K. R. Senwar, N. Nagesh, S. Shrivastava, V. G. M. Naidu, M. Sathish, and A. Kamal. 2016. Design and synthesis of C3-tethered 1, 2, 3-triazolo-β-carboline derivatives: Anticancer activity, DNA-binding ability, viscosity and molecular modeling studies. *Bioorg. Chem.* 64: 42–50.

Shen Y. C., C. Y. Chen, P. W. Hsieh, C. Y. Duh, Y. M. Lin, and C. L. Ko. 2005. The preparation and evaluation of 1-substituted 1, 2, 3, 4-tetrahydro- and 3, 4-dihydroβ-carboline derivatives as potential antitumor agents. *Chem. Pharm. Bull.* 53: 32–36.

Shi B., R. Cao, W. Fan, L. Guo, Q. Ma, X. Chen, G. Zhang, L. Qiu, and H. Song. 2013. Design, synthesis and in vitro and in vivo antitumor activities of novel bivalent β-carbolines. *Eur. J. Med. Chem.* 60: 10–22.

Skropeta D., and L. Wei. 2014. Recent advances in deep-sea natural products. *Nat. Prod. Rep.* 31:999–1025.

Smirnova O. B., T. V. Golovko, and V. G. Granik. 2011. Carbolines. Part 2: Comparison of some of the properties of α-, α-, and δ-carbolines (review). *Pharm. Chem. J.* 45:389–400.

Smith K. L., G. K. Ford, D. S. Jessop, and D. P. Finn. 2013. Behavioural, neurochemical and neuroendocrine effects of the endogenous β-carbolineharmane in fear-conditioned rats. *J. Psychopharmacol.* 27: 162–170.

Sobhani A. M., S. A. Ebrahimi, and M. Mahmoudian. 2002. An in vitro evaluation of human DNA topoisomerase I inhibition by *PeganumharmalaL.* seeds extract and its β-carboline alkaloids. *J. Pharm. Pharmaceut. Sci.* 5: 19–23.

Spindler A., K. Stefan, and M. Wiese. 2016. Synthesis and investigation of tetrahydro-β-carboline derivatives as inhibitors of the breast cancer resistance protein (ABCG2). *J. Med. Chem.* 59:6121–6135.

Srivastava S. K., A. Agarwal, P. M. Chauhan, S. K. Agarwal, A. P. Bhaduri, S. N. Singh, N. Fatima, and R. K. Chatterjee. 1999. Potent 1,3-disubstituted-9H-pyrido [3,4-b]indoles as new lead compounds in antifilarial chemotherapy. *Bioorg. Med. Chem.* 7: 1223–1236.

Stohler R., H. Rommelspacher, and D. Ladewig. 1995. The role of β-carbolines (harman/norharman) in heroin addicts. *Eur. Psychiatr.* 10: 56–58.

Suarez-Jimenez G. M., A. Burgos-Hernandez, and J. M. Ezquerra-Brauer. 2012. Bioactive peptides and depsipeptides with anticancer potential: Sources from marine animals. *Mar. Drugs* 10: 963–986.

Sun R., R. Liu, C. Zhou, Z. Ren, L. Guo, Q. Ma, W. Fan, L. Qiu, H. Yu, G. Shao, and R. Cao. 2015. Synthesis and biological evaluation of piperazine group-linked bivalent β-carbolines as potential antitumor agents. *MedChemComm* 6: 2170–2174.

Szakacs G., J. K. Paterson, J. A. Ludwig, C. Booth-Genthe, and M. M. Gottesman. 2006. Targeting multidrug resistance in cancer. *Nat. Rev. Drug Discov.* 5: 219–234.

Tanaka N., R. Momose, A. Takahashi-Nakaguchi, T. Gonoi, J. Fromont, and J. I. Kobayashi. 2014. Hyrtimomines, indole alkaloids from Okinawan marine sponges *Hyrtiosspp. Tetrahedron* 70: 832–837.

Till M., and M. R. Prinsep. 2009. 5-Bromo-8-methoxy-1-methyl-β-carboline, an alkaloid from the New Zealand marine bryozoan *Pterocellavesiculosa. J. Nat. Prod.* 72: 796–798.

Tokala R., S. Thatikonda, S. Sana, P. Regur, C. Godugu, and N. Shankaraiah. 2018. Synthesis and in vitro cytotoxicity evaluation of β-carboline-linked 2,4-thiazolidinedione hybrids: Potential DNA intercalation and apoptosis-inducing studies. *New J. Chem.*, 42: 16226–16236.

Toshima K., Y. Okuno, Y. Nakajima, and S. Matsumura. 2002. β-Carbolinee Carbohydrate hybrids: Molecular design, chemical synthesis and evaluation of novel DNA photocleavers. *Bioorg. Med. Chem. Lett.* 12: 671–673.

Tsuchiya H. 2016. Anesthetic effects changeable in habitual drinkers: Mechanistic drug interactions with neuro-active indoleamine-aldehyde condensation products associated with alcoholic beverage consumption. *Med. Hypotheses* 92: 62–66.

Wang K. B., C. M. Yuan, C. M. Xue, D. H. Li, Y. K. Jing, H. P. He, X. J. Hao, Y. T. Di, Z. L. Li, H. M. Hua. 2014. A. and B. Pegaharmalines, Two novel β-carboline alkaloids with unprecedented carbon skeletons from *Peganumharmala. RSC Adv.* 4: 53725–53729.

Wang L., C. Dong, X. Li, W. Han, and X. Su. 2017. Anticancer potential of bioactive peptides from animal sources. *Oncol. Rep.* 38: 637–651.

Wesson, K. J., M. T. Hamann, and A. Keenamide. 1996. A bioactive cyclic peptide from the marine mollusk Pleurobranchusforskalii. *J. Nat. Prod.* 59: 629–631.

Xiao S., W. Lin, C. Wang, and M. Yang. 2001. Synthesis and biological evaluation of DNA targeting flexible side-chain substituted β-carboline derivatives. *Bioorg. Med. Chem. Lett.* 11: 437–441.

Xin B., W. Tang, Y. Wang, G. Lin, H. Liu, Y. Jiao, Y. Zhu, H. Yuan, Y. Chen, and T. Lu. 2012. Design, synthesis and biological evaluation of β-carboline derivatives as novel inhibitors targeting B-Raf kinase. *Bioorg. Med. Chem. Lett.* 22: 4783–4786.

Yang J. M., Y. H. Zhu, S. Chen, X. Lu, Y. M. Wu, F. E. Ma, L. P. Li, Y. Yang, Z. H. Shi, K. Y. Huang, X. Hong, and P. Jiang. 2018. A β-carboline derivative-based nickel (II) complex as a potential antitumor agent: Synthesis, characterization, and cytotoxicity. *Med. Chem. Commun.* 9: 100–107.

Yu X., W. Lin, J. Li, and M. Yang. 2004. Synthesis and biological evaluation of novel β-carboline derivatives as Tat-TAR interaction inhibitors. *Bioorg. Med. Chem. Lett.* 14: 3127–3130.

Zhang M., E. J. Park, T. P. Kondratyuk, J. M. Pezzuto, and D. Sun. 2018. Synthesis and structure–activity relationships of tetrahydro-β-carboline derivatives as anticancer and cancer-chemopreventive Agents. *Anticancer Research* 38: 4425–4433.

Zhao M., L. Bi, W. Wang, C. Wang, M. Baudy-Floc'h, J. Ju, and S. Peng. 2006. Synthesis and cytotoxic activities of β-carboline amino acid ester conjugates. *Bioorg. Med.Chem.* 14: 6998–7010.

Zheng C., Y. Fang, W. Tong, G. Li, H. Wu, W. Zhou, Q. Lin, F. Yang, Z. Yang, P. Wang, Y. Peng, X. Pang, Z. Yi, J. Luo, M. Liu, and Y. Chen. 2014. Synthesis and biological evaluation of novel tetrahydro- β-carboline derivatives as antitumor growth and metastasis agents through inhibiting the transforming growth factor-β signaling pathway. *J. Med. Chem.* 57: 600–612.

Zhu S., X. Chen, W. Chen, Q. Ma, M. Li, W. Fan, J. Zhang, and L. Guo. 2022. Multicomponent synthesis of novel β-carboline fused imidazolium derivatives via the Mannich reaction: Cytotoxicity, molecular docking, and mechanistic studies as angiogenesis inhibitors. *New J. Chem.* 46: 4427–4435.

2 β-Carbolines as Antibacterial Agents
Synthesis and Biological Study

Shivendra Singh and Shivangi Sharma

2.1 INTRODUCTION

The pharmacological activities of the carboline family of heterocyclic compounds are unparalleled. A number of new antibiotics have been developed by the pharmaceutical industry, but this has only served to increase microbial resistance (William et al.1998). Traditional medicines have long been promoted by the World Health Organization (WHO) as safe treatments for bacterial and nonbacterial diseases (Céliz et al. 2011). As a result, novel approaches to microbial control must be considered (Ramesh & Hyma 1981). Due to the extensive number of applications in both the biological and pharmaceutical fields, the group of nitrogen-containing heterocycles that contains quinolines and carbolines is the most significant (Sharma et al. 2023; Sharma and Singh 2022; Sharma and Singh 2023; Sharma et al 2023; Sharma et al. 2023; Sharma et al. 2023 Sharma et al. 2023, Kathal et al, 2023). Anti-HIV (Laine et al. 2014; Venkataramana Reddy et al. 2018), antimicrobial (Zhang et al. 2015), antimalarial (Quintana et al. 2016), antileishmanial (Lunagariya et al. 2016), antifungal (Olmedo et al. 2017), and antitumoral (Samundeeswari et al. 2017; Luo et al. 2021) properties have been found in C-1 substituted carbolines (Figure 2.1). The potent antitumor and antimicrobial activities of β-carboline alkaloids have recently sparked interest in them.

Carbolines are a diverse class of nitrogen-containing cyclic alkaloids, and good places to look for them include insects, plants, microorganisms, marine organisms, mammalian tissues, and physiological fluids (Cao et al. 2007). Carbolines are a type of secondary heterocyclic amine that typically consist of an indole ring that has been fused to a pyridine ring. It has been hypothesized that the amino acids tryptamine and tryptophan are involved in the production of beta-carboline in some capacity (Maresh et al. 2008). Nearly 140 distinct β-carboline with a wide range of structures have been discovered in nature thus far. Traditional medicines in many countries make use of the β-carboline harman, while P. harmala species are rich in the more potent β-carboline norharman (Moloudizargari et al. 2013; Stoic 1999). Inhibition of microbial growth on antibiotic-resistant strains has been linked to the presence of multiple carbolines,

DOI: 10.1201/9781351058032-2

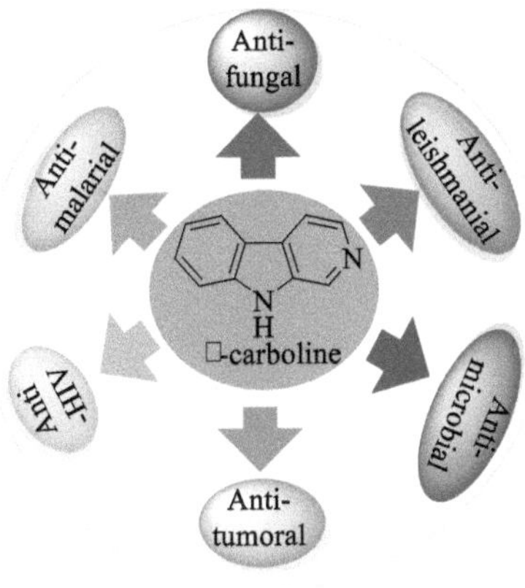

FIGURE 2.1 General Activities of β-carbolines.

which paves the way for the development of new antimicrobials based on carboline (Shin et al. 2010). The cyanobacterium N. harveyana also produce β-carboline norharmane in quantitative amount (Volk 2005). Nostocarboline was isolated from Nostoc species (Becher et al. 2005). An ascidian, Didemnum sp., was used to isolate a rare N-N-coupled β-carboline dimer (Kearns and Rideout 2008; Koehn and Carter 2005; Mishra and Tiwari 2011; Chen et al. 2022).

2.2 OVERVIEW OF BETA-CARBOLINE ALKALOIDS AND THEIR ANTIBACTERIAL PROPERTIES

Marine tunicates of the genus Eudistoma have been the subject of extensive chemical and biological study. Many β-carboline containing alkaloids have been isolated from various Eudistoma species since Wang and his colleagues first reported their structures (Figure 2.2). Biosynthesis of these β-carbolines is thought to involve the coupling of tryptophan with a second amino acid, as demonstrated by in vivo antimicrobial studies with *E. olivaceum*. New natural metabolites with a β-carboline base have been isolated from the tunicate Eudistoma. These metabolites have been given the names Eudistomins Y 1-7. These new metabolites were previously isolated from marine metabolites, and the benzoyl group is attached to the β-carboline nucleus at the C-1 position. Eudistomins Y 1-7 have the potential to inhibit the growth of bacteria; however, only Eudistomin Y 6 displays a trace amount of antimicrobial activity against Gram-positive microbes S. epidermis and B. subtilis at concentrations of 100 μM without causing cytotoxicity (Wang et al. 2008).

Natural β-Carbolines come from a wide variety of sources, including plants (Zhou et al. 1998), fish (Cabrera and Seldes 1999), insects (Kotanen et al. 2003), and mammals, and are synthesized with varying degrees of aromaticity (Beck and Lundman 1983). Interacting with benzodiazepine receptors (Lippke et al. 1983), intercalating into DNA (Csányi et al. 2000), and inhibiting CDK (Song et al. 2002) and topoisomerase (Deveau et al. 2001) are just a few examples of the interesting

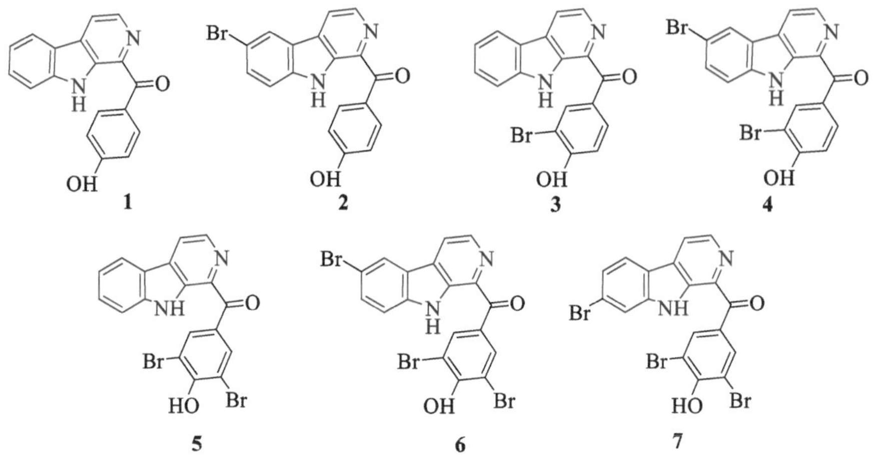

FIGURE 2.2 Structures of Eudistomins-Y and its analogues.

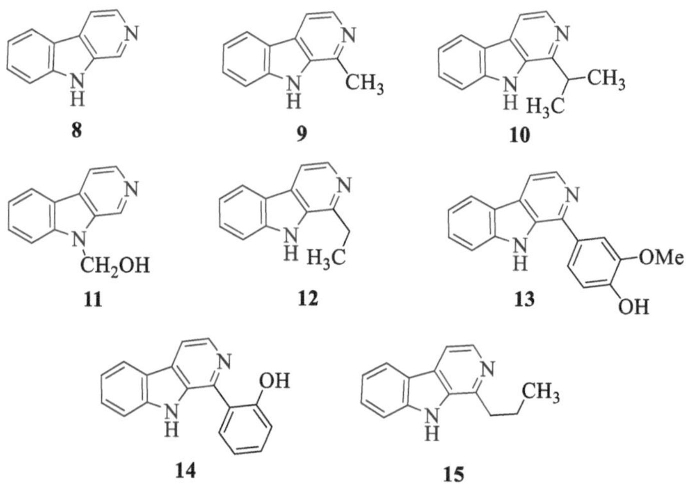

FIGURE 2.3 β-carboline core with diverse substituents.

pharmaceutical properties of most β-carbolines that have garnered a lot of attention in recent years (Deveau et al. 2001). Antibacterial activity was high for both Gram-positive and Gram-negative bacteria for compounds 8 and 9. With inhibition zones of 21 mm and 19 mm against B. cereus, compounds 10 and 11 displayed potent antimicrobial activity. Against both Gram-positive and Gram-negative bacteria, only compounds 12, 13, and 14 showed any significant activity. The carbon chain length resulted in a decline in activity when compared to saturated aliphatic aldehyde derivatives (8 and 10). Isometric β-carboline with a branched side chain is more active than the unbranched counterpart because its activity at position 10 was higher than that at position 15 (Figure 2.3) (Li et al. 2010).

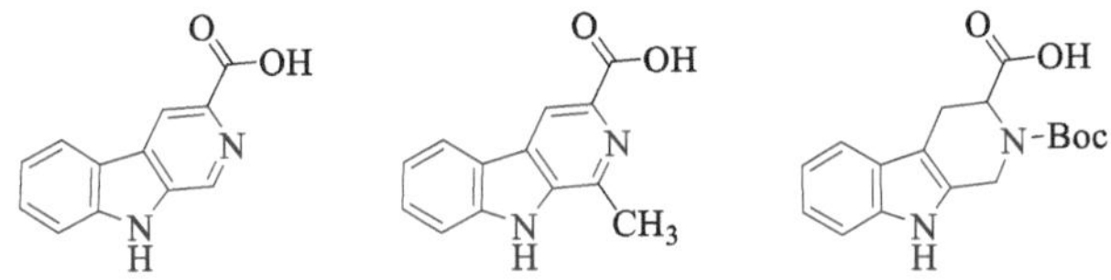

16 **17** **18**

FIGURE 2.4 Structures of Neamine-carboline conjugates.

To date, aminoglycosides have been the only class of antibiotics that can kill both Gram-negative and Gram-positive bacteria in aerobic environments. Aminoglycosides have been shown to readily bind with RNA targets. In addition to this includes both r-RNA and t-RNA, m-RNA (Woodcock et al. 1991; Fourmy et al. 1996; Recht et al. 1996; Spahn and Prescott 1996; Carter et al. 2000; Ye and Zhang 2002). Because aminoglycosides are not selective toward a variety of RNA targets, they are toxic and cause unwanted side effects. As a result, they are not the best choice for use as antibiotics (Woodcock et al. 1991). By coupling neamine and β-carboline-3-carboxylic acids using an aliphatic diamine as a linker, three distinct neamine-carboline conjugates were able to be successfully synthesized in high yields. Some of the conjugates studied had higher binding affinities than neamine for both 16S r-RNA and 18S r-RNA, as determined by SPR (surface plasmon resonance) experiments. Results from testing synthetic compounds' antibacterial activity in vitro showed that some of them were more effective than neamine. Current experimental data suggest that synthetic neamine-carboline conjugates may serve as useful pharmaceuticals. Their antimicrobial efficacy was then evaluated in vitro using a reference strain of Pseudomonas aeruginosa. Results showed that compounds 16, 17, and 18 had higher antibacterial activities than neamine. Specifically, compound 17 showed the most promise (Figure 2.4) (Wu et al. 2010).

A dimeric analogue of N^2-Bn salt showed high toxicity against many microorganisms, demonstrating the inhibitory effects of β-carbolines. These Acinetobacter baumannii bacteria were killed by the salt:E. coli ATCC 25923, Candida albicans, and GAI 07545 (Figure 2.5). For S. aureus specifically, MICs (minimum inhibitory concentrations) of N^2-Bn salt ranged from 0.01 to 0.05 mol/mL. The study found that the MIC for ciprofloxacin against S. aureus GAI 10152 and GAI 10153 was 0.002 mol/mL.

Tobacco's natural β-carboline, harmane 23, exhibits weak antibacterial activity. Kobayashi et al. found that xestomanzamine A 24, which they isolated from the Okinawan marine sponge Xestospongia sp., was cytotoxic against KB cell lines (Kobayashi et al. 1995; Tsukiyama et al. 2002). Antimicrobial and low in phototoxicity, Eudistomin T 25 is isolated from the Eudistoma olivaceum. Intriguing cytotoxic 1-benzylidine-N^2-benzylated-carbolinium bromides 26 and 27 were synthesized by Cao's team (Venkataramana Reddy et al. 2018; Cao et al. 2007). The two most common members of this class, harman (H) and norharman (NH), share structural similarities with nonpolar heterocyclic aromatic amines that are byproducts of the pyrolysis of proteins and amino acids (Figure 2.6).

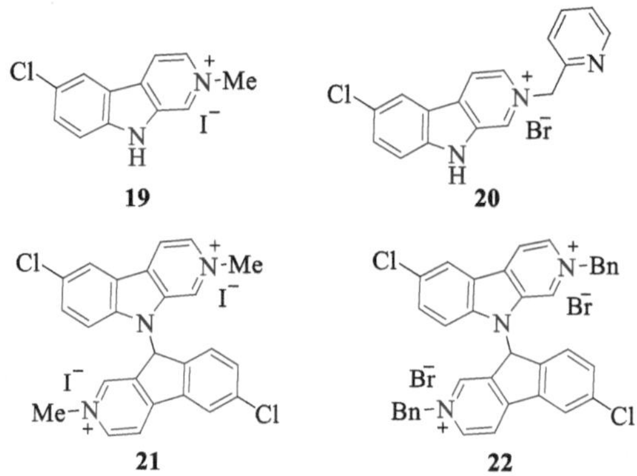

FIGURE 2.5 Structures of Beta-carboline having anti-microbial activity.

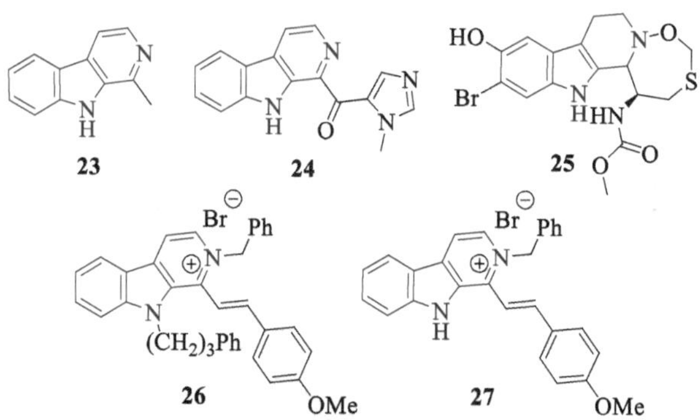

FIGURE 2.6 Pharmacologically interesting β-carbolines.

Harmane is selective for the imidazoline receptor over the 2-adrenoceptor, with an IC_{50} of 30 nM vs. 18 nM, respectively. Further, the harmane is an extremely effective and selective inhibitor of monoamine oxidase (MAO). Harmane's comutagenicity and ability to facilitate mutation are two of its more intriguing properties. The data shows that this β-carboline is not cytotoxic to African green monkey kidney cells, but it is effective against the human immunodeficiency virus (HIV-1). Xestomanzamine A's IC_{50} values for its antifungal activity against Cryptococcus neoformans are just 3.5 μg/mL. Ascidian Eudistoma olivaceum was the source of a naturally occurring β-carboline derivative. Recent research suggests that Eudistomin F may be able to bind with DNA molecules due to its cytotoxic effects. In addition, Eudistomin F was effective against Gram-Positive bacteria. This is because DNA (Deoxy ribose nucleic acid) gyrase is disrupted by Eudistomin F, leading to cell death in the bacteria.

Carboline derivatives with a benzylidine at position 1 were studied for *in vitro* cytotoxicity on a variety of human cell lines. The most intriguing cytotoxic activities were those of N^2-benzylated β-carbolinium bromates. Both compounds were found to be the most effective against 10 different human tumour cell lines, with IC_{50} values of less than 5 μM. These results demonstrated that the N^2-benzyl substituent on the beta-carboline ring plays a significant role in regulating cytotoxic activities.

From a botanical standpoint, harmaline 28, harmalol 29, and other carbolines make up the vast majority of the pharmacologically active alkaloids found in P. harmala seeds (2% to 6% total) (Nenaah 2010). Fluorescent indole alkaloid harmaline is chemically related to beta-carbolines and harmala alkaloids. To produce melatonin, the body's primary sleep-regulating hormone and potent antioxidant, harmala stimulates the anabolic metabolism of serotonin. This describes the use of harmaline and/or other beta-carbolines as a treatment for substance abuse. At noncytotoxic concentrations, harmaline was found to inhibit the immediate early transcription of Herpes Simplex Virus 1 and 2 (HSV-1 and HSV-2). Histamine N-methyltransferase is an enzyme that harmaline blocks. The cardiovascular effects of harmalol include vasorelaxation, angiogenesis inhibition, bradycardia (lower systemic arterial blood pressure), and total peripheral vascular resistance (higher pulse pressure). Harmalol's hypotensive effects are not mediated by the stimulation of cholinergic, beta-adrenergic, or histamine (H) receptors, which is both interesting and promising (Figure 2.7).

Depending on the saturation level and the location of the N atom in the C ring, carbolines are categorized as α-, β-, γ-, or δ-carbolines from a more global structural perspective (Smirnova et al. 2011; Singh and Batra 2012). There is a broad spectrum of biological behaviours among members of this family (Piechowska et al. 2019; Dai et al. 2018; Cao et al. 2007). The fascinating structural diversity and therapeutic potential of β-carboline-containing natural products and their synthetic derivatives has attracted the attention of a number of researchers (Maity et al. 2019; Domínguez and Pérez-Castells 2011). Numerous pharmaceuticals, including vinpocetine, cipargamin, vinamine, brovincamine, yohimbine, tadalafil, abecarnil, reserpine, and lurbinectedin 30–38, contain this one-of-a-kind chemical (Figure 2.8). The leaves of the plant Vinca minor (lesser periwinkle) contain the indole alkaloid vincamine, a monoterpenoid carbazole alkaloid that makes up 25–65% of the plant's total indole alkaloids. In Europe, vincamine is prescribed for the treatment of vascular and degenerative dementias. Vinpocetine is a vincamine derivative that has been synthesized for use in medicine and nutrition. Antiaddictive and antidiabetic properties of vincamine derivatives are another area of investigation. Common dosage forms include extended-release tablets. Potentially nootropic, it has also been investigated as a possible anticancer drug. The vinca alkaloid vincamine is the starting point for the synthetic compound vinpocetine, also known as ethyl apovincaminate. Vinpocetine has been used for more than 30 years to treat cerebrovascular disorders like stroke and dementia in many Asian and European countries. There are numerous nootropic dietary supplements on the market that contain vinpocetine.

28 **29**

FIGURE 2.7 Major β-carboline alkaloids of P. harmala.

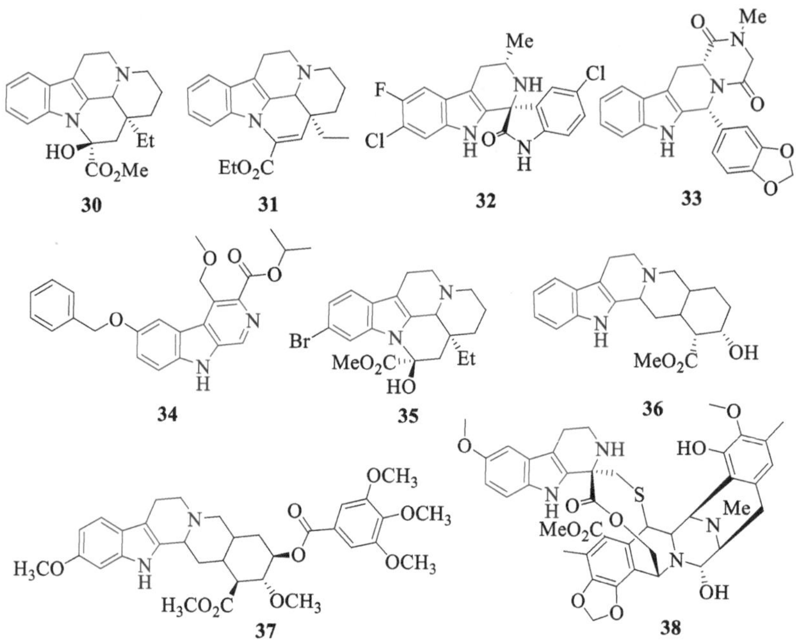

FIGURE 2.8 Structures of beta-carboline possess different biological activities.

The synthetic antimalarial drug ciprofloxacin is a member of the spiroindolone family. Cipargamin is a synthetic antimalarial molecule that belongs to the spiroindolone class. In 2006, a high-throughput phenotypic screen of natural products led to the discovery of this compound as a potent inhibitor of Plasmodium falciparum growth. Known by its brand name Cialis, the β-carboline Cialis is effective against pulmonary arterial hypertension (PAH), benign prostatic hyperplasia (BPH), and erectile dysfunction (ED). The average onset time is 30 minutes, and the maximum duration is 36 hours. Starting with (D)-tryptophan methyl ester and piperonal, tadalafil can be synthesized via a Pictet-Spengler reaction. The diketopiperazine ring is then finished off via condensations with chloroacetyl chloride and methylamine.

Abecarnil is a β-carboline anxiolytic medication. It is in a class of drugs called nonbenzodiazepines, which share the benzodiazepine's effects without the same chemical structure. It functions as a partial agonist at the benzodiazepine site of the GABA receptor. Abecarnil was originally developed as an antianxiety drug, but it has yet to be commercially developed for human use, instead being primarily used for research into the development of other new sedative and anxiolytic drugs. In Japan, brovincamine fumarate was prescribed to people with normal-tension glaucoma so as to increase blood flow to the brain, boost metabolism, and decrease platelet aggregation via the cyclic AMP (Adenosine-3',5'-cyclic monophosphate) pathway. Brovincamine is effective because it blocks calcium channels. Yohimbine is an α_2-adrenergic receptor antagonist that has been used in several studies. It is a veterinary medicine for canine and cervid sedation. Some mammals, including humans, find yohimbine to be sexually stimulating. It has been used to treat erectile dysfunction, but the reported clinical benefits have been minimal, and the PDE5 inhibitor drug class has largely replaced it.

In combination with a thiazide diuretic or vasodilator, it is used to treat hypertension. Combining reserpine and a thiazide diuretic has been shown to reduce hypertension-related mortality in large clinical trials. The ability of reserpine to deplete catecholamines from peripheral sympathetic nerve endings is a major contributor to its antihypertensive effects. Small cell lung cancer is treated with a drug called Zepzelca. Some of the most common unwanted effects include leukopenia, anaemia, lymphopenia, low albumin, neutropenia, fatigue, vomiting, cough, low magnesium, low appetite, high creatinine, high alanine aminotransferase, high aspartate aminotransferase, nausea, constipation, dyspnea, musculoskeletal pain, high glucose, low sodium, and diarrhoea.

Through its pyridine ring and indole moiety, nor-harmane belongs to the class of indole alkaloids (Banoth et al. 2020). There are beta-carbolines in plants, animals, foods, bacteria, and insects (Hamann et al. 2008). Many of these alkaloids have been isolated from marine sources like soft hydroids, corals, and bryozoans, and land plants belonging to the families Caryophyllaceae, Rutaceae, Amaranthaceae, Zygophyllaceae, Simaroubaceae, and Rubiaceae (França et al. 2006; Rajesh and Murugan 2019). There are a few illustrations in Figure 2.9.

Canthi-6-one 39 is of the type 6H-indolo[3,2,1-de][1,5]naphthyridine and is classified as a substituted indole alkaloid. In addition to its role as an antimycobacterial agent, it also plays the role of a metabolite. It is a hetero-tetracyclic organic compound known as an indole alkaloid. Aerobic and nonmotile, genus members are responsible for diseases like tuberculosis and leprosy (Dejos et al. 2014). The Beta-carboline-1-proponoic acid 40 alkaloid known as beta-carboline-1-propionic acid has been shown to have anti-Leishmania and anti-Leishmaniatic properties. Furthermore, apoptotic Leishmania amazonensis (19.5%) and L. infantum (40.4%) promastigotes were assessed after 5 hours of incubation with the alkaloid. It was also found that NO production was suppressed in infected macrophages due to β-carboline-1-propionic

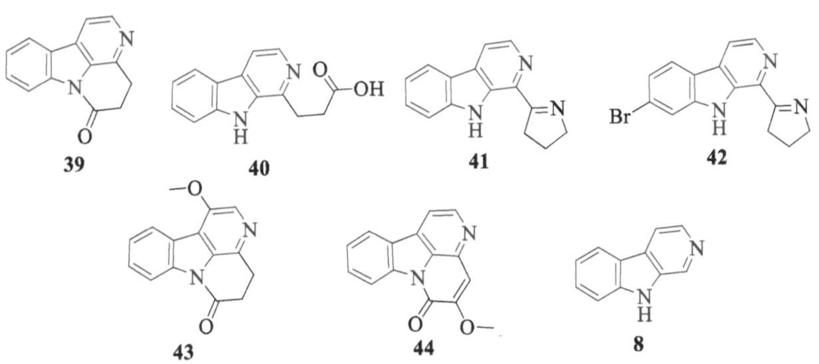

FIGURE 2.9 Structures of beta-carboline derivatives extracts from plants.

acid treatment, suggesting that intracellular amastigote elimination is not reliant on nitrosative stress (Gabriel et al. 2019).

Both Eudistomin I 41 and Eudistomin-H 42 showed potent antimicrobial activity against a wide variety of bacteria and viruses. The zone of inhibition for Eudistomin I, at 100 μg/disc, was 3–7 mm, and for Eudistomin-H, it was 7 mm. The 0.49 μg/ml concentration of Eudistomin-H, also isolated from the same ascidian, shows promise as an anticancer agent. These results provide supporting evidence for further investigation into Eudistomin-H and its derivative Eudistomin I as potential anticancer and antibacterial agents (Rajesh and Murugan 2019). The phytochemistry of L. floridana was the first to report 1-methoxycanthinone 43 and its isomer 5-methoxycanthinone 44. The discovery of simaroubolides in the Leitneriaceae family is significant because this class of compound had been previously isolated from the Simaroubaceae family. The 1-methoxycanthinone isolated from Leitneria floridana showed promising anti-HIV activity and growth inhibition across multiple human tumour cell lines in in vitro biological testing (Xu et al. 2000). Over a hundred different alkaloids and synthetic compounds can be traced back to β-carboline, also known as 9H-pyrido[3,4-b]indole. The Norharmane 8 substituent determines the properties of these chemicals. Natural beta-carbolines typically have a neuroprotective effect but may also have other effects. Analogues of beta-carboline that were created in a lab have recently been shown to have neuroprotective, cognitive-enhancing, and anticancer effects (Casal 2015).

Plakortamine-A, Plakortamine-B, Plakortamine-C, Hyrtioerectin-A, Plakortamine-D, Fascalpysin, and Manzamine-A are all examples of β-carboline compounds that are used as pharmaceuticals. The examples of pharmacologically active, marine alkaloids having beta-carbolines include 6-Deoxymanzamine-X, 8-Hydroxymanzamine-A, Methoxymanzamine-A, and Thorectandramine.

Plakortamine-A 45, B 46, C 47 & D 48 has been synthesized using total synthesis approach. β-Carboline was extracted from the deep-sea Palauan sponge Plakortis nigra. Several of the metabolites were found to be cytotoxic to HCT-116 (human colon tumour) cells (Sandler et al. 2002). In 1986, the novel antitumor alkaloid manzamine A was isolated from the Okinawa sponge; its IC_{50} value against P388 mouse leukaemia cells was determined to be 0.07 μg/ml (Figure 2.10).

The unique bioactivity and structural complexity of manzamine A made its total synthesis an alluring but challenging endeavour. Manzamine A's total synthesis was

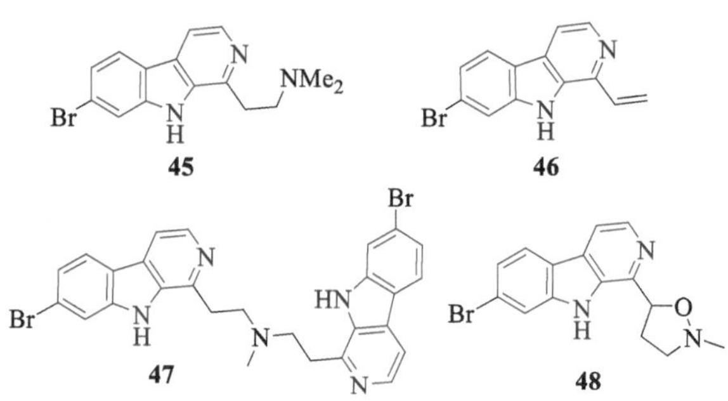

FIGURE 2.10 Different structure of Plakortamine.

first reported in 1998 by Winkler and Axten. After being isolated for over a decade, the antimalarial properties of manzamine A were finally uncovered. Both the P. falciparum D6 and W2 strains were found to be sensitive to manzamine A's antimalarial effects in vitro, with IC_{50} values of 4.5 and 8.0 ng/ml (Ashok et al. 2015). There are a plethora of Cdk inhibitors available, both in the form of biological proteins and small molecules. All Cdks can be inhibited to varying degrees by small molecule Cdk inhibitors; there are no Cdk_4-specific inhibitors. Researchers discovered that the red pigment fascaplysin 50, which is present in marine sponges, inhibits Cdk_4 specifically in vitro. Antimicrobial-based compound were first isolated from the sponge *Fascaplysinopsis* sp. The authors resynthesized fascaplysin and used it to test its effects as a Cdk4 inhibitor on cell lines derived from osteosarcoma, colon carcinoma, and normal diploid fibroblasts (Soni et al. 2000).

The marine sponge Hyrtios erectus from the Red Sea yielded a new class of alkaloids called hyrtioerectines A 51. Carbon-bond-linked moieties in hyrtioerectine A include 6-hydroxy β-carboline and 6-hydroxyindole. Using Fukuzawa's assay, we determined that compound 51 was moderately cytotoxic to HeLa cells; its IC_{50} value was determined to be 10 μg/mL (Youssef 2005). Antimalarial activity of 8-methoxymanzamine-A 52 was high, on par with that of 8-hydroxymanzamine-A but with higher cytotoxicity. The lowest antimalarial potency and absence of cytotoxicity was observed for 8,12-diacetylmanzamine-A (IC_{50}:1300 and 1200 ng/ml against D6 and W2 strains, respectively) (Ashok et al. 2015). An unidentified sponge from the Indo-Pacific region was the source of the antimicrobial compounds 8-hydroxymanzamine-A 53 and the manzamine dimer neo-kauluamine. Both compounds belong to the manzamine family. The in vitro activity of manzamines against AIDS-opportunistic infectious diseases, such as tuberculosis and toxoplasmosis, as well as their in vivo activity as antimalarial agents with superior activity to that of the clinically used drugs artemisinin and chloroquine, are some of the topics that are discussed in this chapter (Ashok et al. 2015). Edrada and colleagues were able to extract new manzamine alkaloids, 6-deoxymanzamine-X 54, from the Philippine marine sponge Xestospongia ashmorica in the year 1996. These alkaloids exhibited low cytotoxicity when tested on L5178y cell lines (Ashok et al. 2015). Recent research has shown that the substance known as thorectandramine 55, which is a novel hexacyclic quaternary β-carbolinium derivative, possesses both antiproliferative and cytotoxic activity. This compound was recently isolated from a Palauan collection of the sponge known as Thorectandra sp. (Charan 2002). The chemical literature contains only a small number of studies on the genus Thorectandra, most of which detail the isolation of various sesterterpenes and sterols (Figure 2.11).

2.3 CHARACTERISTICS OF BETA-CARBOLINES

The bioactive components of raw materials and foods have been the subject of extensive research in recent decades (Pfau & Skog 2004; Herraiz 2002, Laviță et al. 2016; De Meester 1995). Both harman and harmine, which differ from NH in that they have different substituents on their rings, have been the subject of extensive research. Carbolines' biological and pharmacological activities span a wide spectrum (including antitumor, antiviral, and antimicrobial activity) (Herraiz 2012; Li

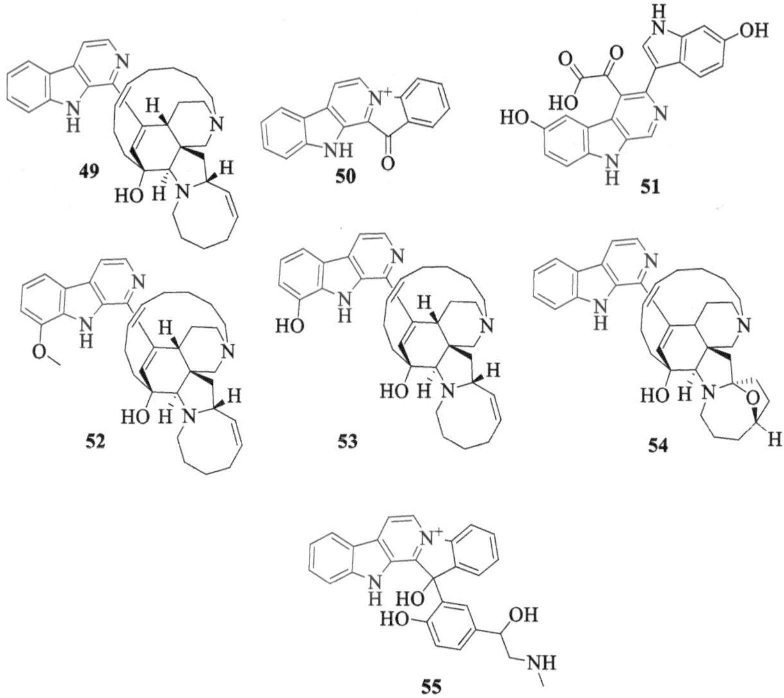

FIGURE 2.11 Different derivatives of carboline alkaloids.

et al. 2016). The β-carbolines H and NH are found naturally in mammalian tissues and body fluids (such as urine and platelets), and they are also stored in the organs of living organisms. They can be produced endogenously in these environments (Zheng et al. 2000). H is particularly interesting because it is stored in large quantities in the brain. Due to the high lipophilicity of H, it is able to spread much more quickly in the tissues of the brain and the peripheral nervous system, which results in levels that are approximately 55 times higher there than in blood plasma. Carbolines are eliminated in urine and faeces, with the majority of H passing out through the bowel movements.

2.4 OCCURRENCE OF BETA-CARBOLINES

The Nitrariaceae succulent P. harmala was the first plant to produce β-carboline compounds. From hallucinogenic plants like B. caapi and T. terrestris, harman can be extracted. Smoke from cigarettes, tobacco leaves, and tobacco chewing all test positive for beta-carbolines (Alves et al. 2010; Poindexter Jr and Carpenter 1962). NH and H have been discovered in processed and refrigerated foods by multiple investigations (Wojtowicz et al. 2015; Totsuka et al. 1999; Allen and Holmstedt 1980). Cookies, coffee (Herraiz 2004; Schultes 1982; Tsuchiya et al. 1999; Casal 2015; Alves et al. 2007; Zawirska-Wojtasiak et al. 2018) (particularly between the ages of 78 and 83), bread, soy, barley, maize, coffee substitutes, and fermented alcoholic beverages are all examples of plant-derived foods that contain these compounds (Tsuchiya et al. 1996; Adachi et al. 1991). H and NH have also been found in foods like meat, fruit, fish, vegetables, and fruit juices (Adachi et al. 1991). Many scientists have conjectured that consumption of items containing NH and H in mammalian tissues may be connected to increased levels of β-carbolines in biological fluids and human tissues. Meat and fish have been reported to contain β-carboline molecules, although the concentration varies depending on the kind of meat and heat treatment. Beef, chicken, and mutton were all found to contain H and NH after being flame-grilled, deep-fried, or baked (Totsuka et al. 1999; Ogawa et al.1993). Herraiz discovered that H and NH levels in raw beef and pork fillets were below the detection limit (Herraiz and Galisteo et al. 2003). After being heated, the amount of carbolines in meat varied depending on the cooking technique used. Well-done best meat had the highest average value, with H ranging from 20.7 to 32.1 ng/g and NH from 36.4 to 128.1 ng/g. Sausage was found to have NH levels between 0 and 3.7 ng/g, with undetectable being the most common. Carbolines were nowhere to be found in raw fish. After processing, NH concentrations in cooked fish ranged from 6.4% to 48.1% by weight, and H concentrations from 0.7% to 4.3% by weight. No matter the method of preparation or the type of fish, NH concentrations were always higher than H concentrations.

2.5 NATURAL PRODUCTS HAVING A SIMPLE B-CARBOLINE SKELETON

Norharman and 1- or 6-substituted β-carboline were synthesized in a single reaction, thanks to work by Wang and coworkers (Scheme 2.1). The designed synthesis begins with racemic tryptophan and other amino acids, and then moves through a series of reactions and chemical transformations, including a decarboxylation, a deamination, a Pictet-Spengler reaction, and an oxidation. Moreover, a number of alkaloids and synthetic analogues were obtained by replacing specific amino acids in the original product (Wang et al. 2018).

The high yield enantioselective synthesis of (S)-eleagnine(-)-tetrahydroharmane was reported by Lood et al (Scheme 2.2). Their synthesis begins with a chirally correct methyl ester of a modified amino acid. The amide of morpholine was synthesized by reacting morpholine with t-BuMgCl. The amide was lithiated laterally, and then o-toluidine that was Boc-protected was added to get the ketone. After reacting the ketone with KHMDS (potassium bis(trimethylsilyl)amide), a mixture of diastereomeric indoline-lactones was obtained by quenching the resulting potassium enolate with ethyl bromoacetate. This diastereomeric mixture was treated with sulfuric acid in

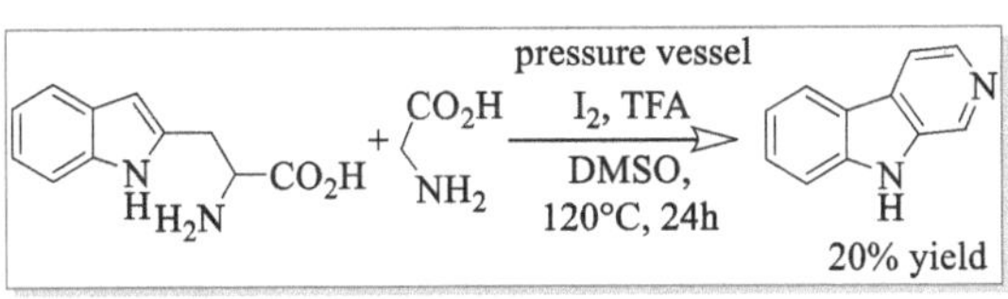

SCHEME 2.1 Acid catalyzed one-pot synthesis of norharman.

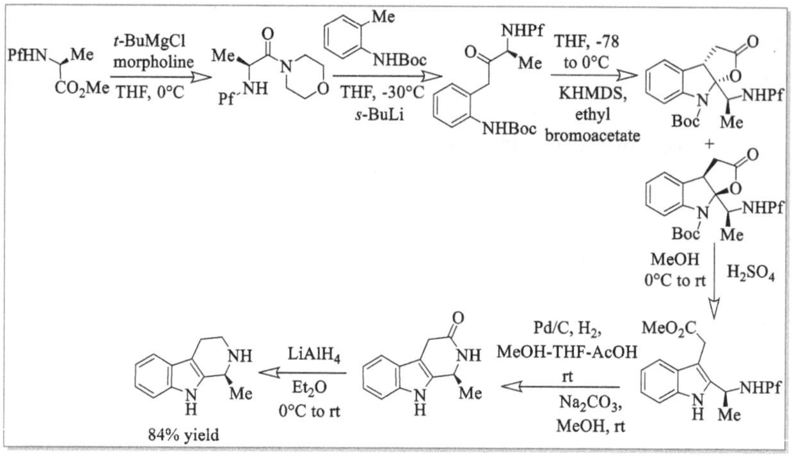

SCHEME 2.2 Synthesis of tetrahydroharmane.

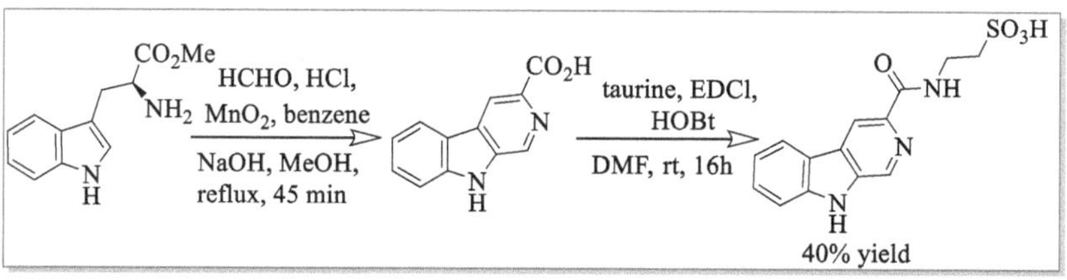

SCHEME 2.3 Synthesis of β-carboline derivative.

MeOH to facilitate sequential lactone ring opening and Boc-cleavage to form indole. The Pf-protecting group (phenylfluorenyl group) was hydrogenolyzed, and the free amine was treated with Na_2CO_3 base to produce lactam. Finally, an enantiomeric excess of (S)-eleagnine was obtained through gram-scale synthesis using a $LiAlH_4$ reducing agent (Lood et al. 2015).

Stolonine C, a new β-carboline alkaloid, was discovered and synthesized as a result of chemical research into the marine tunicate C. stolonifera (Scheme 2.3). The Pictet-Spengler reaction of formaldehyde with L-tryptophan methyl ester is the first step in this synthesis, followed by C-ring dehydrogenation and ester group hydrolysis. Taurine was used to convert the intermediate into an amide, which led to the final product (Tran et al. 2015).

Tryptamine and indole-3-carbonitrile are used in a two-step alternative synthesis of the final product, which also involves dehydrogenation of an intermediate with carbonized palladium (Scheme 2.4) (Pakhare and Kusurkar 2015). To prepare substituted 3,4-dihydro-carbolines in the presence of catalyst, Brányi-Balogh and colleagues showed a microwave-assisted one-pot protocol. From tryptamine and indole-3-carboxylic acid, 50% isoeudistomin U alkaloid was synthesized using the new method (Scheme 2.5) (Ábrányi-Balogh et al. 2016).

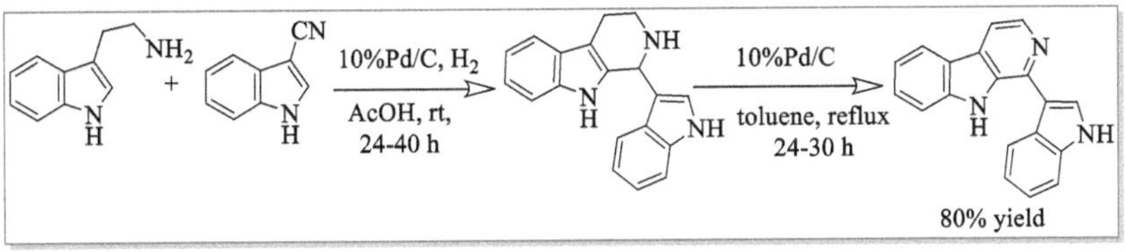

SCHEME 2.4 Two-step synthesis of β-carboline using tryptamine and indole-3-carbonitrile.

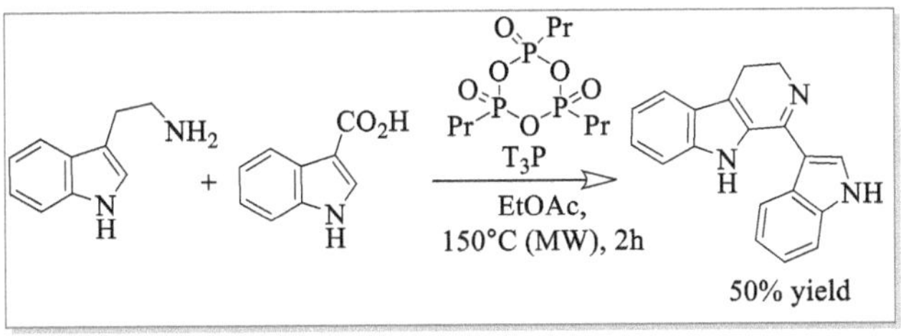

SCHEME 2.5 Microwave-assisted protocol for the preparation of 3,4-dihydro-carbolines.

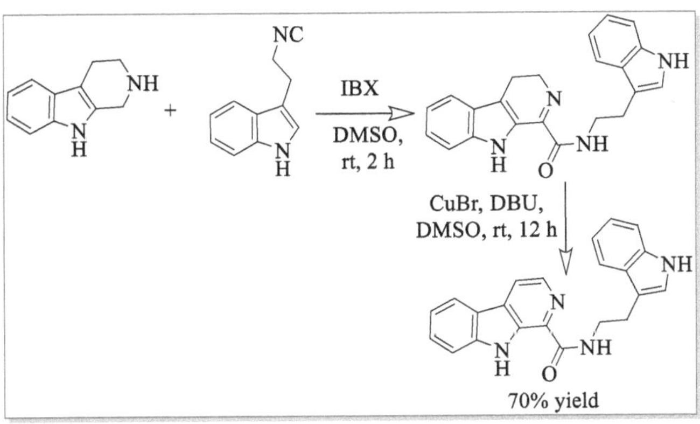

SCHEME 2.6 Efficient method for preparing substituted β-carboline.

Alangiobussine and alangiobussinine were synthesized via the IBX (2-Iodoxybenzoic acid)-mediated total synthesis pathway by oxidatively adding isocyanide to tryptoline. The reaction of tetrahydroisoquinolines and tryptolines with isocyanides is a general method (Scheme 2.6) for preparing imino-carboxamides. The natural product alangiobussine was successfully synthesized on a gram-scale using the IBX-mediated oxidative addition technique. It was subsequently dehydrogenated in the presence of CuBr and DBU (1,8-Diazabicyclo(5.4.0)undec-7-ene) to yield alangiobussinine (Ambule et al. 2019).

Komavine, a spiro-THC derivative, and several analogues with good functional group tolerance were synthesized by Wang et al. using a new Pd-catalyzed cross-annulation of phenols and tryptamine (Scheme 2.7). This alternative protocol for

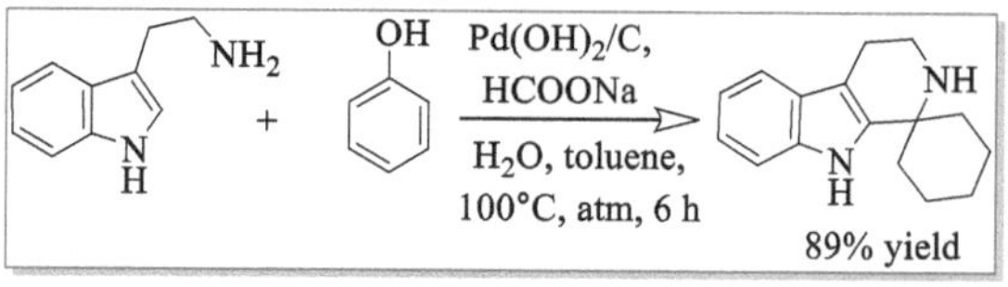

SCHEME 2.7 Pd-catalyzed cross-annulation reaction for the synthesis spiro-THC derivative.

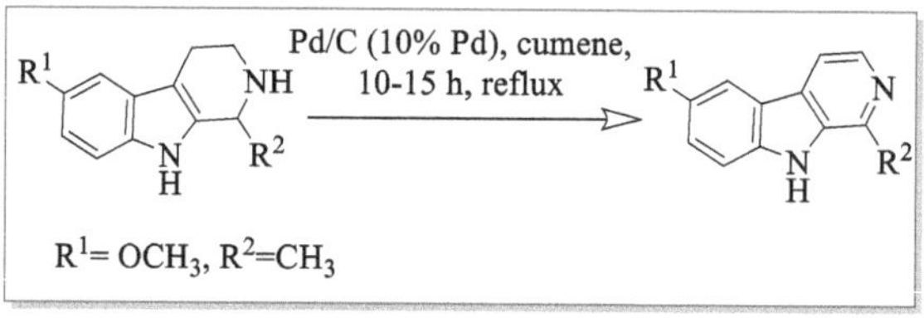

SCHEME 2.8 Synthesis of marine natural product having β-carboline framework.

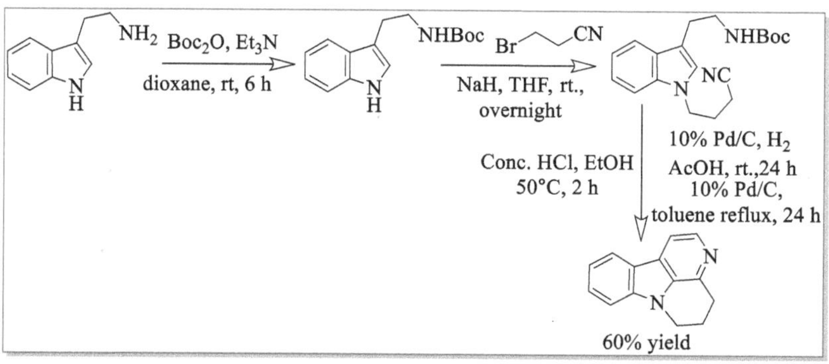

SCHEME 2.9 Pictet-Spengler reductive reaction approach for the synthesis of β-carboline.

synthesizing THC (tetrahydrocoumarin) derivatives involved cleaving the C-O bond in phenol, activating the C-H bond in tryptamine, and forming the C-N/C-C bond. The yield was moderate to excellent. This work demonstrates a feasible strategy for transforming the model monomers of phenolic lignin into compounds with commercial value (Wang et al. 2019).

The synthesis of shishijimicin A, a marine natural substance with exceptional bioactivity activity, was broken down into 48 steps by Schott et al. (Schemes 2.8). By using their approach, the natural product is split into three significant intermediates. Using methods from the literature and functional group transformations, one of them, the β-carboline framework, was built in eight stages to prepare the protected and iodine-functionalized critical intermediate in position 1 (Schott et al. 2006).

Canthine, the framework of biologically active canthine-6-ones, and harmicine were successfully developed using the Pictet-Spengler reductive reaction (Scheme

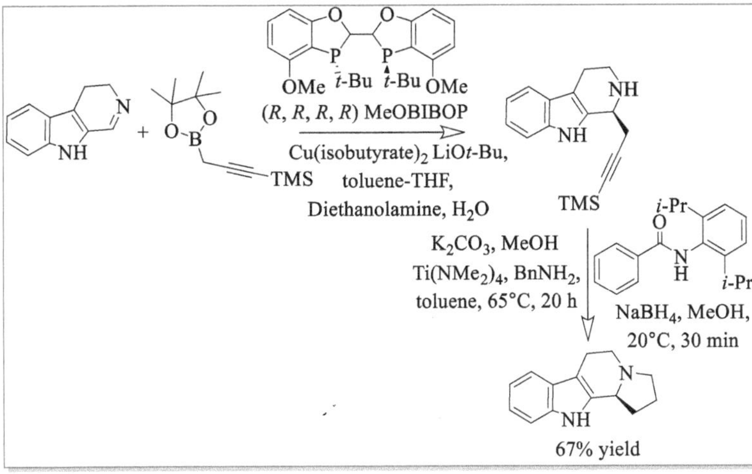

SCHEME 2.10 Metal mediated approach for the synthesis of β-carbolines.

2.9). With 10% Pd/C and H2, nitrile was utilized in place of the often-used aldehydes in the reductive intramolecular cyclization technique. The final product was formed by further dehydrogenation of the C-ring with a 10% Pd/C catalyst (Pakhare and Kusurkar 2015).

Fandrick et al. invented the process of propargylation of cyclic aldimines by copper catalysis (Scheme 2.10). With a high yield and stereoselectivity, (S)-harmicine was synthesized by using a coupling agent, copper(II)isobutyrate, and a chiral ligand. The ring closure and subsequent C=N bond reduction was catalyzed by titanium, and the resulting compound was deprotected and transformed into the final compound (Fandrick et al. 2016).

3.10 SYNTHETIC ASPECTS OF BETA-CARBOLINES HAVING ANTIBACTERIAL PROPERTY

Pictet-Spengler reaction of tryptamine with pyruvaldehyde in acid was the first step in the development of acetyl-carboline, which was then aromatized in situ using 10% palladium on carbon. To get N-boc-1-acetyl-carboline, di-tert-butyl dicarbonate was used to shield the N-H group of 1-acetyl-carboline from oxidation in the presence of dimethyl aminopropional (DMAP) (Scheme 2.11). In addition, Scheme 2.11 gave good-to-excellent yields of β-carboline chalcone analogues after reacting with respective aldehydes in the presence of a 20% NaOH solution in aqueous ethanol for 16 hours. Bromide salts (the final product) were obtained from the N^2-alkylation of three β-carboline chalcones with various alkylbromides with a yield of over 80% (Venkataramana Reddy et al. 2018).

Using Neamine derivatives rich in amino groups, the coupling reaction was executed (Scheme 2.12). The substance was synthesized by reacting amines with acids. Similarly, tetrahydro-2H-pyran-3,4-diyl dibenzoate was obtained by

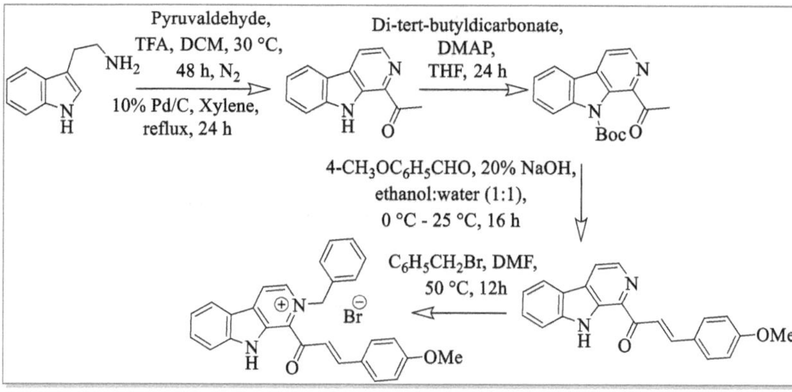

SCHEME 2.11 Preparation of bromide salts of β-carbolines.

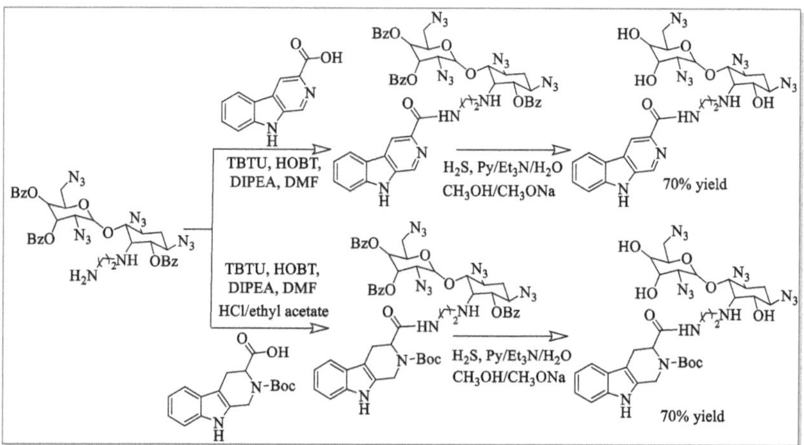

SCHEME 2.12 Coupling approach for the synthesis of β-carboline derivatives.

condensing the derivative of carboline carboxylic acid with amino-containing neamine derivatives at room temperature under acidic conditions. Carboxylic acids are insoluble in carboline, so the coupling reaction was performed in a DMF (Dimethylformamide) solution. N,N-diisopropylethylamine (DIPEA) greatly facilitated product purification by promoting efficient amide-bond formation when combined with HOBT (hydroxybenzotriazole) and TBTU (2-(1H-benzotriazole-1-yl)-1,1,3,3-tetramethylaminium tetrafluoroborate). The reaction was also concluded rapidly after DIPEA addition was completed. The desired compounds were obtained after being saponified with sodium methoxide in methanol and subsequently reduced with hydrogen sulphide (Wu et al. 2010).

The general approach to synthesizing new 1,2,3-triazole-ringed β-carboline derivatives is depicted in Scheme 2.13 (Salehi et al. 2016). L-tryptophan methyl ester and O-propargylated benzaldehydes underwent a Pictet-Spengler reaction, which

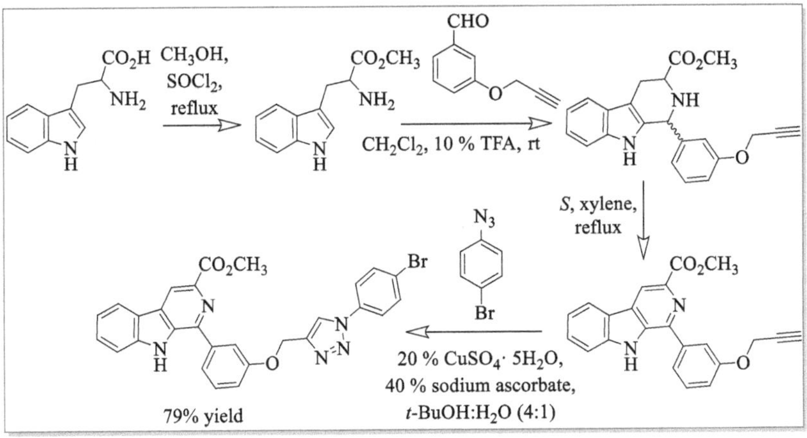

SCHEME 2.13 Routes for the synthesis of 1,2,3,4-tetrahydro-β-carboline compounds.

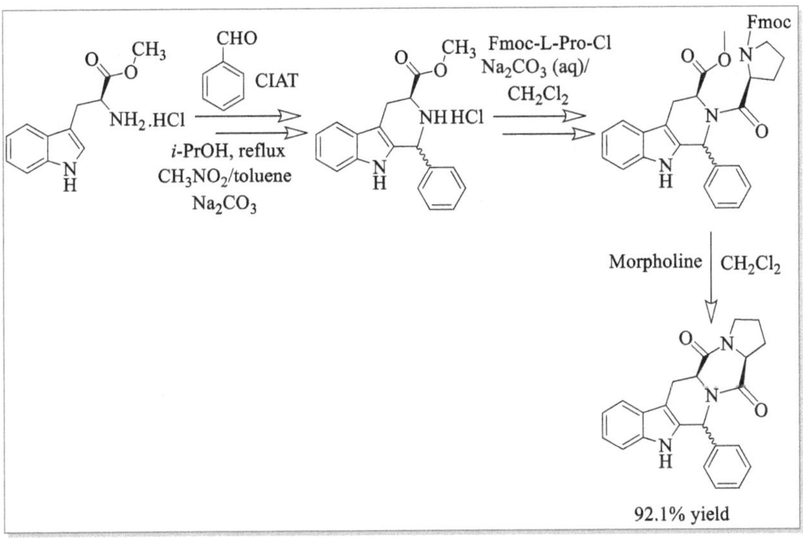

SCHEME 2.14 Stereoselective synthesis of tetrahydro-β-carboline diketopiperazines.

produced a racemic mixture of tetrahydro-carboline compounds. Researchers looked into using sulphur, $KMnO_4$, and IBX as oxidizing reagents to transform 1,2,3,4-tetrahydro-carboline derivatives into their β-carboline counterparts. Other derivatives were oxidized using the same technique.

From L-tryptophan methyl ester hydrochloride and aldehydes, single isomers of tetrahydro-carboline diketopiperazines were made using a four-step reaction that included the Pictet-Spengler reaction, crystallization-induced asymmetric transformations (CIAT), the Schotten-Baumann reaction, and intramolecular ester amidation (Scheme 2.14) (Ma et al.2013).

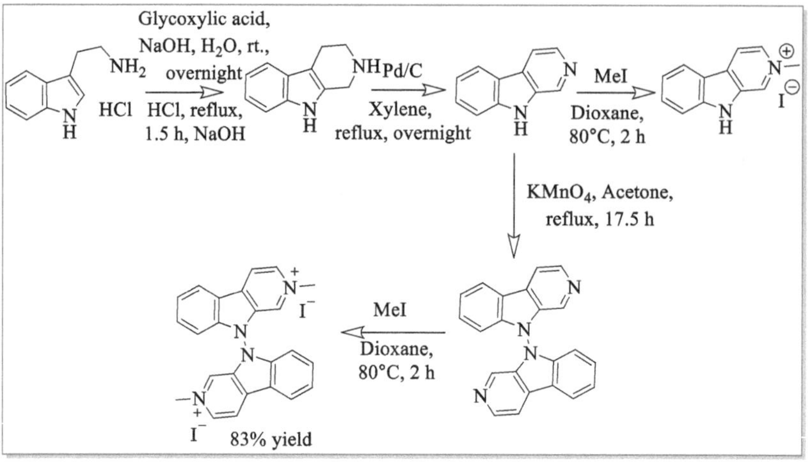

SCHEME 2.15 Synthesis of β-carboline dimers and their N2-alkylated analogues.

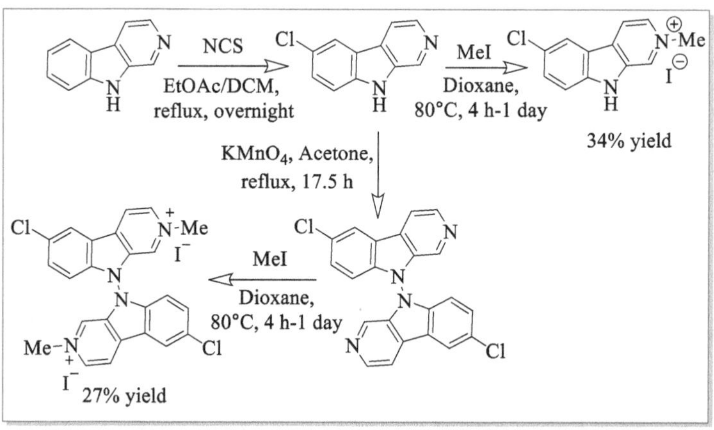

SCHEME 2.16 Synthesis of β-carboline dimers via Oxidative approach.

The biological effects of beta-carbolines, a class of indole alkaloids, are quite varied. The purpose of this research was to determine the structural basis for the inhibitory effects of β-carboline derivatives in vitro against pathogenic microorganisms of clinical significance. Therefore, a number of β-carboline dimers and N^2-alkylated analogues were synthesized and tested for antibacterial activity. Dimeric 6-chlorocarboline N^2-benzylated salt showed strong activity against S. aureus at MICs of 0.010.05 mol/mL. Our results demonstrate that N^1N^1-dimerization and N^2-benzylation significantly elevate compounds' antimicrobial activities (Scheme 2.15 & 2.16) (Suzuki et al. 2018).

Schema 2.17 shows the steps needed to synthesize β-carboline-oxazol-5-ones. The technique developed by the Savariz group was applied to create substituted β-carboline-3-carboxylic acids using commercial L-tryptophan. Analogs of tetrahydro-carbolines were produced by Pictet-Spengler condensation of L-tryptophan methyl

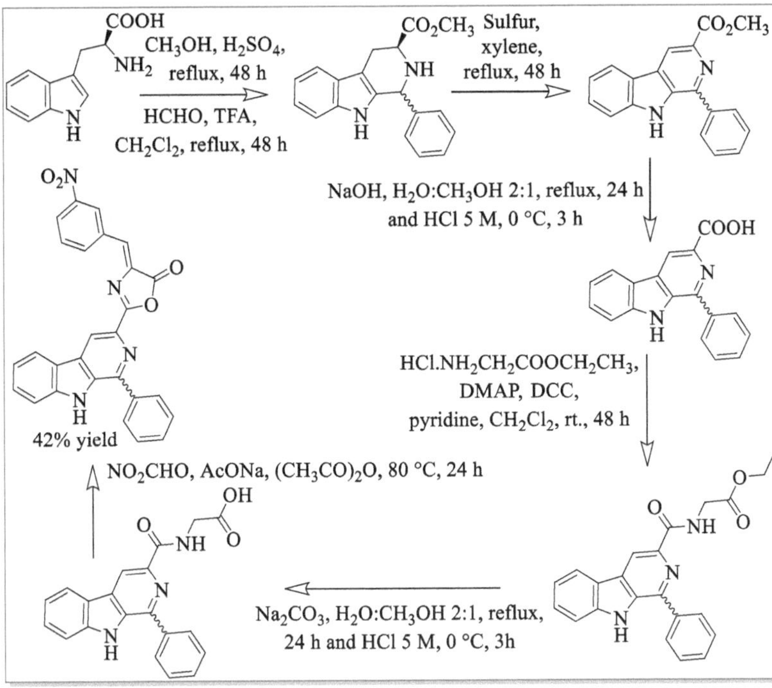

SCHEME 2.17 Synthesis of β-carboline derivatives using L-tryptophan.

ester with benzaldehyde, 4-methoxybenzaldehyde, and 4-hydroxybenzaldehyde. The β-carboline-4H-oxazol-5 ones were made by the Erlenmeyer Plöchl reaction, which is the most widely used technique for producing oxazolones. Analogues of β-carboline-3-carboxylic acids were created by activating β-carboline-3-carboxylic acids in pyridine with DMAP (dimethylaminopyridine) and DCC (N,N'-dicyclohexylcarbodiimide), and then treating the resultant solution with glycine ethyl ester hydrochloride (Savariz et al. 2012).

In order to investigate new antibacterial frameworks and assess antiphytopathogen efficacy in vitro and in vivo, the author synthesized a number of unique THC derivatives. Target compounds showed excellent action against three rogue phytopathogens, namely X. axonopodis pv. citri, P. syringae pv. actinidiae (Psa), and X. oryzae pv. oryzae, as evidenced by their respective EC_{50} values of 2.06 (I23), 2.39 (II9), and 1.69. These results were superior to those seen with 1,2,3,4-THC, the parent structure, and with the control group. In vivo tests showed that at 200 g/mL, they are highly effective at preventing the spread of both rice bacterial blight and kiwifruit (Scheme 2.18) (Liu et al. 2020).

To produce new beta-carboline derivatives, a six-step reaction procedure was used (Scheme 2.19). Initially, L-amine tryptophan's group was shielded with BOC anhydride, and subsequent spectroscopic analyses verified the product's structure. The next step involved esterifying the carboxyl group of BOC tryptophan with 10-undecenol. The esterified product was deprotected using acid to yield a free amine.

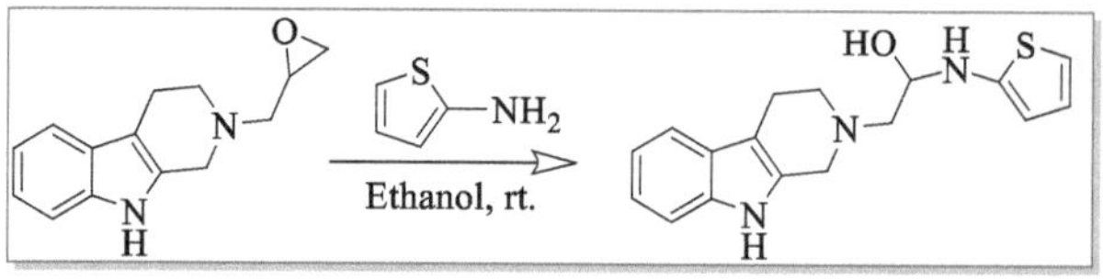

SCHEME 2.18 Synthesis of tetrahydro-β-carboline derivatives via epoxide ring opening.

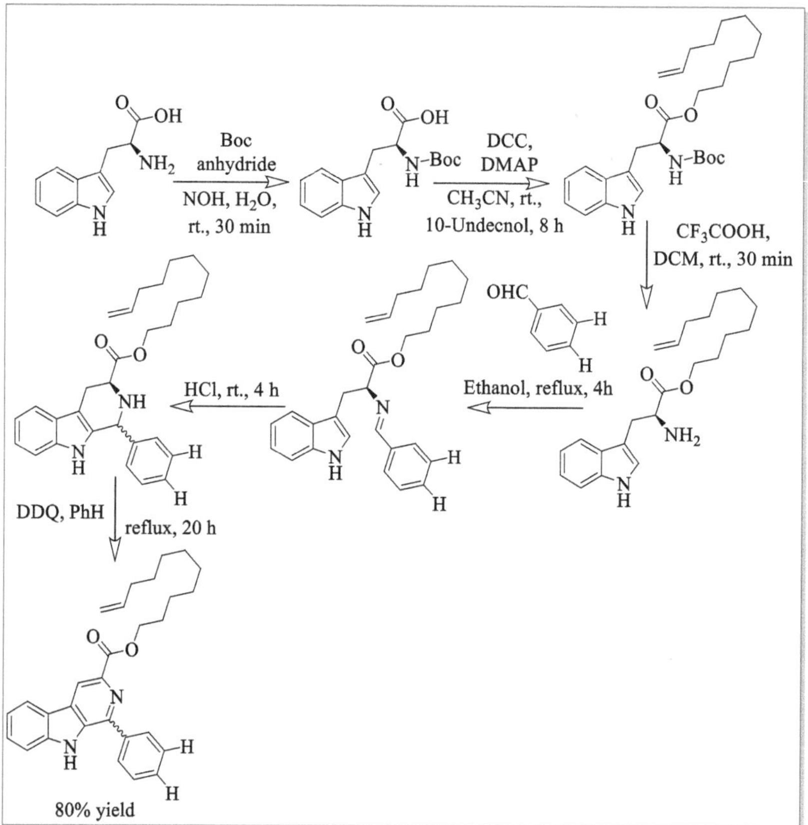

SCHEME 2.19 Synthesis of substituted β-carboline derivatives using L-tryptophan.

Following that, the amino group was condensed with various types of aldehydes to produce Schiff bases. Then, the resulting products were made by cyclizing Schiff bases (Kontham et al. 2021).

Scheme 2.20 depicts the process for creating maleimide, phthalimide, and succinimide-substituted phenylimide-β-carboline derivatives, as described by Ortiz et al (Lopes-Ortiz et al. 2020). Compound precursors were obtained via the Pictet-Spengler reaction of L-tryptophan methyl ester with aromatic aldehydes containing electron-donating and electron-withdrawing groups; subsequent xylene oxidation

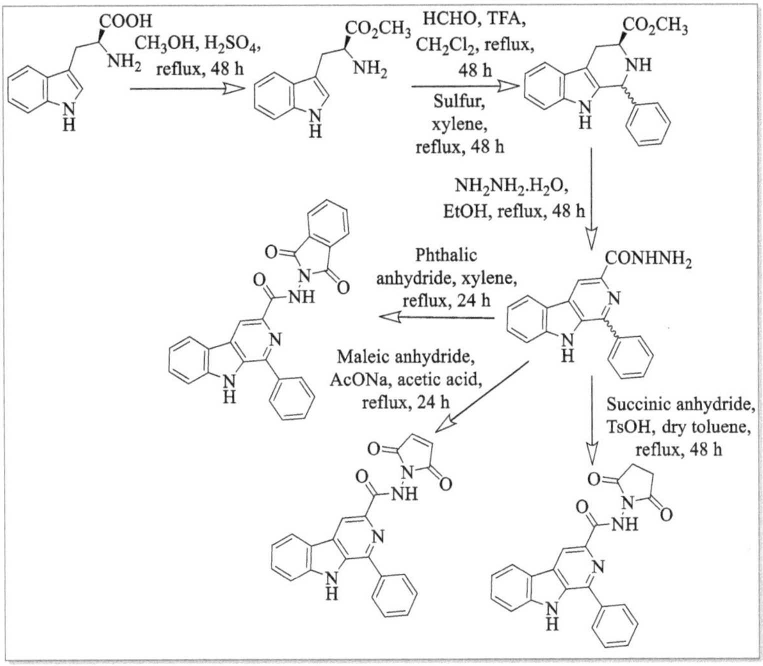

SCHEME 2.20 Different approaches for the synthesis of β-carbolines.

of THC obtained with sulphur yielded the corresponding methyl β-carboline-3-carboxylates. By reacting carbohydrazides with phthalic anhydride, β-carboline-3-N-(1,3-dioxo-1,3-dihydro-isoindol-2-yl)-carboxamides were formed. The phenyl and 4-fluorophenyl groups at position 1 of the β-carboline nucleus were prepared by treating the carbohydrazide with maleic anhydride in acetic acid and anhydrous sodium acetate, respectively. An initial attempt to synthesize β-carboline-3-N-(2,5-dioxo-pyrrolidin-1-yl)-carboxamides by reacting carbohydrazides with succinic anhydride under anhydrous toluene reflux is described by Brosse et al. All carbohydrazides, however, were only found to generate noncyclized intermediates under these conditions. The final compound was synthesized by reacting the appropriate starting carbohydrazides with succinic anhydride in the presence of a catalytic amount of p-toluene sulfonic acid (Brosse et al. 2003).

3.7 APPLICATIONS OF B-CARBOLINES

The effects of natural beta-carbolines are primarily neurological, but they may also have antioxidant effects (Liao et al. 2020). Recently, β-carbolines derivatives developed in a laboratory have been shown to have neuroprotective (Gulyaeva and Aniol 2012), cognitive enhancing, and anticancer properties (Aaghaz et al. 2021). Amazonian peoples used a hallucinogen made from the liana Banisteriopsis caapi, which was first described as such in the middle of the nineteenth century. In the early twentieth century, European pharmacists discovered that harmine was an active ingredient (Djamshidian et al. 2016). Interest in this finding's therapeutic potential has prompted additional study. Harmine's pharmacological effects can be traced back to its ability to inhibit monoamine oxidase (MAO). Banisteriopsis caapi and Peganum harmala extracts have been shown to trigger dopamine release in the striatum in both in vitro and in vivo rodent studies, respectively (Samoylenko et al. 2010; Schwarz et al. 2003). Moreover, in mice treated with MPTP (1-methyl-4-phenyl-1,2,3,6-tetrahydropyridine), harmine increases the number of viable dopaminergic neurons (Barallobre et al. 2014).

3.8 CONCLUSION

β-carbolines are a significant group of nitrogen-containing heterocycles because of their wide range of biological and pharmaceutical uses. There is a wide range of therapeutic applications for the β-carboline family of heterocyclic natural products. Due to their value, many scientists have already made contributions to the bioactivity and preparation of these intriguing frameworks. The antibacterial properties and overall structure of β-carbolines skeletons have been discussed in this chapter. This chapter provided a detailed discussion, supported by examples, of the nature and prevalence of β-carbolines frameworks. We also described the origins, structures, properties, and potential bioactivities of several naturally occurring products with a β-carbolines skeleton. In addition, the synthetic aspect and several important applications were covered.

ACKNOWLEDGMENTS

The authors are thankful to the Department of Applied Chemistry, Amity School of Engineering & Technology (ASET), Gwalior, for all the necessary facilities.

REFERENCES

Aaghaz, S., Sharma, K., Jain, R., & Kamal, A. 2021. β-Carbolines as potential anticancer agents. In *European Journal of Medicinal Chemistry*, Vol. 216. Elsevier Masson s.r.l.

Ábrányi-Balogh, P., Földesi, T., Grün, A., Volk, B., Keglevich, G., & Milen, M. 2016. Synthetic study on the T3P®-promoted one-pot preparation of 1-substituted-3,4-dihydro-β-carbolines by the reaction of tryptamine with carboxylic acids. *Tetrahedron Letters*, 57(18):1953–1957.

Adachi, J., Mizoi, Y., Naito, T., Yamamoto, K., Fujiwara, S., & Ninomiya, I. 1991. Determination of β-carbolines in foodstuffs by high-performance liquid chromatography and high-performance liquid chromatography-mass spectrometry. *Journal of Chromatography A*, 538(2):331–339.

Allen, J. R., & Holmstedt, B. R. 1980. The simple β-carboline alkaloids. *Phytochemistry*, 19(8):1573–1582.

Alves, R. C., Casal, S., & Oliveira, B. P. 2007. Factors influencing the norharman and harman contents in espresso coffee. *Journal of Agricultural and Food Chemistry*, 55(5):1832–1838.

Alves, R. C., Mendes, E., Oliveira, B. P., & Casal, S. 2010. Norharman and harman in instant coffee and coffee substitutes. *Food Chemistry*, 120(4):1238–1241.

Ambule, M. D., Tripathi, S., Ghoshal, A., & Srivastava, A. K. 2019. IBX-mediated oxidative addition of isocyanides to cyclic secondary amines:Total syntheses of alangiobussine and alangiobussinine. *Chemical Communications*, 55(73):10872–10875.

Ashok, P., Lathiya, H., & Murugesan, S. 2015. Manzamine alkaloids as antileishmanial agents:A review. *European Journal of Medicinal Chemistry*, 97(1):928–936.

Banoth, K. K., Faheem, ChandraSekhar, K. V. G., Adinarayana, N., & Murugesan, S. 2020. Recent evolution on synthesis strategies and anti-leishmanial activity of β-carboline derivatives-An update. *Heliyon*, 6(9).

Barallobre, M. J., Perier, C., Bové, J., Laguna, A., Delabar, J. M., Vila, M., & Arbonés, M. L. 2014. DYRK1A promotes dopaminergic neuron survival in the developing brain and in a mouse model of Parkinson's disease. *Cell Death and Disease*, 5(6).

Becher, P. G., Beuchat, J., Gademann, K., & Jüttner, F. 2005. Nostocarboline:Isolation and synthesis of a new cholinesterase inhibitor from Nostoc 78-12A. *Journal of Natural Products*, 68(12):1793–1795.

Beck, O., & Lundman, A. 1983. Occurrence of 6-hydroxy-1-methyl-1, 2, 3, 4-tetrahydro-β-carboline in tissues and body fluids of rat. *Biochemical Pharmacology*, 32(9):1507–1510.

Brosse, N., Pinto, M. F., & Jamart-Grégoire, B. 2003. Preparation of multiply protected Alkylhydrazine derivatives by Mitsunobu and PTC approaches. *European Journal of Organic Chemistry*, 24:4757–4764.

Cabrera, G. M., & Seldes, A. M. 1999. A β-carboline alkaloid from the soft coral Lignopsis spongiosum. *Journal of Natural Products*, 62(5):759–760.

Cao, R., Peng, W., Wang, Z., & Xu, A. 2007. Carboline Alkaloids:Biochemical and pharmacological functions. *Current Medicinal Chemistry*, 14(4):479–500.

Carter, A. P., Clemons, W. M., Brodersen, D. E., Morgan-Warren, R. J., Wimberly, B. T., & Ramakrishnan, V. 2000. Functional insights from the structure of the 30S ribosomal subunit and its interactions with antibiotics. *Nature*, 407(6802):340–348.

Casal, S. 2015. Neuroactive β-Carbolines Norharman and Harman in coffee. *In Coffee in Health and Disease Prevention*: 737–743.

Céliz, G., Daz, M., & Audisio, M. C. 2011. Antibacterial activity of naringin derivatives against pathogenic strains. *Journal of Applied Microbiology,* 111(3):731–738.

Charan, R. D., McKee, T. C., Gustafson, K. R., Pannell, L. K., & Boyd, M. R. 2002. Thorectandramine, a novel β-carboline alkaloid from the marine sponge Thorectandra sp. *Tetrahedron letters*, 43(29):5201–5204.

Chen, B., Zhao, J., Zhang, R., Zhang, L., Zhang, Q., Yang, H., & An, J. 2022. Neuroprotective effects of natural compounds on neurotoxin-induced oxidative stress and cell apoptosis. *Nutritional Neuroscience*, 25(5):1078–1099.

Csányi, D., Hajós, G., Riedl, Z., Timári, G., Bajor, Z., Cochard, F., & Laronze, J. Y. 2000. Synthesis of two new heteroaromatic β-carboline-fused pentacycles. Observation of a new intercalating agent. *Bioorganic & Medicinal Chemistry Letters*, 10(15):1767–1769.

Dai, J., Dan, W., Schneider, U., & Wang, J. 2018. β-Carboline alkaloid monomers and dimers:Occurrence, structural diversity, and biological activities. *European Journal of Medicinal Chemistry*, 157:622–656.

Dejos, C., Voisin, P., Bernard, M., Régnacq, M., & Bergès, T. 2014. Canthin-6-one displays antiproliferative activity and causes accumulation of cancer cells in the G2/M phase. *Journal of Natural Products*, 77(11):2481–2487.

De Meester, C. 1995. Genotoxic potential of β-carbolines:A review. *Mutation Research/Reviews in Genetic Toxicology*, 339(3):139–153.

Deveau, A. M., Labroli, M. A., Dieckhaus, C. M., Barthen, M. T., Smith, K. S., & Macdonald, T. L. 2001. The synthesis of amino-acid functionalized β-carbolines as topoisomerase II inhibitors. *Bioorganic & Medicinal Chemistry Letters*, 11(10):1251–1255.

Djamshidian, A., Bernschneider-Reif, S., Poewe, W., & Lees, A. J. 2016. Banisteriopsis caapi, a Forgotten Potential Therapy for Parkinson's Disease? In *Movement Disorders Clinical Practice* (Vol. 3, Issue 1, pp. 19–26. Wiley-Blackwell.

Domínguez, G., & Pérez-Castells, J. 2011. Chemistry of β-carbolines as synthetic intermediates. *European Journal of Organic Chemistry*, 36:7243–7253.

Fandrick, D. R., Hart, C. A., Okafor, I. S., Mercadante, M. A., Sanyal, S., Masters, J. T., Sarvestani, M., Fandrick, K. R., Stockdill, J. L., Grinberg, N., Gonnella, N., Lee, H., & Senanayake, C. H. 2016. Copper-Catalyzed Asymmetric Propargylation of Cyclic Aldimines. *Organic Letters*, 18(23):6192–6195.

Fourmy, D., Recht, M. I., Blanchard, S. C., & Puglisi, J. D. 1996. Structure of the A site of Escherichia coli 16 S ribosomal RNA complexed with an aminoglycoside antibiotic. *Science*, 274(5291):1367–1371.

França, P. H. B., Barbosa, D. P., da Silva, D. L., Ribeiro, Ê. A. N., Santana, A. E. G., Santos, B. V. O., Barbosa-Filho, J. M., Quintans, J. S. S., Barreto, R. S. S., Quintans, L. J., & de Araújo, J. X. 2006. Indole alkaloids from marine sources as potential leads against infectious diseases. *BioMed Research International*, 2014.

Gabriel, R. S., Amaral, A. C. F., Lima, I. C., Cruz, J. D., Garcia, A. R., Souza, H. A. S., Adade, C. M., Vermelho, A. B., Alviano, C. S., Alviano, D. S., & Rodrigues, I. A. 2019. β-Carboline-1-propionic acid alkaloid:Antileishmanial and cytotoxic effects. *Revista Brasileira de Farmacognosia*, 29(6):755–762.

Gulyaeva, N., & Aniol, V. 2012. Good guys from a shady family. *In Journal of Neurochemistry*, 121, 6:841–842.

Hamann, J., Wernicke, C., Lehmann, J., Reichmann, H., Rommelspacher, H., & Gille, G. 2008. 9-Methyl-β-carboline up-regulates the appearance of differentiated dopaminergic neurones in primary mesencephalic culture. *Neurochemistry International*, 52:688–700.

Herraiz, T. 2002. Identification and occurrence of the bioactive β-carbolines norharman and harman in coffee brews. *Food Additives & Contaminants*, 19(8):748–754.

Herraiz*, T. 2004. Relative exposure to β-carbolines norharman and harman from foods and tobacco smoke. *Food Additives and Contaminants*, 21(11):1041–1050.

Herraiz, T. 2012. β-carbolines as neurotoxins. In *Isoquinolines and Beta-Carbolines As Neurotoxins and Neuroprotectants*, Springer, 77–103.

Herraiz, T., & Galisteo, J. 2003. Tetrahydro-β-carboline alkaloids occur in fruits and fruit juices. Activity as antioxidants and radical scavengers. *Journal of Agricultural and Food Chemistry*, 51(24):7156–7161.

Kathal, R., Sharma, S., Singh, K., & Singh, S. 2023. Simplified procedure for application of DBE, the rule of 13 & Nitrogen rule in structure elucidation of organic compounds. *Letters in Organic Chemistry*, 20(3):276–285.

Kearns, P. S., & Rideout, J. A. 2008. Nonsymmetrical β-carboline dimers from an ascidian, Didemnum sp. *Journal of Natural Products*, 71(7):1280–1282.

Kobayashi, M., Chen, Y. J., Aoki, S., In, Y., Ishida, T., & Kitagawa, I. 1995. Four new β-carboline alkaloids isolated from two Okinawan marine sponges of Xestospongia sp. and Haliclona sp. *Tetrahedron*, 51(13):3727–3736.

Koehn, F. E., & Carter, G. T. 2005. The evolving role of natural products in drug discovery. In *Nature Reviews Drug Discovery*, 4(3):206–220.

Kontham, V., Ippakayala, B., & Madhu, D. 2021. Synthesis of β-carboline fatty alcohol hybrid molecules and characterization of their biological and antioxidant activities. *Arabian Journal of Chemistry*, 14(6).

Kotanen, S., Huybrechts, J., Cerstiaens, A., Zoltan, K., Daloze, D., Baggerman, G., Forgo, P., de Loof, A., & Schoofs, L. 2003. Identification of tryptophan and β-carboline as paralysins in larvae of the yellow mealworm, Tenebrio molitor. *Biochemical and Biophysical Research Communications*, 310(1):64–71.

Laine, A. E., Lood, C., & Koskinen, A. M. P. 2014. Pharmacological importance of optically active tetrahydro-β-carbolines and synthetic approaches to create the C1 stereocenter. *Molecules*. 19(2):1544–1567.

Laviţă, S. I., Aro, R., Kiss, B., Manto, M., & Duez, P. 2016. The role of β-carboline alkaloids in the pathogenesis of essential tremor. *The Cerebellum*, 15(3):276–284.

Li, S., Teng, L., Liu, W., Cheng, X., Jiang, B., Wang, Z., & Wang, C. H. 2016. Pharmacokinetic study of harmane and its 10 metabolites in rat after intravenous and oral administration by UPLC-ESI-MS/MS. *Pharmaceutical Biology*, 54(9):1768–1781.

Li, S., Yang, B., Zhang, Q., Zhang, J., Wang, J., & Wu, W. 2010. Synthesis and bioactivity of β-carboline derivatives. *Natural Product Communications*, 5(10).

Liao, Q., Li, Q., Zhao, Y., Jiang, P., Yan, Y., Sun, H., Liu, W., Feng, F., & Qu, W. 2020. Design, synthesis and biological evaluation of novel carboline-cinnamic acid hybrids as multi-functional agents for treatment of Alzheimer's disease. *Bioorganic Chemistry*, 99.

Lippke, K. P., Schunack, W. G., Wenning, W., & Mueller, W. E. 1983. β-Carbolines as benzodiazepine receptor ligands. 1. Synthesis and benzodiazepine receptor interaction of esters of β-carboline-3-carboxylic acid. *Journal of Medicinal Chemistry*, 26(4):499–503.

Liu, H. W., Ji, Q. T., Ren, G. G., Wang, F., Su, F., Wang, P. Y., Zhou, X., Wu, Z. B., Li, Z., & Yang, S. 2020. Antibacterial functions and proposed modes of action of novel 1,2,3,4-tetrahydro-β-carboline derivatives that possess an attractive 1,3-diaminopropan-2-ol pattern against rice bacterial blight, kiwifruit bacterial canker, and citrus bacterial canker. *Journal of Agricultural and Food Chemistry*, 68(45):12558–12568.

Lood, C. S., Nieger, M., & Koskinen, A. M. P. 2015. Enantiospecific gram scale synthesis of (S)-eleagnine. *Tetrahedron*, 71(30):5019–5024.

Lopes-Ortiz, M. A., Panice, M. R., Borges de Melo, E., Ataide Martins, J. P., Baldin, V. P., Agostinho Pires, C. T., Caleffi-Ferracioli, K. R., Dias Siqueira, V. L., Bertin de Lima Scodro, R., Sarragiotto, M. H., & Cardoso, R. F. 2020. Synthesis and anti-Mycobacterium tuberculosis activity of imide-β-carboline and carbomethoxy-β-carboline derivatives. *European Journal of Medicinal Chemistry*, 187.

Lunagariya, N. A., Gohil, V. M., Kushwah, V., Neelagiri, S., Jain, S., Singh, S., & Bhutani, K. K. 2016. Design, synthesis and biological evaluation of 1,3,6-trisubstituted β-carboline derivatives for cytotoxic and anti-leishmanial potential. *Bioorganic and Medicinal Chemistry Letters*, 26(3):789–794.

Luo, B., & Song, X. 2021. A comprehensive overview of β-carbolines and its derivatives as anticancer agents. *European Journal of Medicinal Chemistry*, 224.

Ma, Y., Wu, H., Zhang, J., & Li, Y. 2013. Enantioselective synthesis and antimicrobial activities of tetrahydro-β-carboline diketopiperazines. Chirality, 25(10):656–662.

Maity, P., Adhikari, D., & Jana, A. K. 2019. An overview on synthetic entries to tetrahydro-β-carbolines. *In Tetrahedron*, 75(8):965–1028. Elsevier Ltd.

Maresh, J. J., Giddings, L. A., Friedrich, A., Loris, E. A., Panjikar, S., Trout, B. L., Stöckigt, J., Peters, B., & O'Connor, S. E. 2008. Strictosidine synthase:Mechanism of a Pictet-Spengler catalyzing enzyme. *Journal of the American Chemical Society*, 130(2):710–723.

Mishra, B. B., & Tiwari, V. K. 2011. Natural products:An evolving role in future drug discovery. *European Journal of Medicinal Chemistry,* 46(10):4769–4807.

Moloudizargari, M., Mikaili, P., Aghajanshakeri, S., Asghari, M., & Shayegh, J. 2013. Pharmacological and Therapeutic effects of Peganum harmala and its main alkaloids. *Pharmacognosy Reviews*, 7(14):199–212.

Nenaah, G. 2010. Antibacterial and antifungal activities of (beta)-carboline alkaloids of Peganum harmala (L) seeds and their combination effects. *Fitoterapia*, 81(7):779–782.

Ogawa, Y., Adachi, J., & Tatsuno, Y. 1993. Accumulation of 1-methyl-tetrahydro-B-carboline-3-carboxylic acid in blood and organs of rat. A possible causative substance of eosinophilia-myalgia syndrome associated with ingestion ofl-tryptophan. *Archives of Toxicology*, 67(4):290–293.

Olmedo, G. M., Cerioni, L., González, M. M., Cabrerizo, F. M., Rapisarda, V. A., & Volentini, S. I. 2017. Antifungal activity of β-carbolines on Penicillium digitatum and Botrytis cinerea. *Food Microbiology*, 62:9–14.

Pakhare, D. S., & Kusurkar, R. S. 2015. Synthesis of tetrahydro-β-carbolines, β-carbolines, and natural products, (±)-harmicine, eudistomin U and canthine by reductive Pictet Spengler cyclization. *Tetrahedron Letters*, 56(44):6012–6015.

Pfau, W., & Skog, K. 2004. Exposure to β-carbolines norharman and harman. *Journal of Chromatography B*, 802(1):115–126.

Piechowska, P., Zawirska-Wojtasiak, R., & Mildner-Szkudlarz, S. 2019. Bioactive β-carbolines in food:A review. *Nutrients*, 11(4).

Poindexter Jr, E. H., & Carpenter, R. D. 1962. The isolation of harmane and norharmane from tobacco and cigarette smoke. *Phytochemistry*, 1(3):215–221.

Quintana, V. M., Piccini, L. E., Panozzo Zénere, J. D., Damonte, E. B., Ponce, M. A., & Castilla, V. 2016. Antiviral activity of natural and synthetic β-carbolines against dengue virus. *Antiviral Research*, 134:26–33.

Rajesh, R. P., & Murugan, A. 2019. Spectroscopic identification of brominated, non-brominated alkaloids and evaluation of antimicrobial activity of Eudistomin-I, Eudistomin H, from green ascidian Eudistoma viride. *Journal of Applied Pharmaceutical Science*, 9(2):116–123.

Ramesh, A., & Hyma, B. 1981. Traditional Indian medicine in practice in an Indian metropolitan city. *Social Science & Medicine. Part D:Medical Geography*, 15(1):69–81.

Recht, M. I., Fourmy, D., Blanchard, S. C., Dahlquist, K. D., & Puglisi, J. D. 1996. RNA sequence determinants for aminoglycoside binding to an A-site rRNA model oligonucleotide. *Journal of Molecular Biology*, 262(4):421–436.

Salehi, P., Babanezhad-Harikandei, K., Bararjanian, M., Al-Harrasi, A., Esmaeili, M. A., & Aliahmadi, A. 2016. Synthesis of novel 1,2,3-triazole tethered 1,3-disubstituted β-carboline derivatives and their cytotoxic and antibacterial activities. *Medicinal Chemistry Research*, 25(9):1895–1907.

Samoylenko, V., Rahman, M. M., Tekwani, B. L., Tripathi, L. M., Wang, Y. H., Khan, S. I., Khan, I. A., Miller, L. S., Joshi, V. C., & Muhammad, I. 2010. Banisteriopsis caapi, a unique combination of MAO inhibitory and antioxidative constituents for the activities relevant to neurodegenerative disorders and Parkinson's disease. *Journal of Ethnopharmacology*, 127(2):357–367.

Samundeeswari, S., Chougala, B., Holiyachi, M., Shastri, L., Kulkarni, M., Dodamani, S., Jalalpur, S., Joshi, S., Dixit, S., Sunagar, V., & Hunnur, R. 2017. Design and synthesis of novel phenyl -1, 4-beta-carboline-hybrid molecules as potential anticancer agents. *European Journal of Medicinal Chemistry*, 128:123–139.

Sandler, J. S., Colin, P. L., Hooper, J. N. A., & Faulkner, D. J. 2002. Cytotoxic β-carbolines and cyclic peroxides from the Palauan sponge Plakortis nigra. *Journal of Natural Products*, 65(9):1258–1261.

Savariz, F. C., Foglio, M. A., de Carvalho, J. E., Ruiz, A. L. T. G., Duarte, M. C. T., da Rosa, M. F., Meyer, E., & Sarragiotto, M. H. 2012. Synthesis and evaluation of new β-carboline-3-(4-benzylidene)- 4H-oxazol-5-one derivatives as antitumor agents. *Molecules*, 17(5):6100–6113.

Schott, Y., Decker, M., Rommelspacher, H., & Lehmann, J. 2006. 6-Hydroxy-and 6-methoxy-β-carbolines as acetyl-and butyrylcholinesterase inhibitors. *Bioorganic & Medicinal Chemistry Letters*, 16(22):5840–5843.

Schultes, R. E. 1982. The beta-carboline hallucinogens of South America. *Journal of Psychoactive Drugs*, 14(3):205–220.

Schwarz, M. J., Houghton, P. J., Rose, S., Jenner, P., & Lees, A. D. 2003. Activities of extract and constituents of Banisteriopsis caapi relevant to parkinsonism. *Pharmacology Biochemistry and Behavior*, 75(3):627–633.

Sharma, S., Singh, K., & Singh, S. 2023. Synthetic strategies for Quinoline based derivatives as potential bioactive heterocycles. *Current Organic Synthesis*, 20.

Sharma, S., & Singh S. 2022. Synthetic Routes to Quinoline-Based Derivatives having Potential Anti-Bacterial and Anti-Fungal Properties. *Current Organic Chemistry*, 26(15).

Sharma, S., & Singh, S. 2023. Molecular docking study for binding affinity of 2H-thiopyrano [2, 3-b] quinoline derivatives against CB1a. *Interdisciplinary Perspectives on Infectious Diseases*, 2023.

Sharma, S., & Singh, S. 2023. The biological and pharmacological potentials of Indole-based heterocycles. *Letters in Organic Chemistry*, 20(8):711–729.

Sharma, S., Das, A. K., & Singh, S. 2023. Bioactive Indole Heterocycles and their synthetic routes:A comprehensive review. *Current Trends in Biotechnology and Pharmacy*, 17(3s):1141–1158.

Sharma, S., Monga, Y., Gupta, A., & Singh, S. 2023. 2-Oxindole and related heterocycles:synthetic methodologies for their natural products and related derivatives. *RSC Advances*, 13(21):14249–14267.

Sharma, S., Monga, Y., Gupta, A., & Singh, S. 2023. In silico screening and molecular docking study of quinoline based compounds with Human kallikrein 7 in complex with 1, 4-diazepane-7-one 1-acetamide derivative receptor target for potential antibacterials. *Journal of Molecular Chemistry*, 3(1):585–585.

Shin, H. J., Lee, H. S., & Lee, D. S. 2010. The synergistic antibacterial activity of 1-Acetyl-β-Carboline and β-Lactams against Methicillin-Resistant Staphylococcus aureus (MRSA). *Journal of Microbiology and Biotechnology*, 20(3):501–505.

Singh, V., & Batra, S. 2012. 1-Formyl-9H-β-Carboline:A useful scaffold for synthesizing substituted-and fused β-Carbolines. *Current Organic Synthesis*, 9(4):513–528.

Smirnova, O. B., Golovko, T. V., & Granik, V. G. 2011. Carbolines. Part 2:Comparison of some of the properties of α-, γ-, and δ-carbolines. *Pharmaceutical Chemistry Journal*, 45(7):389–400.

Song, Y., Wang, J., Teng, S. F., Kesuma, D., Deng, Y., Duan, J., & Sim, M. M. 2002. β-Carbolines as specific inhibitors of cyclin-Dependent kinases. *Bioorganic & Medicinal Chemistry Letters*, 12(7):1129–1132.

Soni, R., Muller, L., Furet, P., Schoepfer, J., Stephan, C., Zumstein-Mecker, S., Fretz, H., & Chaudhuri, B. 2000. Inhibition of cyclin-dependent kinase 4 (Cdk4) by fascaplysin, a marine natural product. *Biochemical and Biophysical Research Communications*, 275(3):877–884.

Spahn, C. M. T., & Prescott, C. D. 1996. Throwing a spanner in the works:Antibiotics and the translation apparatus. *Journal of molecular medicine*, 74(8):423–439.

Stoic, S. 1999. Indole derivatives as neuroprotectants. *Life Sciences*, 65(18-19):1943–1950.

Suzuki, K., Nomura, I., Ninomiya, M., Tanaka, K., & Koketsu, M. 2018. Synthesis and antimicrobial activity of β-carboline derivatives with N 2 -alkyl modifications. *Bioorganic and Medicinal Chemistry Letters*, 28(17):2976–2978.

Totsuka, Y., Ushiyama, H., Ishihara, J., Sinha, R., Goto, S., Sugimura, T., & Wakabayashi, K. 1999. Quantification of the co-mutagenic β-carbolines, norharman and harman, in cigarette smoke condensates and cooked foods. *Cancer Letters*, 143(2):139–143.

Tran, T. D., Pham, N. B., Ekins, M., Hooper, J. N. A., & Quinn, R. J. 2015. Isolation and total synthesis of stolonines A-C, unique taurine amides from the Australian marine tunicate Cnemidocarpa stolonifera. *Marine Drugs*, 13(7):4556–4576.

Tsuchiya, H., Sato, M., Hayashi, H., Kato, H., Kureshiro, H., & Hayashi, T. 1996. Simultaneous determination of tetrahydro-β-carbolines and β-carbolines. *Chromatographia*, 43(7):419–425.

Tsuchiya, H., Shimizu, H., & Iinuma, M. 1999. Beta-carboline alkaloids in crude drugs. *Chemical and Pharmaceutical Bulletin*, 47(3):440–443.

Tsukiyama, R. I., Katsura, H., Tokuriki, N., & Kobayashi, M. 2002. Antibacterial activity of licochalcone A against spore-forming bacteria. *Antimicrobial Agents and Chemotherapy*, 46(5):1226–1230.

Venkataramana Reddy, P. O., Hridhay, M., Nikhil, K., Khan, S., Jha, P. N., Shah, K., & Kumar, D. 2018. Synthesis and investigations into the anticancer and antibacterial activity studies of β-carboline chalcones and their bromide salts. *Bioorganic and Medicinal Chemistry Letters*, 28(8):1278–1282.

Volk, R. B. 2005. Screening of microalgal culture media for the presence of algicidal compounds and isolation and identification of two bioactive metabolites, excreted by the cyanobacteria Nostoc insulare and Nodularia harveyana. *Journal of Applied Phycology*, 17(4):339–347.

Wang, W., Nam, S. J., Lee, B. C., & Kang, H. 2008. β-carboline alkaloids from a Korean tunicate Eudistoma sp. *Journal of Natural Products*, 71(2):163–166.

Wang, Z., Niu, J., Zeng, H., & Li, C. J. 2019. Construction of spirocyclic tetrahydro-β-carbolines via cross-annulation of phenols with tryptamines in water. *Organic Letters*, 21(17):7033–7037.

Wang, Z. X., Xiang, J. C., Cheng, Y., Ma, J. T., Wu, Y. D., & Wu, A. X. 2018. Direct Biomimetic synthesis of β-Carboline Alkaloids from two amino acids. *Journal of Organic Chemistry*, 83(19):12247–12254.

Williams, R. J., & Heymann, D. L. 1998. Containment of antibiotic resistance. *Science*, 279(5354):1153–1154.

Wojtowicz, E., Zawirska-Wojtasiak, R., Przygoński, K., & Mildner-Szkudlarz, S. 2015. Bioactive β-carbolines norharman and harman in traditional and novel raw materials for chicory coffee. *Food Chemistry*, 175:280–283.

Woodcock, J., Moazed, D., Cannon, M., Davies, J., & Noller, H. F. 1991. Interaction of antibiotics with A- and P-site-specific bases in 16S ribosomal RNA. *EMBO Journal*, 10(10):3099–3104.

Wu, S., Fu, Y., Yan, R., Wu, Y., Lei, X., & Ye, X. S. 2010. Synthesis of neamine-carboline conjugates for RNA binding and their antibacterial activities. *Tetrahedron*, 66(19):3433–3440.

Xu, Z., Chang, F. R., Wang, H. K., Kashiwada, Y., McPhail, A. T., Bastow, K. F., Tachibana, Y., Cosentino, M., & Lee, K. H. 2000. Anti-HIV agents 451 and antitumor agents 205.2 Two new sesquiterpenes, leitneridanins A and B, and the cytotoxic and anti-HIV principles from Leitneria floridana. *Journal of Natural Products*, 63(12):1712–1715.

Ye, X. S., & Zhang, L. H. 2002. Aminoglycoside mimetics as small-molecule drugs targeting RNA. *Current Medicinal Chemistry*, 9(9):929–939.

Youssef, D. T. A. 2005. Hyrtioerectines A-C, cytotoxic alkaloids from the red sea sponge Hyrtios erectus. *Journal of Natural Products*, 68(9):1416–1419.

Zawirska-Wojtasiak, R., Piechowska, P., Wojtowicz, E., Przygoński, K., & Mildner-Szkudlarz, S. 2018. Bioactivity of selected materials for coffee substitute. *PLoS One*, 13(11):e0206762.

Zhang, J., Li, L., Dan, W., Li, J., Zhang, Q., Bai, H., & Wang, J. 2015. Synthesis and Antimicrobial activities of 3-Methyl-β-Carboline derivatives. *Natural Product Communications*, 10(6).

Zheng, W., Wang, S., Barnes, L. F., Guan, Y., & Louis, E. D. 2000. Determination of harmane and harmine in human blood using reversed-phased high-performance liquid chromatography and fluorescence detection. *Analytical Biochemistry*, 279(2):125–129.

Zhou, T. S., Ye, W. C., Wang, Z. T., Che, C. T., Zhou, R. H., Xu, G. J., & Xu, L. S. 1998. β-Carboline alkaloids from Hypodematium squamuloso-pilosum. *Phytochemistry*, 49(6):1807–1809.

3 β-Carbolines as Antifungal Agents
Synthesis and Biological Study

Dattatraya Pansare, Rohini Shelke, Rajita Ingle, Shankar Thopate, Mubarak Shaikh, Aniket Sarkate, and Rajendra P. Pawar

3.1 INTRODUCTION

In recent years, fungous contaminations have increased, resulting in approximately 1.4 million deaths annually among immune compromised patients worldwide (Felix et al. 2017). β-Carbolines, which are fused indole-pyridine based tricyclic heterocyclic compounds, can be found in various plant kingdoms (Youssef 2001; Chen et al. 2010). Compounds containing β-Carbolines retain a comprehensive variety of biotic actions. These activities includes antitumoral effect (Kumar et al. 2017), DNA intercalation (Sobhani et al. 2002), CDK inhibition (Song et al. 2002), embarrassment of DNA topoisomerase-I and II (Nil et al. 2003), palliative effects (Lippke et al. 1983), and antimicrobial (Savariz et al. 2012), anti-inflammatory (Chen et al. 2010), antithrombin (Liu et al. 2010), antiviral (Hudson et al. 1986), anticonvulsant (Braestrup et al. 1980), antileishmanial effects (Barea et al. 2018), anti-HIV activity (Ashok et al. 2015), and antimalarial properties (Pierrot et al. 2019). Nitrogen-containing hetero-aromatic structures play a crucial role as intermediates in the synthesis of promising antifungal candidates (Song et al. 2014a). The most prevalent fungal infections of pathogens in humans, counting *Candida albicans* and non-albicans species *Cryptococcus neoformans (C. neoformans)*, and *Candida glabrata (C. glabrata)*, are responsible for approximately 75% of all fungal contagions and this study reported that these are the quarter leading source of nosocomial diseases (Richardson et al. 2010). Despite the discovery of numerous antifungal agents, pathogenic fungi continuously develop resistance to these drugs (Sanglard et al. 2016). β-Carboline, or norharmane, represents a notable class of alkaloids derived from indole characterized by a tricyclic pyrido[3,4-b]indole ring arrangement. This class of compounds encompasses around 64 distinct β-carboline alkaloids, distributed over a minimum of eight plant families. The initial members of this alkaloid class are identified as harmine (1) and harmaline (2) as depicted in Figure 3.1. The first discovery of harmaline, a prototypical β-carboline alkaloid, took place in 1841 when it was isolated

 DOI: 10.1201/9781351058032-3

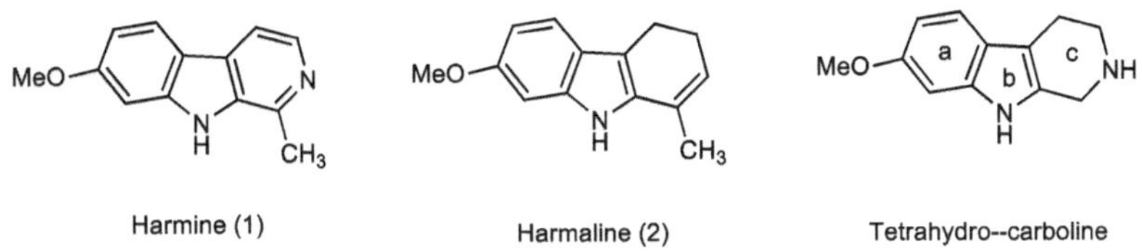

FIGURE 3.1 Harmine (1), Harmaline (2) and Tetrahydro-β-carboline.

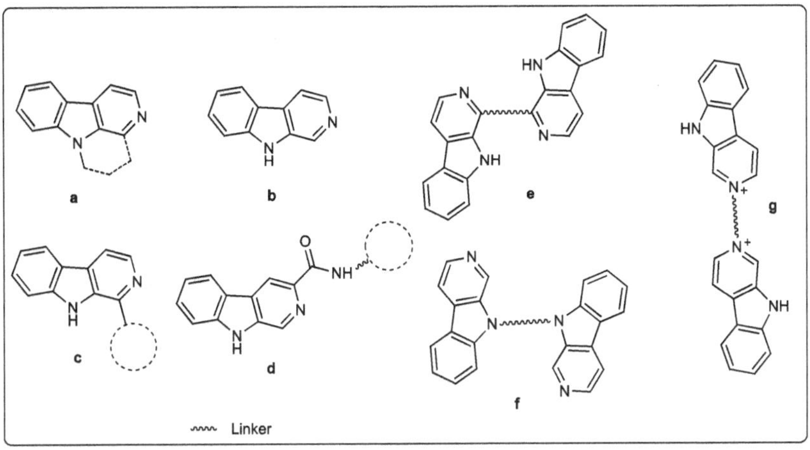

FIGURE 3.2 Representative examples of different β-carbolines and their derivatives (a-g).

from Peganum Harmala (Syrian Rue) [1c]. Subsequently, in 1847, Fritzsche also successfully isolated hormine from the seeds of the same plant (Fritzsche et al. 1847).

Despite being entirely aromatic, β-carboline compounds, including partially saturated derivatives such as 1,2,3,4-tetrahydro-β-carbolines (THBCs), and 3,4-dihydro-β-carboline, are renowned for their normal incidence and diverse biological activities (Figure 3.1). THBCs, a class of compounds, exist in numerous humble and multifaceted forms, both regular and artificial (Grate et al. 1996; Hesse et al. 2002). In THBCs, as shown in Figure 3.1, three rings are referred to as a, b, and c-rings.

β-carbolines, a class of heterocyclic compounds, has emerged as an important class of medicinal compounds owing to their diverse living actions. β-carbolines are characterized by a tricyclic arrangement containing of a pyridine ring merged to an indole moiety. They exhibit an extensive variety of pharmacological possessions like antifungal activities. In this introduction, we focus specifically on the synthesis of β-carboline and their antifungal activity.

Among the various naturally occurring carbolines including α-, β-, γ-, and δ-carbolines, β-carbolines are becoming more and more significant because of their versatile submissions (Figure 3.2) (Dai et al. 2018).

Carbolines are recognized as noteworthy alkaloids with biological significance. Tetrahydro-β-carboline, characterized by its inflexible indole ring and adaptable piperidine structure, displays substantial reactivity with diverse receptors and plays a dual function as an acceptor and donor of 'H' in the realm of drug development. Both naturally occurring and synthetic derivatives of tetrahydro-β-carboline demonstrated effectiveness against HPV, TMV, malaria, fungal infections, thrombosis, and leishmaniasis in a study carried out by Wang et al. (2021).

3.2 SYNTHESIS OF β-CARBOLINES WITH A MARINE ALKALOID FOCUS

Marine alkaloids are widely recognized for their broad spectrum of biological activities. Notably, a marine alkaloid based on β-carboline, containing a pityriacitrin scaffold, exhibits antifungal properties (Mayser et al. 2002; Zhang et al. 2011; Li et al. 2012; Irlinger et al. 2005; Mexia et al. 2015; Tan et al. 2015;. Considering its biological significance, Xu and colleagues developed new derivatives **(5a–d)** as shown in Scheme 3.1 (Xu et al. 2019).

A series of such complexes, β-carboline derivatives centered around pityriacitrin were prepared through the linkage of carboxylic acid (compound 4) using a 1-hydroxybenzotriazole (HOBt)-based approach. Compound 4 was generated by reacting the methyl ester of L-tryptophan (compound 1) and 3-acetyl indole with iodine/DMSO in a Pictet–Spengler reaction to yield 1-(3-indoloyl-β-carboline-3-carboxylic

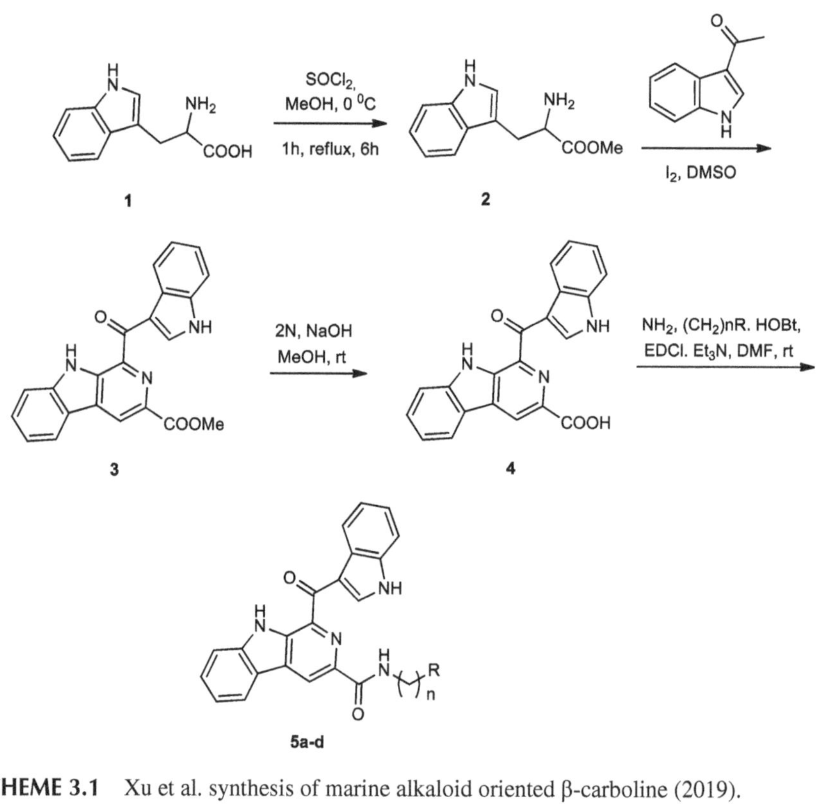

SCHEME 3.1 Xu et al. synthesis of marine alkaloid oriented β-carboline (2019).

acid) methyl ester (compound 2). The subsequent de-esterification process led to the production of the desired compounds.

3.3 β-CARBOLINE PRODUCTS AS ANTIFUNGAL MEDIATORS

Multiple 1,4-disubstituted 1,2,3-triazoles were synthesized by Huo et al. (2018), including those discussed in previous studies (Ruddarraju et al. 2016; Kant et al. 2016; Huo et al. 2017) and their harmonic derivatives. Their primary emphasis was on 9-fluoro-substituted hormones, as explored in related research (Pommier et al. 2010; Nenaah 2011; Abbasipour et al. 2010). The antifungal properties of the 1,2,3-triazoles 1a-c, as illustrated in Figure 3.3, were evaluated in relation to a number of phytopathogenic species, such as *Botrytis cinerea*, *Fusarium oxysporum*, *R. solani*, and sunflower sclerotinia rot. Interestingly, compound 1b showed improved fungicidal action against sunflower sclerotinia rot (SCR) and rice spikelet rot (RSR) in comparison to the commercial fungicides carbendazim and azoxystrobin.

In 2018, Zhang and colleagues devised a novel synthetic route for an antifungal compound targeting rice sheath blight caused by the R. solani. They successfully produced a benzoyl urea derivative, N-[1-(3,4,5-trimethoxyphenyl)-9H-pyrido[3,4-b]indole-3-carbomoyl]-2,6-difluorobenzamide, through their innovative synthesis method at the C3-position 2 (Figure 3.4), which exhibited superior fungicidal activity

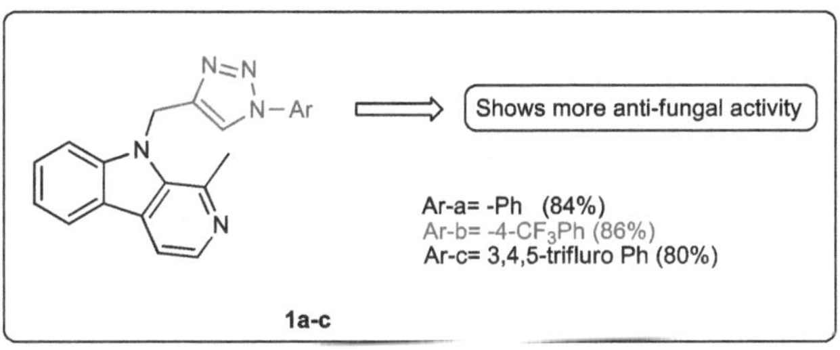

FIGURE 3.3 1,4-disubstituted-1,2,3-triazoles.

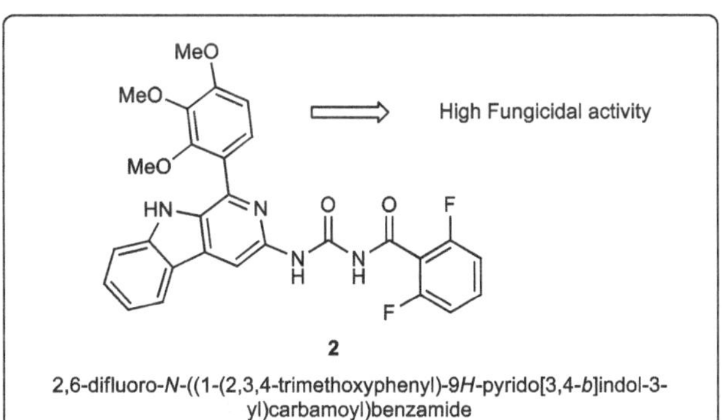

FIGURE3.4 2,6-difluoro-N-((1-(2,3,4-trimethoxyphenyl)-9H-pyrido[3,4-b]indol-3-yl)carbamoyl) benzamide.

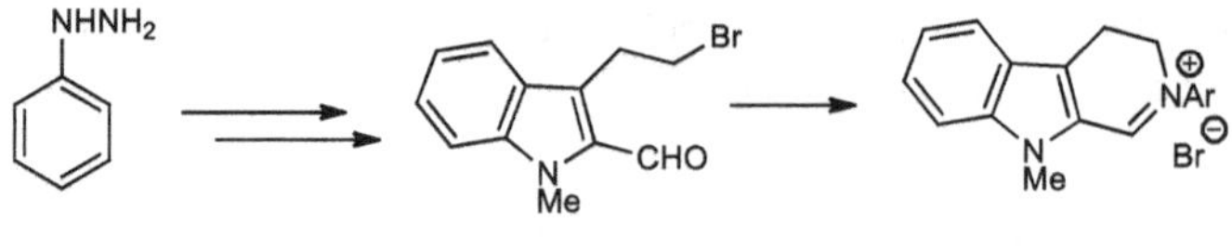

1 2 3

SCHEME 3.2 Zhou's strategy.

compared to benzoylthiourea. This increased activity can be attributed to its higher electronegativity, affecting receptor binding. However, the compound's fungicidal activity decreased, likely due to this same reason. Compound 2 displayed excellent fungicidal activity in isolated leaf assays and greenhouse experiments, with activity twice that of validamycin A.

In 2016, Zhou (as shown in Scheme 3.4) presented a study on the antifungal characteristics of 2-aryl-3,4-dihydro-β-carbolin-2-ium salts in collaboration with Hou et al. The research revealed that, when tested against various fungal strains, 10 of these compounds exhibited an average inhibition rate exceeding 80% at a concentration of 150 μM, with the majority of EC50 values falling within the range of 2.0 to 30.0 μM.

Schemes 3.3–3.5 were used to outline synthesis routes for β-carbolines that are 1-substituted. Scheme 3.3 shows the synthesis procedure for producing derivatives (4a–4l) of methyl 1-substituted β-carboline-3-carboxylate. This process initiated with L-tryptophan and related aldehydes through a Pictet–Spengler condensation. Subsequently, a conversion to methyl occurred under acidic conditions, using thionyl chloride and methanol, yielding 1-substituted-2,3,4,9-tetrahydro-β-carboline-3-carboxylate (3a-3l). One more step involves oxidizing 1-substituted-β-carboline-3-carboxylate derivatives (4a–4l) in dimethylformamide (DMF) using potassium permanganate. The application of sodium hydroxide-mediated hydrolysis led to the formation of 1-substituted-β-carboline-3-carboxylic acids **(5c-5l)** (Ashok et al. 2018; Shi et al. 2013). Furthermore, a straightforward synthesis under mild reaction conditions was carried out for the synthesis of methyl 1-substituted-β-carboline-3-carboxylates (4m-4p) via an adjustable one-step approach utilizing the Minisci reaction (Lin et al. 2011).

Scheme 3.4 shows the successful synthesis of β-carboline-1-amides (compounds 4q–4r) through the use of 1-hydroxybenzotriazole (HOBt) as the coupling reagent and N-(3-dimethylaminopropyl)–N'-ethylcarbodiimide hydrochloride (EDCl) using common acid intermediate 3 and various amines. In order to explore the impact of the carbamoyl group's location and the degree of unsaturation in the C ring (either pyridine or tetrahydropyridine) on the antifungal characteristics, ammonia was used to ammonolyze compounds 6, 7 (as per Scheme 5), and 8a-8j (outlined in Scheme 3), along with compounds 8q and 8r (from Scheme 4). Additionally, the structural and stereochemical features of the derivatives of β-carboline were confirmed by means of single-crystal X-ray analyses of **6, 4k**, and **8e**, complemented by MASS, ^{1}H NMR, and ^{13}C NMR spectra.

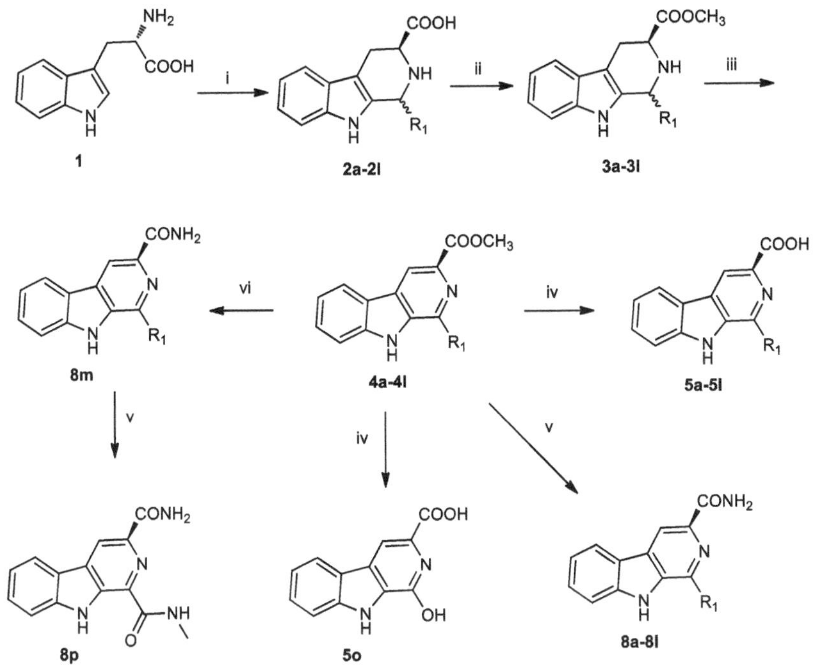

SCHEME 3.3 Synthesis of 1-substituted β-carbolines. i) Aldehyde, H_2SO_4, H_2O, rt, 24 h, 70–90%; ii) $SOCl_2$, MeOH, 0 °C-rt, 12 h, 71–91%; iii) KMnO4, DMF, 0 °C-rt, 24 h, 51–79%; iv) NaOH, MeOH, water, reflux, 6 h, 79–95%; v) $NH_3.H_2O$, MeOH, reflux, 24 h, 57–81%; vi) Reagent, H_2SO_4, 30% H_2O_2, $FeSO_4$, 0 °C-rt, 1-3 h, 45–65%.

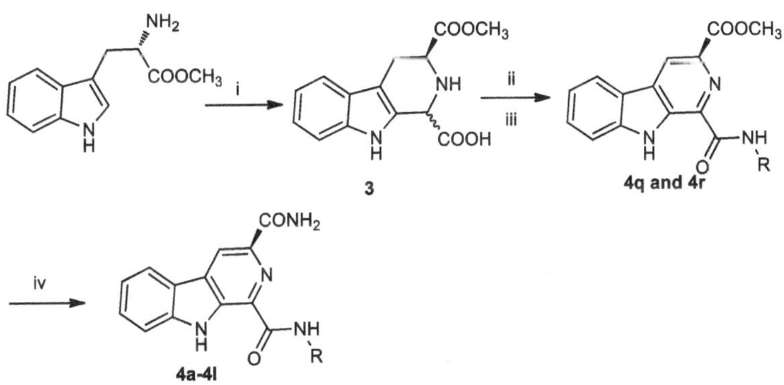

SCHEME 3.4 Synthesis of β-carboline-1-amides from L-tryptophan. i) Glyoxylic acid, H_2SO_4, H_2O, rt, 24 h, 93%; ii) RNH_2, EDCI, HOBt, triethylamine, CH_2Cl_2, 0 °C-rt, 12 h, 75–78%; iii) $KMnO_4$, DMF, 0 °C-rt, 24 h, 61–75%; iv) $NH_3.H_2O$, MeOH, reflux, 24 h, 70–74%.

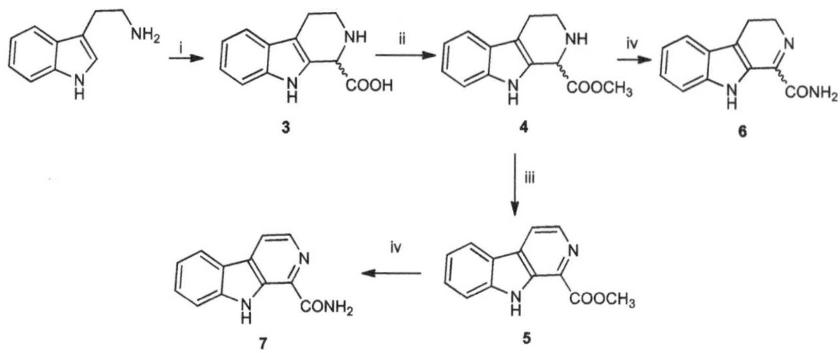

SCHEME 3.5 Synthesis of β-carboline-1-amides from tryptamine. i) Glyoxylic acid, H_2SO_4, H_2O, rt, 24 h, 90%; ii) $SOCl_2$, MeOH, 0 °C-rt, 12 h, 95%; iii) $KMnO_4$, DMF, 0 °C-rt, 24 h, 75%; iv) $NH_3.H_2O$, MeOH, reflux, 24 h, 76–82%.

3.4 ANTIFUNGAL ACTIONS OF β-CARBOLINE PRODUCTS

Two compounds, harmane and tetrahydroharmane, at a 50 µg/mL dose have demonstrated efficient inhibition of gamma-glutamyl transpeptidase (GGT). To enhance their potential as antifungal agents, various modifications were introduced at the first and third positions of the fundamental β-carboline structure **4a**. The newly engineered carbolines demonstrated varying degrees of antifungal efficacy against against nine diverse species of plant fungi like *Botrytis cinerea*, *Gaeumannomyces graminis var. Tritice*, *Sclerotinia sclerotiorum*, *Fusarium oxysporum*, *Thanataphorus cucumeris*, *Fusarium graminearum*, *Phytophthora capsici*, *Colletotrichum orbicularis*, and *Fusarium moniliforme* are among the nine diverse species of plant fungi.

A series of innovative hybrid aryl-1,2,3-triazole-β-carboline were synthesized and designed according to Scheme 3.6. The antifungal potential of these compounds was assessed in vitro using the mycelia growth inhibition assay at a concentration of 50 µg/mL against Rhizoctorzia solani, Fusarium oxysporum, Botrytis cinerea Pers., sunflower sclerotinia rot, and rape sclerotinia rot. The assessment revealed that compounds **3a, 3b**, and **3c** displayed significant antifungal action against Sclerotinia rot of sunflowers. Huo et al. (2018) found that these results provide an intriguing basis for the development of potent, novel antifungal agents.

A different investigation sought to explore the potential medicinal properties of β-carboline alkaloid compounds that inhibit C. gattii and Cryptococcus neoformans. Minimum inhibitory concentrations (MICs) were established in compliance with the directives provided by the Clinical and Laboratory Standards Institute for these alkaloids and derivatives VNI and VGI of *C. neoformans* and *C. gattii*. One particular active compound underwent further evaluation for its cytotoxicity against various genotypes of Candida albicans ATCC 36232, Candida neoformans (VNII, VNIII, and VNIV), Candida gattii (VGI, VGIII, and VGIV), MRC-5, and its effects on human fibroblast cell lines. The study also delved into the impact on fungal cells, including examination of the cell wall, ergosterol leakage, and nucleic acid changes (Cruz et al. 2019).

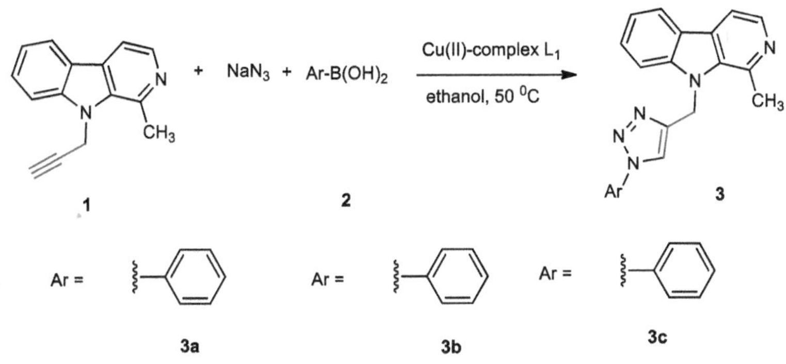

SCHEME 3.6 Synthesis of the 9-(1,2,3-triazolyl)-β-carboline hybrids **3a-c**.

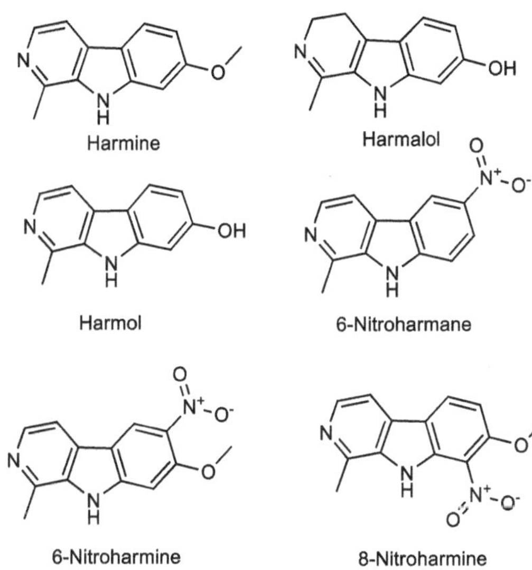

FIGURE 3.5 Structures of β-carboline compounds.

3.5 FUNGAL ACTIVITIES

Significant fungicidal effects against a range of fungi were also demonstrated by these alkaloids and their metabolites. The fungicidal efficacy was substantially impacted by the kind and positioning of substituents on the benzene ring (A ring) when the core structure was β-carboline, with the C ring being pyridine. Among these compounds, Compound 5 demonstrated outstanding fungicidal activity against the majority of the fungi tested, while Compound 4 exhibited moderate effectiveness. Compound 8, which featured a pivaloyloxy group, showed superior fungicidal activity when compared to the acetoxy-containing Compound 9. Additionally, some compounds demonstrated specialized fungicidal efficacy against particular fungi. For instance, Hermane exhibited over 90% efficacy in resistance to Cercospora arachidicola

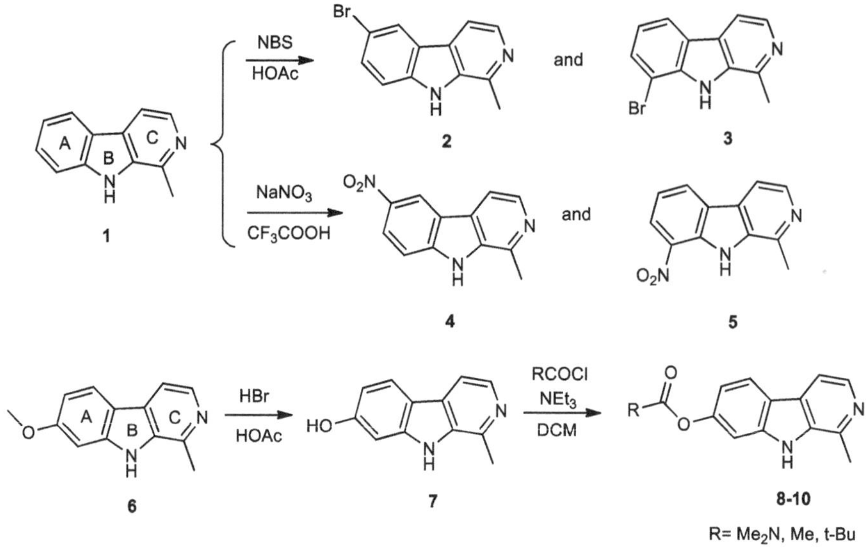

SCHEME 3.7 Synthesis of compounds.

hori, and Compound 2 displayed over 90% efficacy against Physalospora piricola, surpassing its effectiveness against other fungi (Song et al. 2014b).

3.6 CONCLUSION

In conclusion, the synthesis of β-carbolines offers a versatile approach to access these structurally diverse derivatives with extensive variety of biological activities. Their antifungal properties make them particularly attractive as potential therapeutic agents against fungal infections. Additional investigation and advancement in this area offer potential for uncovering innovative β-carboline derivatives that exhibit enhanced antifungal properties, thereby playing a role in the advancement of novel antifungal medications. The appliance of antifungal action of β-carbolines is believed to involve interference with fungal cell membrane integrity and function. Additionally, β-carbolines have been reported to inhibit crucial enzymes and processes involved in fungal cell wall synthesis and maintenance, further compromising fungal cell viability. Furthermore, β-carbolines have demonstrated synergistic effects when used in combination with existing antifungal drugs, enhancing their efficacy and reducing the development of drug resistance. This makes them promising candidates for combination therapy against fungal infections.

REFERENCES

Abbasipour, H., M. Mahmoudv, F. Rastegar, M. Basij. 2010. Insecticidal activity of Peganum harmala seed extract against the diamond back moth. *Plutellaxylostella, Bull. Insectol.* 63: 259–63.

Ashok, P., S. Chander, J. Balzarini, C. Pannecouque, S. Murugesan. 2015. Design, synthesis of new β-carboline derivatives and their selective anti-HIV-2 activity. *Bioorg. Med. Chem. Lett.* 25: 1232–35.

Ashok, P., S. Chander, T.K. Smith, M. Sankaranarayanan. 2018. Design, synthesis and biological evaluation of piperazinyl-b-carbolinederivatives as anti-leishmanial agents. *Eur. J. Med. Chem.* 150: 559–66.

Barea, P., V.A. Barbosa, D.L. Bidoia, J.C. de Paula, T.F. Stefanello, W.F. da Costa, C.V. Nakamura, M.H. Sarragiotto. 2018. Synthesis, antileishmanial activity and mechanism of action studies of novel β-carboline-1,3,5-triazine hybrids. *Eur. J. Med. Chem.* 150: 579–90.

Braestrup, C., M. Nielsen, C.E. Olsen. 1980. Urinary and brain beta-carbo- line-3-carboxylates as potent inhibitors of brain benzodiazepine receptors. *Proc. Natl. Acad. Sci. U.S.A.* 77: 2288–92.

Chen, Y.F., P.C. Kuo, H.H. Chan, I.J. Kuo, F.W. Lin, C.R. Su, M.L. Yang, D.T. Li, T.S. Wu. 2010. β-Carboline alkaloids from Stellaria dichotoma var. lanceolata and their anti-inflammatory activity. *J. Nat. Prod.* 73: 1993–98.

Cruz, K.S., E.S. Lima, M.J. Amazonas da Silva, E.S. de Souza, A. Montoia, Adrian M. Pohlit, J.V. Braga de Souza. 2019. Screening and antifungal activity of a β-carboline derivative against cryptococcus neoformans and C. gattii. *Hindawi, Intern. J. Micro.* 2019: 1–8. https://doi.org/10.1155/2019/7157845

Dai, J., W. Dan, U. Schneider, J. Wang. 2018. β-carboline alkaloid monomers and dimers: Occurrence, structural diversity, and biological activities. *Eur. J. Med. Chem.* 157: 622–56.

Felix, B., G. Sara, O.O. Rita, W.D. David. 2017. Global and multi-national prevalence of fungal diseases-estimate precision. *J. Fungi. (Basel).* 3: 57.

Fritzsche, J. 1847. Bestandtheile der Samen von Perganum hamala. *Justus Liebigs Ann. Chem.* 64: 360–64.

Grate, J.W., G.C. Frye. 1996. New metal oxide sensors. In: H. Baltes, W. G€opel, J. Hesse (Eds.). *Sensors Update*, Wiley-VCH, Weinheim. 2: 10–20.

Hesse, M. 2002. Alkaloids: Nature's curse or blessing. In: P. M. Wallimann (Ed.). *Alkaloids: Nature's Curse or Blessing*, Wiley-VCH, Zürich, Switzerland: 14–27.

Hou, Z., L.F. Zhu, X.C. Yu, M.Q. Sun, F. Miao, L. Zhou. 2016. Design, Synthesis, and structureeactivity relationship of new 2-aryl-3,4-dihydro-b-carbolin-2-ium salts as antifungal agents. *J. Agric. Food Chem.* 64: 2847–54.

Hudson, J.B., E.A. Graham, G.H.N. Towers. 1986. Antiviral effects of harmine, a photoactive β-carboline alkaloid. *Photochem. Photobiol.* 43: 21–26.

Huo, J.P., H. Hu, M. Zhang, X. Hu, M. Chen, D. Chen, J. Liu, G. Xiao, Y. Wang., Z. Wen. 2017. A mini review of the synthesis of poly-1,2,3-triazole-based functional materials. *RSC Adv.* 7: 2281–87.

Huo, X.Y., L. Guo, X.F. Chen, Y.T. Zhou, J. Zhang, X.Q. Han, B. Dai. 2018. Design, synthesis and antifungal activity of novel aryl-1,2,3-triazole-β-carboline hybrids. *Molecules.* 23: 1344.

Irlinger, B., A. Bartsch, H.J. Kramer, P. Mayser, W. Steglich. 2005. New tryptophan metabolites from cultures of the lipophilic yeast Malassezia furfur. *Helv. Chim. Acta.* 88: 1472–85.

Kant, R., D. Kumar, D. Agarwal, R.D. Gupta, R. Tilak, S.K. Awasthi, A. Agarwal. 2016. Synthesis of newer 1,2,3-triazole linked chalcone and flavone hybrid compounds and evaluation of their antimicrobial and cytotoxic activities. *Eur. J. Med. Chem.* 113: 34–49.

Kumar, S., A. Singh, K. Kumar, V. Kumar. 2017. Recent insights into synthetic β-carbolines with anti-cancer activities. *Eur. J. Med. Chem.* 142: 48–73.

Li, S., Y. Zhang, Y. Li, X. Li, L. Kong, C. Tan, S. Li, Y. Di, H. He, X. Hao. 2012. β-carboline alkaloids from the leaves of Trigonostemonlii Y.T. Chang. *Bioorg. Med. Chem. Lett.* 22: 2296–99.

Lin, G., Y. Wang, Q. Zhou, W. Tang, J. Wang, T. Lu. 2011. A facile synthesis of 1-substituted β-carboline derivatives via minisci-reaction. *Synth. Commun.* 41: 3541–50.

Lippke, K.P., W.G. Schunack, W. Wenning, W.E. Mueller. 1983. Conjugates of cate- cholam ines. 1. N-Alkyl-functionalized carboxylic acid congeners and amides related to isoproterenol. *J. Med. Chem.* 26: 499–503.

Liu, J., X. Jiang, M. Zhao, X. Zhang, M. Zheng, L. Peng, S. Peng. 2010. A class of 3S-2-aminoacyltetrahydro- β-carboline-3-carboxylic acids: Their facile synthesis, inhibition for platelet activation, and high in vivo anti-thrombotic potency. *J. Med. Chem.* 53: 3106–16.

Mayser, P., U. Schaefer, H.J. Krämer, B. Irlinger, W. Steglich. 2002. Patyriacitrin –an ultraviolet-absorbing indole alkaloid from the yeast Malassezia furfur. *Arch. Dermatol. Res.* 294: 131–34.

Mexia, N., G. Gaitanis, A. Velegraki, A. Soshilov, M.S. Denison, P. Magiatis. 2015. Pityriazepin and other potent AhR ligands isolated from Malassezia furfur yeast. *Arch. Biochem. Biophys.* 571: 16–20.

Nenaah, G. 2011. Toxicity and growth inhibitory activities of methanol extract and the beta-carboline alkaloids of Peganum harmala L. against two coleopteran stored-grain pests. *J. Stored Prod. Res.* 47: 255–26.

Nil, H. 2003. Possibility of the involvement of 9H-pyrido[3,4-b]indole (norharman) in carcinogenesis via inhibition of cytochrome P450-related activities and intercalation to DNA. *Mutation Res.* 541: 123–36.

Pierrot, D., V. Sinou, S.S. Bun, D. Parzy, N. Taudon, J. Rodriguez, E. Ollivier, D. Bonne. 2019. Design and synthesis of simplified speciophylline analogues and β-carbolines as active molecules against plasmodium falciparum. *Drug Dev. Res.* 80: 133–37.

Pommier, Y., E. Leo, H. Zhang, C. Marchand. 2010. DNA topoisomerases and their poisoning by anticancer and antibacterial drugs. *Chem. Biol.* 17: 421.

Richardson, M., F. Lass. 2010. Changing epidemiology of systemic fungal infections. *Clin. Microbiol. Infect.* 14: 5.

Ruddarraju, R.R., A.C. Murugulla, R. Kotla, M.C.B. Tirumalasetty, R. Wudayagiri, S. Donthabakthuni, R. Maroju, K. Baburao, L.S. Parasa. 2016. Design, synthesis, anticancer, antimicrobial activities and moleculardocking studies of theophylline containing acetylenes and theophylline containing 1,2,3-triazoles with variant nucleoside derivatives. *Eur. J. Med. Chem.* 123: 379–96.

Sanglard, D. 2016. Emerging threats in antifungal-resistant fungal pathogens. *Front Med (Lausanne).* 3: 1.

Savariz, F.C., M.A. Foglio, J.E. de Carvalho, A.L. Ruiz, M.C. Duarte, M.F. da Rosa, E. Meyer, M.H. Sarragiotto. 2012. Synthesis and evaluation of new β-Carboline-3-(4-benzylidene)-4H-oxazol-5-one derivatives as antitumor agents. *Molecules.* 17: 6100–13.

Shi, B., R. Cao, W. Fan, L. Guo, Q. Ma, X. Chen, G. Zhang, L. Qiu, H. Song. 2013. Design, synthesis and in vitro and in vivo antitumor activities of novel bivalent bcarbolines. *Eur. J. Med. Chem.* 60: 10–22.

Sobhani, A.M., S.A. Ebrahimi, M. Mahmoudian. 2002. An in vitro evaluation of human DNA topoisomerase I inhibition by peganum harmala L. seeds extract and its beta-carboline alkaloids. *J. Pharm. Pharm. Sci.* 5: 19–23.

Song, H., Y. Liu, Y. Liu, L. Wang, Q. Wang. 2014. Carboline,dihydro- β-carboline, tetra hydro-β-carboline alkaloids, and their derivatives. *J. Agric. Food Chem.* 62: 1010–18.

Song, H., Y. Liu, Y. Liu, L. Wang, Q. Wang. 2014. Synthesis and antiviral and fungicidal activity evaluation of β-carboline, dihydro-β-carboline, tetrahydro-β-carboline alkaloids, and their derivatives. *J. Agric. Food Chem.* 62: 1010–18.

Song, Y., J. Wang, S.F. Teng, D. Kesuma, Y. Deng, J. Duan, J.H. Wang, R.Z. Qi, M.M. Sim. 2002. Beta-carbolines as specific inhibitors of cyclin-dependent kinases. *Bioorg. Med. Chem. Lett.* 12: 1129–32.

Tan, C., Y. Zhang, N. Kong, Y. Di, X. Hao. 2015. Further alkaloids from the leaves of Trigonostemonlii. *Helv. Chim. Acta* 98: 72–77.

Wang, J., F. Gong, T. Liang, Z. Xie, Y.X. Yang, C. Cao, J. Gao, T. Lu, X. Chen. 2021. A review of synthetic bioactive tetrahydro-β-carbolines: A medicinal chemistry perspective. *Eur. J. Med. Chem.* 225: 113815.

Xu, L.S.T., Y. Zhang, K. Wang, Z. Yang, S. Ke. 2019. Synthesis and biological evaluation of marine alkaloid-oriented β-carboline analogues. *Eur. J. Med. Chem.* 168: 293–300.

Youssef, D.T.A. 2001. Alkaloids of the flowers of Hippeastrum vittatum. *J. Nat. Prod.* 64: 839–41.

Zhang, P., X. Sun, B. Xu, K. Bijian, S. Wan, G. Li, M. A-Jamali, T. Jiang. 2011. Total synthesis and bioactivity of the marine alkaloid pityriacitrin and some of its derivatives. *Eur. J. Med. Chem.* 46: 6089–97.

Zhang, Z.L., Y. Zeng, Z.Y. Jiang, B.S. Shu, V. Sethuraman, G.H. Zhong. 2018. Design, synthesis, fungicidal property and QSAR studies of novel β-carbolines con- taining urea, benzoylthiourea and benzoylurea for the control of rice sheath blight. *Pest Manag. Sci.* 74: 1736–46.

4 β-Carbolines as Kinase Inhibitors

Synthesis and Biological Studies

Pankaj Teli, Shivani Soni, Sunita Teli, Dinesh K. Agarwal, Dinesh K. Jangid, and Shikha Agarwal

4.1 INTRODUCTION

Continuously developing new drugs for treating different diseases remains a persistent challenge in the field of medicine. The findings for novel therapeutic agents have led to the discovery of many naturally occurring compounds with biological activities, such as alkaloids. One class of alkaloids that has gained attention in recent years is β-carbolines and their derivatives.

β-carbolines (β-Cs) are a class of naturally occurring alkaloids (Cao *et al.* 2007; Sarkar, Pandya, and Bhadra 2014; Poindexter Jr and Carpenter 1962; Ayipo *et al.* 2021) (Figure 4.1) and synthetic compounds that are widespread in nature. They have a tricyclic pyrido[3,4-b]indole ring in structure (Figure 4.2) and are produced by plants, animals, bacteria, and fungi, and are also synthesized in the laboratory. β-Cs contain a wide range of therapeutical activities, involving antileishmanial(Banoth *et al.* 2020), anti-inflammatory, anticancer (Aaghaz *et al.* 2021; Luo and Song 2021), antioxidant, antidiabetic, anti-HIV (Xu *et al.* 2000), antiviral (Formagio *et al.* 2009), and antimicrobial (Zhang *et al.* 2015) activities (Abinaya *et al.* 2022; Patel *et al.* 2012; Thatikayala *et al.* 2022). A lot of these biological activities are mediated by the ability of β-Cs to inhibit various kinases.

Kinases are a diverse group of enzymes that play a crucial part in many physiological activities like signal transduction and cell cycle progression and also in gene expression (Malumbres and Barbacid 2007; Roux and Blenis 2004). These enzymes transfer phosphate groups from ATP to specific substrates, thereby modulating their activity and function (Bossemeyer 1995). Diseases like cancer, autoimmune disorders, and some neurodegenerative diseases have been aroused due to irregularities in several kinase activities. Therefore, kinases have become attractive targets for drug development.

In this chapter, we explore the role of β-Cs and their derivatives as inhibitors of various kinases. We discuss the different kinases that have been targeted by β-Cs,

DOI: 10.1201/9781351058032-4

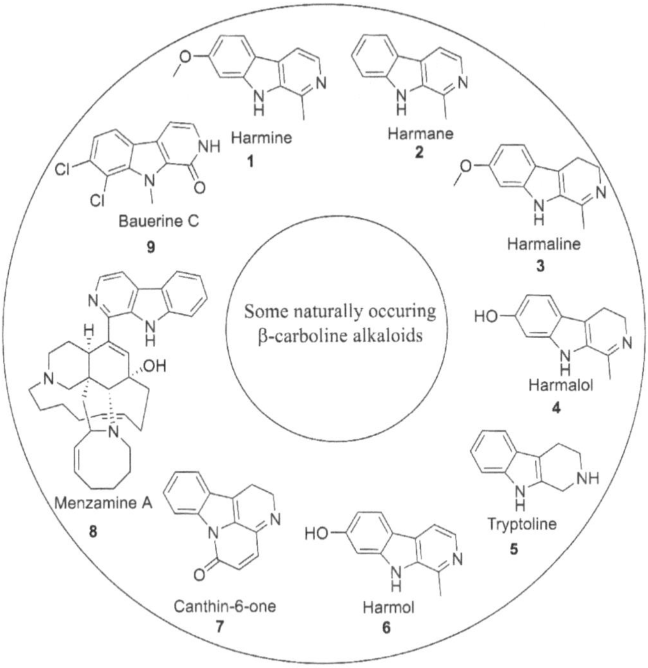

FIGURE 4.1 Various naturally occurring β-carboline alkaloids.

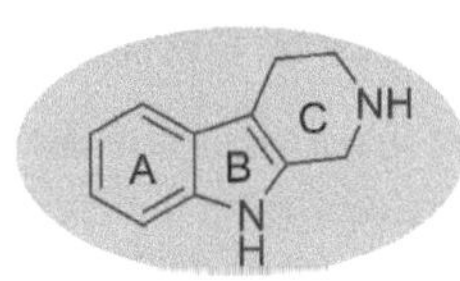

FIGURE 4.2 General structure of β-carboline.

including cyclin-dependent kinases (CDKs), IkappaB kinase (IKK), mitogen-activated protein kinase-activated protein kinase 2 (MAPKAPK2), haspin, dual-specificity tyrosine phosphorylation-regulated kinase 1A (DYRK1A), protein kinase B (PKB or Akt), CDC-like kinase (CLK), glycogen synthase kinase (GSK) (Figure 4.3), and so on. We also examine the biological activities of these kinase inhibitors, their mechanisms of action, and their potential as therapeutic interventions.

4.2 β-CARBOLINES AS DIFFERENT KINASE INHIBITORS

In recent times, β-carboline derivatives have been employed against several kinases such as CDK, MAPKAPK, DYRK, and so on and these molecules have shown significant kinase inhibitory activities and the activities are discussed here on the basis of different kinases.

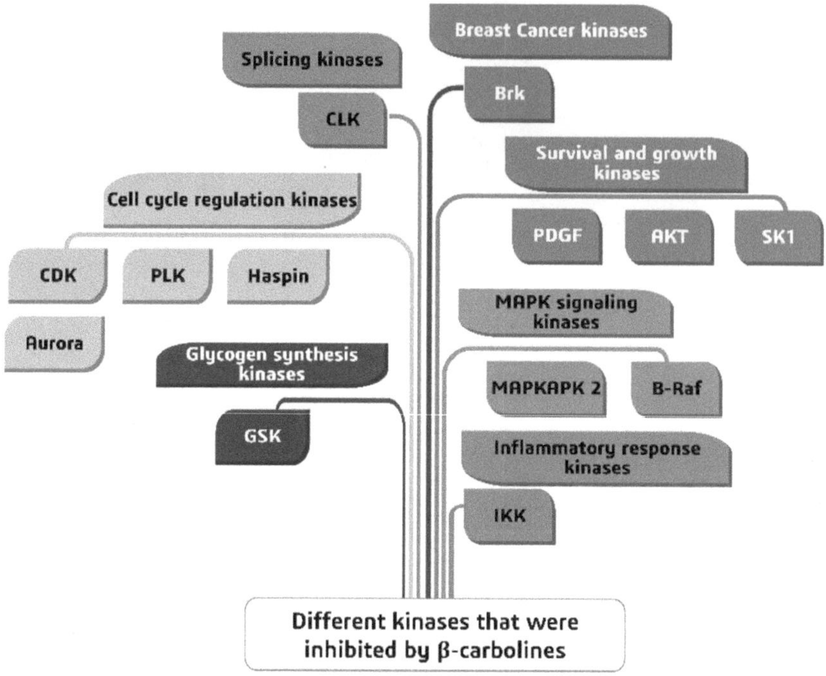

FIGURE 4.3 Several kinases that were inhibited by β-carboline derivatives.

4.2.1 β-Carbolines as Inhibitors of Cyclin-dependent Kinases (CDKs)

Cyclin-regulated kinases (CDKs) are a group of enzymes that primarily function in regulating different cellular processes (i.e., cell cycle progression, transcriptional regulation, metabolic pathways, and apoptosis). These enzymes play a vital role in regulating cell proliferation by managing cell cycle checkpoints and gene expression reacting to internal and external signals (Ding *et al.* 2020). Thus, inhibition of CDKs has been shown to be a favourable strategy for cancer therapy. Several β-C derivatives with significant CDK inhibitory activity are illustrated in Figure 4.4 and Figure 4.5.

A recent investigation has pinpointed harmine (1), a compound present in certain herbal remedies, as a distinctive inhibitor of Cdk5/p25, Cdk1/cyclin B, and Cdk2/cyclin A and with low micromolar IC_{50} values. Harmine showed negligible influence on other kinases and proved to be a competitive inhibitor of CDKs with ATP-Mg^{2+}. This indicates that it is bound to the ATP Mg^{2+} binding pocket of CDKs. It showed a potent inhibitory impact on the growth and multiplication of cancer cells, while showing no significant influence on quiescent fibroblasts, indicating its promising selectivity for targeting cancer cells. Moreover, it was found to hinder carcinoma cell DNA replication. Therefore, the outcomes of the study suggest that harmine's inhibitory action is specific to CDKs and cell division (Song *et al.* 2004).

A study demonstrated that β-Cs inhibit CDK2 and CDK5, with specific inhibition towards CDK2 observed in certain C1-substituted β-Cs such as 10, 11, and 12, while

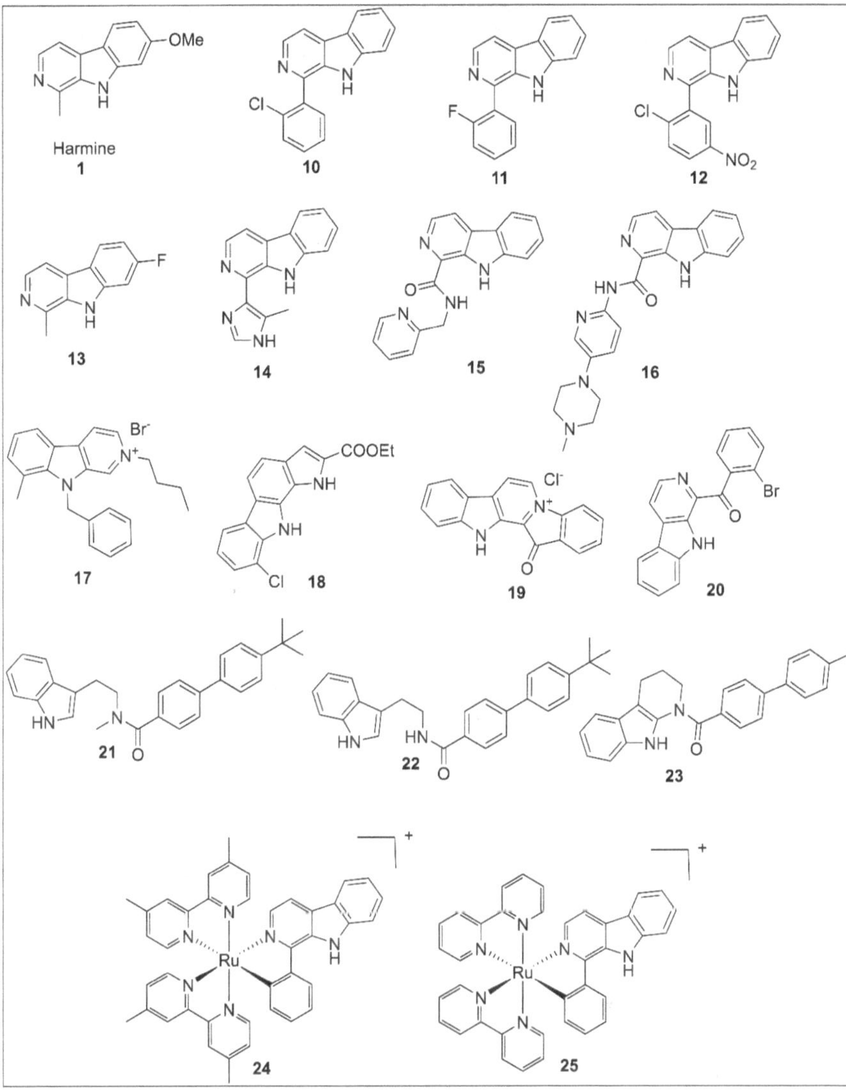

FIGURE 4.4 Several β-carboline derivatives as a potential CDK inhibitors.

1, 13, and 14 have shown high potency and selectivity as inhibitors of CDKs. The inhibitory activity of β-Cs on CDKs is influenced by two significant factors – the aromaticity index exhibited by the tricyclic system and the location of functional groups (Song *et al.* 2002).

A library of novel β-C compounds was developed, with compounds 15 and 16 exhibiting the most potent CDK4 enzymatic inhibition and antiproliferative activities. Furthermore, they demonstrated noteworthy inhibition of oncogenesis in HCT116 tumor heterograft models while exhibiting minimal toxicity in in vivo and in vitro

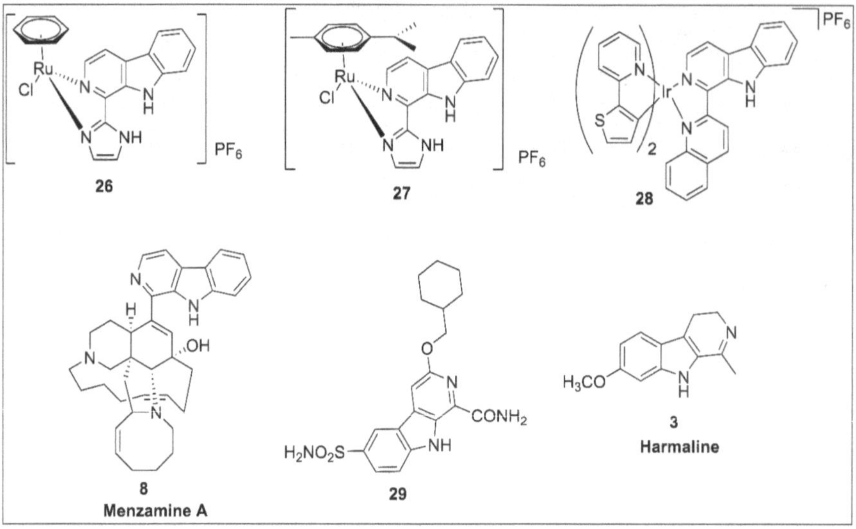

FIGURE 4.5 Various β-carbolines with significant CDK inhibitory activity.

trials. The CDK4 inhibitory activity of compounds 15 and 16 was demonstrated with IC_{50} values of 0.38 μM and 6.51 μM, respectively. The anti-HCT116 activity of 15 and 16 was observed to be potent in in vitro investigations, including mitigation of colony formation, intrusion and migration, initiation of apoptosis, and cell cycle blockage in G1 phase. According to the studies conducted in silico, it has been illustrated that they display favorable biological effects and possess acceptable physicochemical properties and ADME (absorption, distribution, metabolism, excretion) predictions. These findings suggest that compounds 15 and 16 could be promising candidates for further advancement and preclinical investigation as anticancer drugs that target CDK4 (Li *et al.* 2022).

A study using budding yeast as a model system found that 17, a β-C derivative, inhibited yeast growth and blocked the cell cycle initiation by inhibiting CDK activity. Laboratory-based studies have indicated that β-C drugs suppressed the kinase activity of Cdk2/Cyclin A. This suggests that the CDK inhibition mediated by β-C drugs could be responsible for their potential antitumor effects (Li *et al.* 2007).

Inhibitory activity against CDK1/cyclinB was assessed for a group of compounds known as pyrrolo[2,3-a] carbazole analogs. The compound 18 was discovered to be the most powerful inhibitor, with an IC_{50} in the low micromolar range and producing 90% inhibition at higher doses. Two potential binding modes of 18 in the ATP-binding cleft were revealed through computational modelling of the CDK1-18 binding interaction. These modes include interactions with enzyme residues Thr14, Lys130, and Asp146. Based on the results, it appears that compound 18 has the potential to develop as a CDK1/cyclin B inhibitor (Fousteris *et al.* 2008).

An anticancer agent fascaplysin (19) based β-C was examined for biological activities and it was found that they specifically inhibit CDK4 more effectively than CDK2. The most successful compound in the series was 20, which had an IC_{50} (CDK4-cyclin

D1) of 11 μmol, indicating strong inhibitory activity against CDK4-cyclin D1 (García *et al.* 2006).

A set of tryptamine and tetrahydro-β-C biphenyl derivatives were produced with the use of Suzuki-Miyaura reaction, showcasing flexibility of this crucial procedure. The outcome of this study indicated that most of the compounds displayed notable selectivity for CDK4/cyclin D1 as opposed to CDK2/cyclin A. Moreover, the para-biphenyl variants of the tryptamine and tetrahydro-β-C compounds exhibited better biological activity when compared to their ortho- and meta-biphenyl analogs. This observation suggested that the inhibitor's biphenyl moiety tends to align in a preferred orientation in the "Phe 93 pocket" of the enzyme. The most potent compounds identified, namely 21, 22, and 23, displayed IC_{50} values of 9.2 μM, 11.2 μM, and 9 μM, respectively. These biphenyls with parasubstituents contain t-butyl or methyl in the 4th position. The potent binding mode of these compounds was attributed to their strong lipophilic component and the suggested p-stacking interaction of the inhibitors that occurs between them and the "Phe 93 pocket," which might be the reason for the observed inhibition properties of the inhibitors against CDK4/cyclin D1 (Jenkins *et al.* 2008).

In a study, two new cyclometalated Ru(II) complexes, 24 and 25, were developed and their antiproliferative activities were evaluated. The complexes showed lower IC_{50} values compared to other noncyclometalated Ru(II)-β-C complexes and cisplatin. The interaction of Ru(II) polypyridyl moieties with 1-Ph-β-C resulted in a higher degree of anticancer activity. The hydrophobicity of 24 and 25 was found to be closely associated with their ability to enter cells and their efficacy in inducing cytotoxicity in tumor cells. Studies on absorption and distribution indicated that the complexes were primarily conveyed into the cells via the clathrin-mediated endocytosis route and accumulated within the cell nuclei. It was discovered that both efficiently triggered cell death in HeLa cells by causing mitochondrial impairment, accumulating intracellular reactive oxygen species as ROS, and inducing DNA damage through ROS. Compound 24 prompted cell cycle inhibition at the G0/G1 checkpoint by preventing the CDK2/4 kinases through downregulation of the expression of CDC25A. The studies uncovered valuable information for formulating and creating novel cyclometalated Ru(II) compounds with powerful anticancer properties (Chen *et al.* 2017).

A wide range of Ru (II)-containing arene entities with a β-C derivative, having potential CDK inhibitory qualities, were synthesized and investigated for their anticancer activity. Stability studies revealed that the hydrolysis behavior of these complexes was significantly affected by the nature of β-C ligand, and hydrolysis was restrained at high chloride concentrations. These complexes displayed high potency against several cancer cell lines, particularly 26 and 27, which showed less cytotoxicity against normal lung fibroblasts. The mechanism investigations revealed that these compounds triggered cell death by disrupting mitochondrial function through a pathway that involved the generation of ROS. Moreover, these complexes inhibited CDK1 activity and activated G2M phase arrest, suppressing the level of expression of CDK1 and cyclin B1, an essential regulator gene at the G2/M boundary. The effectiveness of these complexes as antitumor was observed to have a positive correlation with their ability to inhibit CDK1/cyclin B. Additionally, the alliance of Ru (II)-arene complexes with β-C alkaloids appears to be a promising approach for developing targeted antineoplastic agents that can overcome cisplatin resistance. According to a study on the structure-relationship of these complexes, it was found that their stability in water and capacity for hydrogen bonding with protein targets are crucial factors (He *et al.* 2013).

A study revealed that two cyclometalated Ir (III) complexes inhibited a higher level of anticancer activity against various cancer cell lines. These complexes demonstrated an anticancer mechanism that was distinct from that of cisplatin. In contrast to cisplatin, which stimulates apoptotic cell death in A549 cells and initiates autophagy as a cytoprotective reaction, the mechanism of action of compound 28 involved autophagy-mediated apoptosis that was caspase-independent and ROS-mediated, without inducing apoptosis. The activation of autophagy is triggered by mTOR signaling by compound 28. Furthermore, compound 28 exhibited a strong inhibitory effect on CDK2 and mild activity against mTOR and CDK1. The distinct mechanism of action exhibited by these cyclometalated Ir (III) complexes has been found to make them highly effective against A549 cells that are resistant to cisplatin. This unique antitumor mechanism suggested that these complexes have the ability to act as therapeutics for the cancer treatment that are apoptosis deficient and/or resistant to cisplatin (He *et al.* 2014).

Based on the research findings, it is evident that marine sponges produce several secondary metabolites as defense mechanisms against predation. Among these, menzamine A (8) is a pentacyclic alkaloid fused with β-C, which exhibits anticancer activity against pancreatic cancer. Nevertheless, its cytotoxicity against other tumors and the mechanism by which it exerts its antineoplastic effects have yet to be fully understood. This study demonstrated that manzamine A was effective against colorectal cancer (CRC) cell lines by reducing cell proliferation. To understand the molecular regulation of manzamine A gene expression analysis was performed by using microarray technology. The analysis revealed dysregulation in pathways related to the steps: first cell cycle, second DNA repair, then mRNA metabolism, and lastly apoptosis. Moreover, manzamine A has been demonstrated to induce the arrest of cell cycle at the G0/G1 phase by inhibiting cyclin-dependent kinases through p53/p21/p27, and to trigger caspase-mediated apoptotic cell death by depleting mitochondrial transmembrane potential. In addition, it suppressed the epithelial-mesenchymal transition process and mesenchymal transcription elements (e.g., snail, slug and twist), leading to the acquisition of an epithelial-like phenotype and the inhibition of migration.

In summary, manzamine A exhibited potential as an effective anticancer drug against metastatic CRC. Based on the findings of this research, it showed that manzamine A has the potential to be used as a foundation for creating a cancer-fighting medication that could be utilized in the treatment of metastatic CRC. Therefore, the results offered a groundwork for additional studies aimed at exploring the potential of manzamine A and its derivatives of cancer treatments (Lin *et al.* 2018).

An array of 3-substituted 6-aminosulfonyl-β-Cs was produced, and their mode of interaction to cyclin-dependent kinase 2 was examined through molecular docking and fluorescence measurements to determine their binding mode. The investigation demonstrated that replacing the 3-cyclohexylmethoxy group with a different moiety resulted in an enhancement of interaction between hydrophobic residues and the intense hydrophobic pocket of CDK2, thereby leading to an increase in the affinity. Compound 29, which contained 3-(cyclohexyl methoxy) group, exhibited the strongest binding capacity to CDK2. The outcome indicated that compound 29 has the potential to function as a CDK2 inhibitor, and additional assessments regarding its antitumor efficacy are necessary to ascertain the likelihood of these novel compounds being employed as potential CDK2 inhibitors (Wang *et al.* 2012).

The β-C alkaloid, Harmaline (HAR) (3), is a naturally occurring compound found in Peganum harmala seeds, and has demonstrated significant antitumor effects. A study assessed the possible antitumor impacts of HAR on stomach cancer by conducting experiments in both in vitro and in vivo, with a particular emphasis on the supression of CDK. The outcomes revealed that HAR successfully repressed the proliferation process of cells in SGC-7901 cancer cell lines, causing G2/M cell cycle arrest and boost in cell death. HAR stimulated the activation of proteins involved in cell cycle regulation, such as p-Cdc2, p21, p-p53, and Cyclin-B, while suppressing the activation of p-Cdc25C. Furthermore, the research indicated that HAR enhanced the expression of Fas/FasL, and the activation of Caspase-8 and Caspase-3, suggesting that Fas/FasL-mediated mechanisms were engaged in the induction of cell death by HAR. The antitumor efficacy of HAR was further validated in vivo using a 15 mg/kg/day dosage that also resulted in cell cycle arrest. Based on this, it was suggested that HAR has the capacity to function as a CDK inhibitor, and that it may serve as a basis for future research into its antitumor activity (Wang, *et al.* 2015). In general, β-C and their derivatives have exhibited promising potential as CDK inhibitors in the management of several diseases, including cancer. However, additional investigation is needed to fully comprehend their mode of operation and to maximize their therapeutic capabilities.

4.2.2 The Role of β-C as Inhibitors of Dual-specificity Tyrosine Phosphorylation-regulated Kinases (DYRKs)

DYRKs refer to a group of protein kinases that are evolutionarily conserved. These kinases possess both tyrosine and serine/threonine kinase activities and are involved in the regulation of many cellular functions, like cell division, differentiation, and cell death (Yuan *et al.* 2022). Among the four members of this family (DYRK1A, DYRK1B, DYRK2, and DYRK3), DYRK1A has been studied thoroughly due to its involvement in the regulation of brain development and its association with the neurodegenerative diseases including Down syndrome and Alzheimer's disease (Becker and Sippl 2011). Different β-C analogs with potential DYRK activity are illustrated in Figure 4.6.

Because of their connection to Down's syndrome and neurodegenerative disorders, DYRKs have received proper attention in the field of medical research. To gain a deeper understanding related to structure-activity relationship, a combination of ligand-based pharmacophore analysis and protein/ligand docking was utilized with MOE (molecular operating environment) software. This approach facilitated the synthesis of some novel β-C derivatives to validate the theoretical framework. As a result, a slightly modified lead compound (30) was identified with

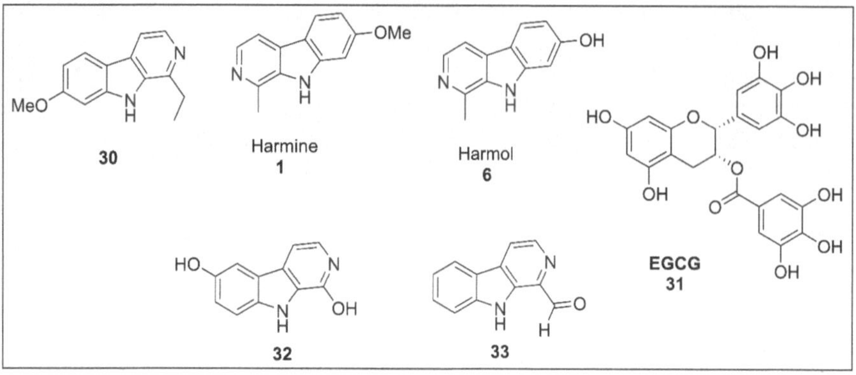

FIGURE 4.6 Different β-carboline derivatives with excellent DYRK inhibitory activity.

potent DYRK1A inhibitory activity (IC_{50} of 130 nM) and notable specificity against monoamine oxidase A (MAO-A), DYRK2, DYRK3, and DYRK4, along with CLK2 (Drung *et al.* 2014).

The potential therapeutic application of harmine was hindered through its unintended impact on MAO-A, which has an impact on various nervous system targets. This study utilized an optimized time-dependent fluorescence energy transfer-based methodology for DYRK1A to screen for potential inhibitors in a high-throughput manner using a 384-well format. The screening discovered harmine (1) and other four related complexes as potent DYRK1A inhibitors. In summary, the study found that harmol (6) showed better selectivity for DYRK1A than MAO-A and demonstrated a better therapeutic window in a glioma tumour xenograft model compared to harmine. The position of modifications in harmine analogs affected the DYRK1A inhibition but not the MAO-A inhibition. These results present insights into the potential development of β-C compounds as therapeutic molecules for the treatment of diseases related to DYRK1A inhibition. Further research is required to explore the curative potential of these entities in vivo and optimize their selectivity and potency (Tarpley *et al.* 2021).

A different research investigation discovered that harmine (1) was identified as a powerful and particular repressor of DYRK1A both in cultured cells and in vitro. Harmine inhibited the substrate phosphorylation by DYRK1A more effectively compared with different kinases in the DYRK family. A greater concentration was required to completely restrain DYRK1A tyrosine autophosphorylation within a bacterial in vitro translation system. Harmine hindered the phosphorylation of the particular substrate from DYRK1A in cultured cells without negative effects on cell viability. Furthermore, harmine was found to interfere with the neuritogenesis in cultured hippocampal neurons, providing evidence for the participation of DYRK1A in the control of neurite formation. These findings suggest that harmine or other beta-carboline alkaloids may have potential therapeutic applications in conditions where DYRK1A is implicated, such as Down syndrome and neurodegenerative diseases (Göckler *et al.* 2009).

Investigations have shown that DYRK1A is involved in phosphorylating a protein called tau, which is associated with Alzheimer's disease. Inhibiting DYRK1A may therefore be a potential strategy for preventing or slowing down the growth of tau pathology in Alzheimer's disease. During a study, harmine and other β-C derivatives were assessed to control tau phosphorylation in cell culture screenings and in-vitro phosphorylation screenings. The findings indicated that these complexes can decrease the expression of all three phosphorylated forms of tau protein and inhibited the phosphorylation of tau protein on serine 396 by DYRK1A. Additionally, certain functional groups that can influence the efficacy of β-C compounds in inhibiting tau phosphorylation were identified by researchers (Frost *et al.* 2011).

Harmine not only acts as an inhibitor of the kinase activity of DYRK1A, but it also inhibits its auto-phosphorylation on tyrosine, thereby preventing the formation of the active enzymes. DYRK1A phosphorylates caspase 9 at an inhibitory site, which can prevent caspase 9 from activating the apoptotic pathway at Thr125, which serves as an inhibitory region and defines a standard for the activation of caspase 9 via basal inhibitory phosphorylation of this protease. The inhibitory phosphorylation of caspase 9 at Thr125 by DYRK1A has great importance in apoptosis regulation during development and pathogenesis. Harmine can block the DYRK1A-dependent phosphorylation of Thr125, indicating its potential role in modulating the apoptotic pathway (Seifert, Allan, and Clarke 2008).

Further investigation into the mechanism of action revealed that ATP-competitive inhibition against DYRK1A is demonstrated by harmine binding to residues inside the binding pocket for ATP and displacing ATP. The substitution of valine with alanine at position 306 (V306A) results in a significant reduction in harmine potency against DYRK1A, as it selectively modifies the affinity of the protein for both harmine and ATP. However, the V306A mutation does not seem to affect DYRK1A activity, with the exception of reducing its ATP binding strength. This decrease can be entirely offset by providing ATP within the range considered physiological. Consequently, harmine continues to be a prospective primary compound for crafting selective ATP-competitive restraints on DYRK1A, exploring the ATP-binding site of DYRK1A (Adayev, Wegiel, and Hwang 2011).

Diabetes, a health condition resulting from an inadequate number of functional insulin-producing β-cells within the pancreas, affects an estimated 380 million people worldwide. In humans, beta cell proliferation occurs during a limited time period around birth. However, after early childhood and during embryonic life, the replication of beta cells is barely detectable. Efforts to expand beta cells in adult humans have been unsuccessful so far, emphasizing the significance of creating therapeutic compounds capable of stimulating the regeneration and expansion of these cells in mature individuals. Through a large-scale screening of small molecules, researchers discovered that derivatives of harmine can serve in the role of a new class of mitogenic compounds for human beta cells. They also pinpointed DYRK1A as the probable target of harmine, while nuclear factors of activated T cells (NFAT) likely act as mediators of human β-cell proliferation and differentiation. Their study showcased harmine's capacity to stimulate beta cell growth, increase islet mass, and enhance glycaemic control across three distinct human islet and mouse in vivo models. These discoveries imply that harmine analogs hold distinctive promise as therapeutic agents for managing diabetes in humans. The researchers intend to further amplify the effectiveness and specificity of these compounds in upcoming investigations (Wang, Alvarez-Perez, *et al.* 2015).

A study has demonstrated that epigallocatechin gallate (EGCG) (31), a flavonoid found in green tea, can modulate the activity of FoxO transcription parameters in human skin fibroblasts cultured in vitro. EGCG is a polyphenol compound found in green tea, which has been shown to have various health benefits. One of the reported effects of EGCG is its ability to stimulate the nuclear accumulation and DNA-binding activity of FoxO transcription factors at a low concentration of 1 μM. At greater concentrations (100 μM) of EGCG, the compound can generate hydrogen peroxide, which then stimulates the activity of phosphoinositide-3-kinase (PI3K)/Akt signaling. The activation of PI3K/Akt signaling results in the suppression of FoxO activity. FoxO phosphorylation prompts its removal from the nucleus and diminishes its ability to bind to DNA. At lower concentrations, harmine not only inhibits FoxO but also induces its nuclear aggregation and enhances DNA binding activity. When Caenorhabditis elegans worms are exposed to EGCG, it elevates the expression of the DAF-16 target gene, sod-3, and triggers nuclear accumulation of the FoxO ortholog, DAF-16. Additionally, the mean and maximal lifespan of *C. elegans* was found to be increased by 20% and 13%, respectively, upon exposure to EGCG, indicating that the effects of EGCG may have significant physiological implications. The results demonstrate that EGCG has insulin-mimetic and insulin-antagonistic effects, depending on its concentration. EGCG enhances FoxO nuclear localization and DNA-binding activity at low concentrations, but at higher concentrations, it triggers a signaling cascade that involves the generation of hydrogen peroxide, leading to the stimulation of PI3K/Akt signaling and inhibition within FoxO activity. In conclusion, the study suggests that EGCG holds potential as a therapeutic remedy for age-related diseases (Bartholome *et al.* 2010).

Researchers have investigated the potential of *Picrasma quassioides*, a Chinese traditional medicine, to treat Alzheimer's disease. They found that the extract from its stems showed neuroprotective properties in cell models and enhanced cognitive abilities and memory in AD mice. Suppressing neuro-inflammation and reducing Aβ1-42 deposition were linked to the anti-AD mechanism. Seven canthin-6-one alkaloids, six β-C alkaloids, and five quassinoids were identified from the extract. The in vitro studies have shown that four β-Cs and six canthin-6-ones have potential neuroprotective activities. Moreover, two of these compounds (32 and 33) were found to have high affinity for both DYRK1A and butyrylcholinesterase (BuChE) according to molecular docking experiments (Guo *et al.* 2019).

These studies highlight the diverse range of potential therapeutic applications of β-Cs and their derivatives, particularly harmine, including the treatment of neurodegenerative disorders, bone disorders, and cancer, as well as their potential use in stem cell research and regenerative medicine. However, to fully optimize the efficacy and safety of β-Cs, and to understand their mechanisms of action, further research is required.

4.2.3 β-Carbolines Acting as Inhibitors of Mitogen-activated Protein Kinase-activated Protein Kinase 2 (MAPKAPK2)

MAPKAPK2 is a type of kinase that phosphorylates serine and threonine residues. It is activated as a downstream effector within the MAPK (mitogen-activated protein kinase) signaling pathway (Pearson *et al.* 2001). The MAPKs are a family of enzymes that are highly conserved across different species, and they perform an essential role in regulating diverse physiological processes within mammalian cells. These processes include cell differentiation, development, proliferation, immune function, cell death, and stress responses (Coulombe and Meloche 2007). Inhibiting MAPKAPK-2 has surfaced as a promising target for drug development aimed at alleviating a variety of conditions, which includes various autoimmune diseases and inflammatory diseases, neurodegenerative disorders, and cancer (Soni, Anand, and Padwad 2019). Several drugs have been developed to target MAPKAPK-2 including monoclonal antibodies, small molecule inhibitors, and siRNAs. Various β-C analogs with potential MAPKAPK2 activity are presented in Figure 4.7.

A group of novel tetrahydro-β-C-1-carboxylic acid analogs has been discovered, and their potential to inhibit MAPKAPK-2 has been investigated. The researchers have also explored the structure-activity relationship (SAR) of these analogs. These observations encompass the importance of an ether functionality at the 6-position of the A-ring, the pivotal role of the carboxylic acid, and the improvement in binding affinity achieved by connecting the C-ring nitrogen with the 4-position. Compound 34 has been found to demonstrate significant selectivity for MAPKAPK-2 as compared to other closely related kinases, such as PRAK and MAPKAPK-3. In U937 cells, the researchers observed that the carboxylic acids were inactive. However, they discovered that compound 35, an ester prodrug derived from compound 34, effectively regulated

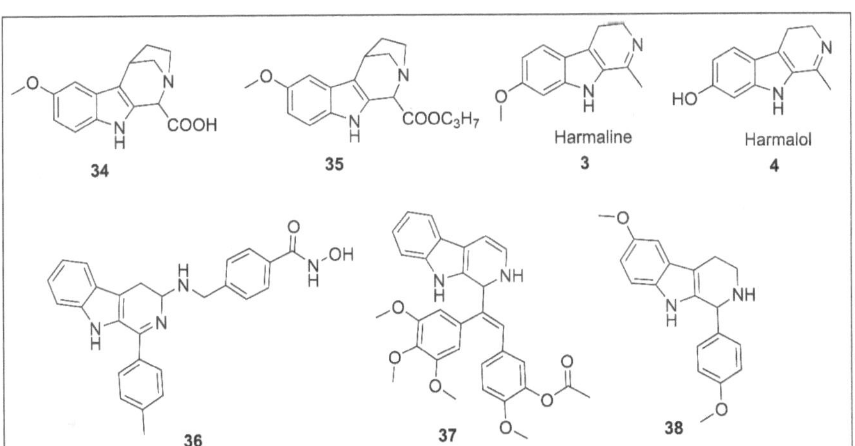

FIGURE 4.7 β-carbolines having significant MAPKAPK-2 inhibitory activities.

the production of TNFα in both U937 cellular assays and a rat LPS model (Trujillo *et al.* 2007).

A collection of highly effective inhibitors of MAPKAPK-2 based on β-C scaffolds has been synthesized and the DELFIA assay revealed that these compounds exhibited remarkable inhibition of MAPKAPK-2, with low IC_{50} values of 10 nM. Based on structural studies using X-ray crystallography, it has been found that these inhibitors bound to a region that is located in close proximity with the hinge region and p-loop of MAPKAPK-2a. These derivatives exhibited moderate selectivity against ERK2 and MAPKAPK-1 (Wu *et al.* 2007).

A study was performed over carboline derivatives using pharmacophore modelling and docking studies to identify potential anti-inflammatory agents that can bind to the MAPKAPK-2 receptor. The study concluded that receptor binding necessitates two hydrogen bond donors, two H-bond acceptors, and one hydrophobic characteristic. The 3D QSAR models generated by the study showed good consistency and correlation with experimental IC_{50} values. CoMSIA and CoMFA contour maps were also developed, which can guide the structural modification of compounds to improve their biological activity. The study's findings can help in understanding the structural features necessary for biological interaction and improving the potency of molecules that inhibit MAPKAPK-2 (Nayana *et al.* 2009).

The production of melanin is controlled by specific enzymes and transcription factors in melanocytes. Harmaline (3) and harmalol (4), β-C alkaloids found in various plants and alcoholic drinks were tested for their effects on melanoma cells. The study results indicated that both harmalol and harmaline increased tyrosinase activity and melanin content in a manner dependent on concentration and time. Furthermore, it was observed that expression of proteins that are associated in melanin production (tyrosinase, TRP-1, and TRP-2) also increased. The treatment with both harmaline and harmalol was observed to activate pathways involving microphthalmia-associated transcription factor, the cAMP response element-binding protein, and the p38 MAPK pathway. This activation ultimately led to an increase in melanogenesis, a process of melanin production (Park *et al.* 2010).

A group of new hydroxamate derivatives based on β-C were generated and tested for their effects in various in vitro assays. They were found to exhibit potent inhibition of HDAC1/3/6 and also showed significant antitumor activity against five different human cancer cell lines. Among the tested β-C derivatives, compound 36 emerged as the most potent, exhibiting the highest anticancer activity with 0.53~1.56 μM IC_{50} values, significantly greater than the harmine (1) (IC_{50} value of 46.7~55.3 μM) and was also observed to be 3 to 10 times lower to SAHA (IC_{50} = 4.48~6.26 μM). The HDAC inhibitory effects of the compound 36 were confirmed by immunoblot analysis, which showed a concentration-dependent inhibition of histone H3 and α-tubulin acetylation. Additionally, compound 36 caused concentration-dependent cell cycle arrest in HepG2 cells at the G2/M phase via the inhibition of cell cycle-related proteins cyclin-B and CDK1. Notably, it also exhibited strong antimetastasis activity by inhibiting the MAPK signaling pathway and decreasing the protein level of MMP2 and MMP9 (Ling *et al.* 2018).

The researchers synthesized a couple of β-C scaffolds using Pictet-Spengler reaction and evaluated their anti-inflammatory activity for human neutrophils. Compound 37 and 38, two compounds synthesized in the study, were found to have notable inhibitory effects on generating superoxide anion induced by N-formyl-Met-Leu-Phe in human neutrophils, without causing cytotoxicity. Specifically, compound 38 selectively hindered fMLF-induced phosphorylation of p38 and intracellular calcium levels, while compound 37 was noted for its inhibition of p38 MAPK phosphorylation and reduction of intracellular calcium levels in fMLF-activated human neutrophils. Molecular docking analysis demonstrated the favorable binding affinity of compound 38 towards p38 MAP. The suggested synthetic strategy could lead to the advancement of novel anti-inflammatory drugs targeting neutrophilic inflammation (Kumar *et al.* 2021).

Overall, recent literature highlights the significant potential of β-Cs and their derivatives as inhibitors of MAPKAPK-2, with applications in treating diverse diseases such as cancer, neurological disorders, and inflammatory conditions. Further research is required to optimize the therapeutic efficacy of these compounds and to evaluate their safety and efficacy in clinical trials.

4.2.4 β-Carbolines as Inhibitors of IκB Kinase (IKK)

IκB kinase (IKK) is a serine-threonine protein kinase pivotal in the activation of the transcription factor NF-κB (Hacker and Karin 2006). The IKK enzyme complex comprises three subunits, IKKγ, IKKα, and IKKβ, collectively known as NEMO (Regnier *et al.* 1997). IKKβ specifically phosphorylates the inhibitor of NF-κB (IκB), triggering its degradation and subsequently activating NF-κB (Karin 1999). Targeting IKK for inhibition has emerged as a promising strategy for treating various cancers and inflammatory conditions. Numerous drugs aimed at IKK inhibition are under development, including small molecule inhibitors and antisense oligonucleotides (Llona-Minguez, Baiget, and Mackay 2013). In preclinical studies, IKK inhibition has demonstrated promising outcomes in the treatment of chronic inflammatory conditions such as rheumatoid arthritis, inflammatory bowel diseases, and certain cancers (Yi *et al.* 2022). Several β-C derivatives with excellent IKK activity are shown in Figure 4.8.

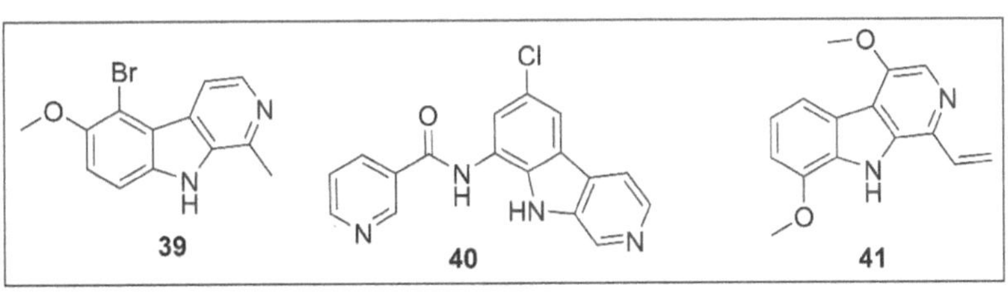

FIGURE 4.8 β-carbolines with potential IKK inhibitory activity.

Scientists have been searching for specific inhibitors of IKK to NF-κB. A natural product derivative called 5-bromo-6-methoxy-β-C (39) was identified as a nonspecific IKK inhibitor in an evaluation of the endogenous IKK complex. This discovery led to the refinement of this derivative, yielding a new category of selective IKK inhibitors with IC50 in the nanomolar range. Among the identified analogs, compound 40 emerged as effective in inhibiting IKK and its downstream signaling pathways both in cellular and in vivo models. Its capability to hinder the phosphorylation of IκBα, subsequently blocking NF-κB activation in cells, along with its ability to curtail TNF-α release in lipopolysaccharide-challenged mice, positions it as a potential candidate for treating diverse cancers and inflammatory conditions. Further investigations are underway to analyze its efficiency and safety in clinical trials (Castro *et al.* 2003).

Control over nuclear factor (NF)-κB transcription factors plays a pivotal role in the pathogenesis of numerous diseases. Research aimed to investigate the impact of a β-C alkaloid (41) on suppressing NF-κB signaling pathways in lipopolysaccharide-stimulated RAW 264.7 cells. The alkaloid displayed a concentration-dependent reduction in the inducible nitric oxide synthase protein level and NOS promoter activities. Furthermore, it diminished nuclear translocation, phosphorylation, and degradation of IκB, NF-κB DNA binding activity, and attenuated IKK activity in LPS-stimulated RAW 264.7 cells. These findings indicate the potential of the β-C alkaloid (41) to inhibit IKK activity, subsequently suppressing the NF-κB signaling pathway in LPS-stimulated RAW 264.7 cells (Woo Yoon *et al.* 2005).

β-Cs and their derivatives have shown promising potential as IKK inhibitors. These compounds could offer a novel approach for developing therapeutic strategies to address diseases associated with NF-κB pathway dysregulation. Further research is necessary to elucidate the exact mechanism of action of these compounds and to refine their pharmacodynamic and pharmacokinetic properties.

4.2.5 β-Carbolines as Inhibitors of CDC-like Kinase 1 (CLK1)

CDC-like kinase 1 known as CLK1 is a protein kinase that plays a vital role in pre-mRNA splicing by phosphorylating serine-arginine-rich proteins. It also participates in governing circadian rhythm and apoptosis (Martín Moyano, Němec, and Paruch 2020). CLK1 has been labelled as a potential therapeutic target for diseases such as viral infections, cancer, and neurological disorders. Inhibition of CLK1 has been shown to decrease tumour growth, enhance the efficacy of chemotherapy, and inhibit viral replication (ElHady *et al.* 2022). Small molecule inhibitors of CLK1 are currently under development and show favorable results as potential therapies for cancer and other diseases. Different β-C analogs with significant CLK1 activity are illustrated in Figure 4.9.

A series of hybrid β-C compounds, amalgamating benzimidazole and benzoxazole fragments, were synthesized and tested for their anticancer potential against four distinct human tumor cell lines. These hybrids exhibited noteworthy anticancer efficacy against A549 (lung), MCF-7 (breast), A2780 (ovarian), and Colo-205 (colon) cell lines. Molecular interaction analysis, focusing on cancer cell migration and specific kinase targets, unveiled their potential binding modes, with certain hybrid β-C compounds demonstrating selectivity towards CDC-like kinases (CLK-1 through

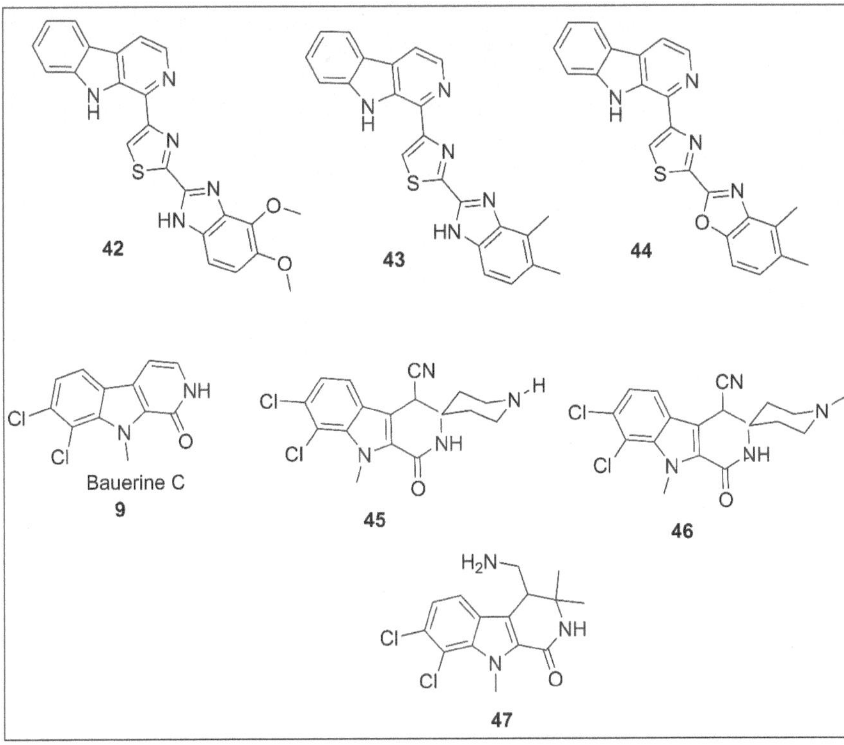

FIGURE 4.9 Various β-carbolines with CLK inhibitory activity.

CLK-4). Combining the thiazole-linked β-C nucleus with the benzimidazole core notably enhanced the antitumor activity against the former three cell lines, while the presence of benzoxazole rendered the β-C moderately active. This rationale was supported by a promising antineoplastic profile observed against MCF-7 breast cancer cell lines (Sireesha *et al.* 2021).

A study introduced a derivative, 7,8-dichloro-1-oxo-β-Cs, as a versatile framework for non-ATP mimetic kinase inhibitors. The primary component of this compound was derived from the natural complex bauerine C (9), and structure-activity relationships were validated. The complexes of 4-cyano-1,2,3,4-tetrahydro-1-oxo-β-C type, including spiropiperidine 45, have been shown to possess significant inhibitory activity against oncogenic DAPK3 and PIM kinases. X-ray crystallography revealed their unusual binding mode and good specificity with scarce off-targets outside and within the AGC (protein kinase A, G ,and C) family. Cellular assays showed that compound 45 exhibited antiproliferative activity at lower micro-molar concentrations and depleted PIM-regulated phosphorylation of downstream effector proteins. The N-methyl analogue 46 demonstrated intriguing binding mode with DAPK3 by forming halogen bonds with the kinase hinge region, contributing to inhibitor selectivity. The 6,7-dichloroindole scaffold can also serve as a template for new CLK inhibitors, and selected compounds such as primary amine 47 and alkaloid 9 could be optimized to generate new targeted inhibitors of DRAK1 and BMP2K (BIKE). Research is currently being conducted to increase the potency and selectivity of these compounds (Huber *et al.* 2012).

Overall, these studies demonstrated the prospects of β-Cs and their derivatives as selective and potent inhibitors of CLK1 and other CLK family members. Additional studies are necessary to enhance the therapeutic capacity of these compounds and assess their effectiveness and safety in preclinical and clinical environments.

4.2.6 β-Carbolines as Inhibitors of Protein Kinase B (PKB or Akt)

Protein kinase B (PKB), also recognized as Akt kinase, is a serine/threonine-specific kinase pivotal in diverse cellular functions. These encompass glucose metabolism, cell proliferation, transcription, apoptosis, and cell migration (Uko et al. 2020). Dysregulation of the Akt pathway has been linked to various diseases, including cardiovascular diseases, cancer, and diabetes. Akt inhibitors have been developed as potential cancer therapeutics, and have shown promising results in preclinical studies (Bhutani, Sheikh, and Niazi 2013). Additionally, inhibition of Akt has been shown to reduce oxidative stress and inflammation in some neurodegenerative diseases like Parkinson's disease and Alzheimer's disease. Figure 4.10 demonstrated several β-C compounds that were active against PKB.

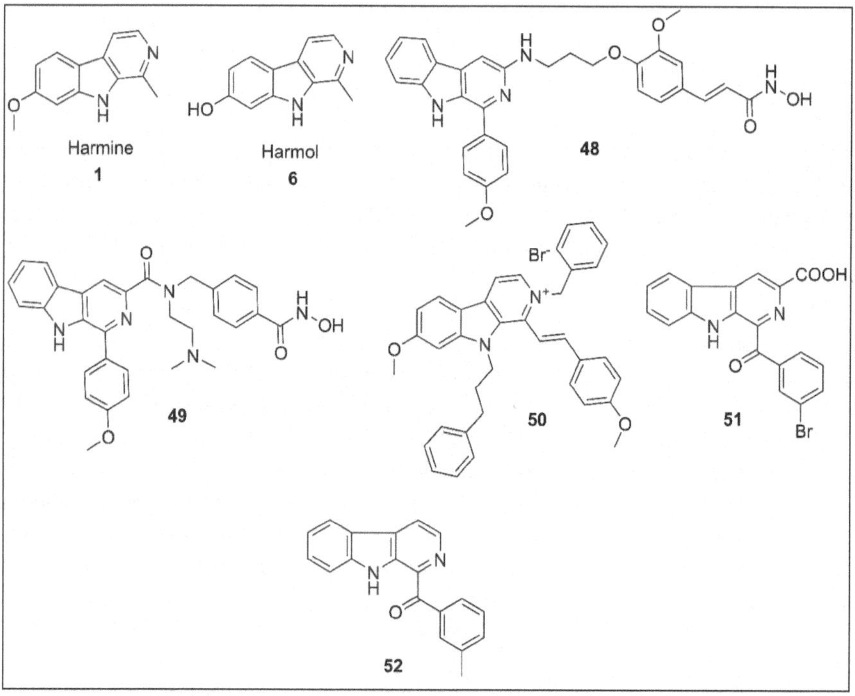

FIGURE 4.10 Different β-carbolines having inhibitory activity against PKB.

The β-C alkaloids have remarkable pharmacological properties and can be used to control plant pathogenic nematodes and pests. However, there is a lack of research on their potential insecticidal properties. A previous study found that these complexes can trigger apoptosis in cancer cell lines, which led to further investigation of its toxicological operations on insects. A recent study focused on examining the cytotoxicity of natural harmala alkaloids (five in number) and found that they have an autophagy-inducing effect on the Spodoptera frugiperda Sf9-cultured cell line. The study unveiled that natural harmala alkaloids demonstrated inhibition of Sf9 cell proliferation, exhibiting dependency on both time and dosage. The study found that two unsaturated β-C alkaloids, harmol (6) and harmine (1), showed more potent autophagy induction activity as evidenced by LysoTracker Red and monodansyl cadaverine staining. The administration of β-C resulted in an elevation in the expression of several autophagy-related genes at the RNA level, accompanied by an increase in the protein expression of Sf-Atg8 post-treatment. Throughout this investigation, notable alterations were observed in the core autophagic signaling pathway, specifically the PI3K/Akt/mTOR pathway, in response to β-C. To verify this, specific activators and inhibitors were employed, and the outcomes suggested that the PI3K/Akt/mTOR pathway primarily regulated the induction of autophagy prompted by harmine in insect cellular models. These discoveries hold potential implications for the utilization of these bioactive components across various domains (Cui *et al.* 2019).

A novel series of β-Cs that contain hydroxamic acid and are linked through a hydroxycinnamic acid moiety were proposed as potential antitumor agents. The aim was to incorporate histone deacetylase (HDAC) inhibition to enhance the potency of naturally occurring β-Cs, which have limited anticancer activities. The in vitro assays conducted on these novel β-Cs demonstrated significant antitumor activities against four human cancer cell lines. From the series, compound 48 showed the highest cytotoxic potency. Compound 48 was observed to enhance the acetylation levels of histone H3 and α-tubulin in tumor cells. It also induced a high rate of apoptosis as compared to the reference HDAC inhibitor SAHA. Additionally, it demonstrated potential autophagic flux activity in Bel7402-cells. Moreover, this compound was found to significantly inhibit the PI3K/Akt/mTOR signaling route, which is a critical path for promoting cellular growth and is often activated in various types of cancers. One of the most remarkable discoveries was its selective inhibition of tumor cell proliferation, particularly the drug-sensitive Bel7402 and the drug-resistant Bel7402/5-FU cells, without exerting a significant effect on normal LO2 cells. Based on the results of the in vitro assays, it can be suggested that the hydroxamic-acid containing β-C analogs have the potential quality to be a more effective contender for carcinoma therapy in humans. The derivatives combine the main structural parts of classical HDACIs, harmine (1) and HCA, leading to improved efficacy and synergistic effects or even drug resistance reversal (Ling *et al.* 2019).

The researchers developed and synthesized a series of HDAC (hydroxamic acid histone deacetylase) inhibitors by incorporating the β-C motif and exploring the impact of C3 amide substitution on HDAC inhibition and antiproliferative activity. Most of these compounds exhibited remarkable HDAC inhibition and antiproliferative activity, demonstrating IC_{50} values in the lower micromolar range.

The most potent compound identified in the series, N-(2(dimethylamino) 1 1 ethyl)-N-(4-(hydroxylcarbamoyl) 3 37 benzyl)-1-(4-methoxyphenyl)-9H pyrido[3,4-b] indole-3-carboxamide (49), displayed an HDAC inhibition IC_{50} value five times lower than that of suberoylanilide hydroxamic acid (SAHA, vorinostat). Moreover, this compound increased the acetylation reaction of histone H3 and α-tubulin. Alongside its HDAC-inhibitory effects, compound 49 induced DNA damage and hindered several signaling pathways commonly activated in cancers, including Stat3, Akt, and ERK. The moderate Caco-2 cell permeability and reasonable solubility of compound 49 suggest that β-C-based HDAC inhibitors hold promise as therapeutic agents for human cancers (Ling *et al.* 2017).

A set of harmine (1) derivatives was synthesized and studied with a benzylidene derivative in the 1-position of the β-C ring, aiming to explore their potential as anticancer agents. The most noteworthy anticancer activities were observed with the N2-benzylated β-C derivatives, and compound 50 exhibited the highest activity against various cancer cell lines. It induced apoptosis in cancer cells with minimum toxicity to normal cells. It also promoted ROS production and inhibited phosphorylation of AKT leading to activation of the mitochondria-dependent cell apoptotic pathway and suppression of phosphorylated AKT. It also showed promising antitumor effects in a nude mice xenograft model and has great potential as active therapeutic agent for cancer treatment (Zhang *et al.* 2016).

A comprehensive array of 1-substituted and 1,3-disubstituted β-C analogs were synthesized in a study employing a modified one-step Pictet-Spengler reaction. The anti-inflammatory potential of these synthesized compounds was assessed by their ability to inhibit the production of NO (nitric oxide) and PGE2 (prostaglandin E2), alongside evaluating their cellular toxicity against four human cancer cell lines. The majority of the examined compounds displayed robust inhibitory activity against the production of NO and PGE2. Notably, compounds 51 and 52 exhibited significant reductions in the expression of inducible cyclooxygenase 2 (COX2) and nitric oxide synthase (iNOS), indicating the capacity of β-C analogs to curtail the production of these molecules at a translational level. Certain β-C derivatives also demonstrated noteworthy inhibition of superoxide anion (O^{2-}) generation or elastase production compared to the reference compound, with compound 51 displaying the highest potency. The synthetic 1-benzoyl-3-carboxy β-C analogs showcased promising potential as anti-inflammatory drugs. Furthermore, it was observed that these analogs hindered N-formyl-L-methionyl-phenylalanine (FMLP)-induced phosphorylation of JNK (c-Jun N-terminal kinase) and AKT (protein kinase B), suggesting their role in reducing human neutrophil functions by inhibiting the activation of AKT and JNK signaling pathways (Yang *et al.* 2011).

In conclusion, β-Cs have shown promising potential as inhibitors of AKT kinase, a critical signaling molecule involved in numerous cellular processes. Research has shown their capability to suppress AKT activation and impede cancer cell growth, rendering them a potential therapeutic target for cancer treatment. Further investigations are warranted to enhance the efficacy and safety of β-Cs as AKT kinase inhibitors. However, their distinct chemical properties and mode of action present a promising avenue for exploration.

4.2.7 β-Carbolines as Inhibitors of Polo-like Kinases (PLKs)

PLKs denote a cluster of serine/threonine protein kinases overseeing several cellular functions, such as mitosis, cell cycle progression, cytokinesis, and meiosis. Within mammals, PLKs serve a pivotal role in overseeing cell division, centriole duplication, and chromosome segregation (Barr, Sillјé, and Nigg 2004). Abnormal expression and activity of Polo-like kinases (PLKs) were linked to several diseases, including cancer. As a result, inhibitors targeting PLKs are being investigated as potential therapeutic agents for cancer treatment. For instance, polo-like kinase 1 (PLK1) inhibitors are being studied for their potential use in non-small cell lung carcinoma therapy (Zhang *et al.* 2021). Additionally, abnormal expression of Polo-like kinase 4 (PLK4) was observed in various human cancers, proposing a potential role for PLK4 inhibition in cancer therapy (Garvey *et al.* 2021). β-C compounds with good PLK inhibitory activity are demonstrated in Figure 4.11.

The inhibitory activity of PLK-1 was examined for a library of β-C/Schiff bases. Compound 53, from all the synthesized compounds, showed the highest potency against colon cancer with a GI_{50} range of 3-45 μM in the NCI-60 panel cell assay. The other compounds also demonstrated strong to mild cytotoxic activity against the panel cell lines. During Kinome scan screening, it was found that it can target PLK-1 specifically at a concentration of 15 μM. Furthermore, it was observed that when administered at varying concentrations, it caused a corresponding halt in the cell cycle progression at S/G2 phase in HCT-116 cells. Further, it triggered programmed cell death by activating procaspase-3 and cleaved PARP (poly(ADP)ribose polymerase. Experiments conducted on DLA and EAC animal models indicated that compound 53 administered at 100 mg/kg/bd. wt dosage, notably extended the average lifespan

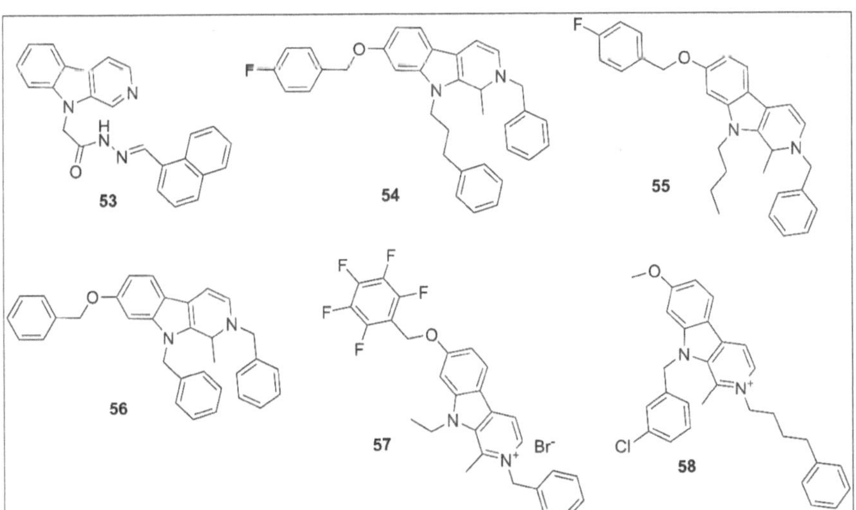

FIGURE 4.11 Several β-carboline derivatives possessing potential PLK inhibitory activity.

of mice with tumors, demonstrating its potential as an effective antitumour agent. Despite observing a reduction in the body mass of tumor-enduring mice in comparison to the tumor-regulated mice, the results from the experiments indicated that compound 53 possessed the potential to act as an effective antitumor agent (Jeyapal *et al.* 2019).

This study aimed to construct and authenticate a model capable of forecasting the efficacy of *β*-C analogs as antitumor agents against HepG2 human cancer cell lines. Using ALMOND software, an alignment-free 3D-QSAR analysis was conducted to compute alignment-independent descriptors (known as GRIND descriptors). After analyzing a training 30-compound set the resulting model achieved a squared correlation coefficient of 0.957 and a standard deviation error of 0.116 in calculations. To validate the model, both leave-one out and leave-multiple out methods were employed, which demonstrated a high level of accuracy in predicting results with a 0.66 score for a test of set consisting ten compounds. In addition, the research study employed molecular docking techniques to explore the approach used to assess the interaction between *β*-C analogs and the active site of the most probable anticancer receptor, the PLK protein. According to the research findings, it was determined that the primary factors contributing to the antitumor effectiveness of the compounds under investigation were hydrophobicity, shape effects, and hydrogen bond interactions. This research study offered valuable insights that can facilitate the development of some novel drugs that target HepG2 cell lines by providing guidance on the design of such drugs (Ghasemi and Davoudian 2014).

Throughout the study, scientists created a range of β-C derivatives and determined that three specific compounds, namely 54, 55, and 56, exhibited strong potential as novel inhibitors of PLK. The compounds synthesized in this research selectively inhibited the PLK1, PLK2, and PLK3 in vitro kinase activity, and exhibited potent antitumor properties against multiple cancer cell lines. Additionally, the compounds induced cell cycle block and cellular death in tumor cells while demonstrating comparably lower levels of toxicity towards noncancer cells. One interesting observation from the study was that the compounds displayed contrasting levels of sensitivity between normal and cancer cells. Based on their findings, the researchers deduced that β-carboline derivatives 54, 55, and 56 exhibited strong potential as novel PLK inhibitors and could be considered as potential candidates for cancer treatment. Further studies will clarify the mechanism contributing to the differences in cell cycle block induced by these compounds (Han *et al.* 2012).

The researchers conducted an analysis of the SARs within a collection of β-C alkaloid analogs through the use of both protein docking and 2D QSAR methodologies. Their goal was to discover the interaction method between PLK1 kinase, and β-C derivatives as well as ascertain the vital substituents dictating the cytotoxic activity of these derivatives. The QSAR models demonstrated strong correlations between the observed and predicted activities of the β-C alkaloid derivatives, highlighting lipophilicity as a pivotal factor influencing their cytotoxicity. Molecular docking studies revealed that ligand 57 displayed elevated binding energy and proficiently interacted with the active sites of the PLK1 kinase through hydrophobic interactions and hydrogen bonding. These findings offer insights that can be leveraged to design novel and more potential β-C derivatives as effective cytotoxic agents. In addition, the authors designed three new compounds that were predicted to have fever binding affinity and enhanced cytotoxic activity. Taken together, these outcomes provide valuable insights that can be leveraged in virtual screening for novel β-C derivatives and facilitate the design of new compounds possessing superior biological activities (Akabli *et al.* 2018).

A β-C derivative, 58, was determined to be more effective in restraining the proliferation of cdc5-2 temperature-sensitive mutant yeast cells, which contain a human PLK1 homologue, than wild-type yeast cells. It was also observed to inhibit purified PLK1 kinase activity in an ATP-competitive manner at low micromolar concentrations, as confirmed by docking investigations based on the crystal structure of PLK1. The β-C derivative demonstrated its ability to inhibit tumor cell proliferation, induce mitotic block, enhance the cyclin B1 accumulation, and stimulate apoptosis by potentially downregulating PLK1. These results provide evidence that the cytotoxicity of beta-carboline derivatives towards tumor cells may be attributed to PLK1 inhibition, which is a novel and significant finding in this field of research (Zhang *et al.* 2009).

In conclusion, β-Cs and their derivatives have shown potential as PLK inhibitors, with preclinical studies showcasing their potential to trigger apoptosis and induce cell cycle arrest in cancer cells. Further research is required to enhance the efficacy and safety profile of β-C derivatives acting as PLK inhibitors and to evaluate their potential as anticancer agents in clinical trials.

4.2.8 β-Carbolines as Inhibitors of Glycogen Synthase Kinase (GSK)

Glycogen synthase kinase labelled as GSK is a serine/threonine protein kinase with diverse functions, notably regulating glycogen metabolism and glucose storage (Buschiazzo *et al.* 2004). Present in two isoforms, GSK-3 is primarily located in neurons within the CNS (central nervous system). It plays a pivotal role in glycogen metabolism and glucose storage, serving as a key mediator in intracellular signaling downstream of receptors crucial for neuronal function. Inhibition of GSK-3 has been investigated as a potential therapeutic strategy for a range of diseases including bipolar disorder, AD, and type 2 diabetes. GSK inhibitors can be of different chemotypes, including non-ATP competitive inhibitors, from natural sources, substrate-competitive inhibitors, and cations and synthetic ATP (Marosi *et al.* 2022). Additionally, GSK-3 is directly regulated by other protein kinases such as AMPK, PKA, and casein kinase 2 (CK2), which led to catalytically inactive and phosphorylated glycogen synthase (Embi, Rylatt, and Cohen 1980; Proud *et al.* 1977). β-C derivatives with significant GSK inhibitory activity are shown in Figure 4.12.

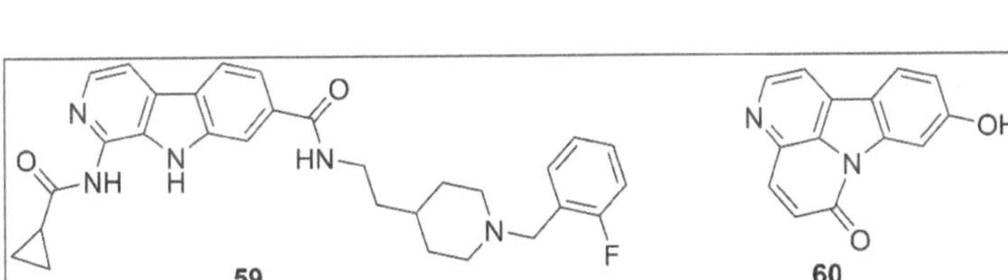

FIGURE 4.12 β-carbolines with significant GSK inhibitory activity.

In a study, the authors identified a novel series of harmine compounds incorporating the N-benzylpiperidine moiety for potential treatment of AD. These derivatives displayed notable activity against AChE and showed selectivity over BChE. Among these compounds, compound 59 exhibited strong anti-AChE activity with 0.27 mM IC_{50} value and selective BChE inhibition with 20.82 mM IC_{50} value. Additionally, it showed moderate GSK-3 (glycogen synthase kinase-3) inhibition with an IC_{50} value of 6.78 mM. Based on ligand-protein docking studies and molecular simulation methods, compound 59 was found to be able to form a stable interaction with both GSK-3b and AChE. Additionally, it displayed great selectivity towards GSK-3b compared to multikinases and showed low cell toxicity towards SH-SY5Y, HL-7702 HEK-293T, and HepG2 cell lines. The authors observed that ZLWH-23 effectively decreased tau hyperphosphorylation at the Ser-396 site in Tau (P301L) 293T cellular framework. Based on these investigations, the authors suggested that derivatives based on harmine carry the potential to be formulated into drug leads for Alzheimer's treatment (Liu *et al.* 2022).

A group of researchers conducted a screening of a repository of plant extracts to identify compounds that could inhibit the transcriptional activity of TCF/β-catenin, which is a critical element of the Wnt signaling route. From the root extract of *E. longifolia*, researchers discovered a compound named 9-hydroxycanthin-6-one (60) that showed inhibitory activity against TCF/β-catenin transcription. The researchers discovered that compound 1 had the ability to inhibit Wnt signaling by activating GSK3β without the need for CK1α, a kinase necessary for the initial phosphorylation of β-catenin. In zebrafish embryos, the compound reduced the expression in both Wnt target genes, and in SW480 cells, it reduced β-catenin levels. Additionally, they found that compound 60 has a unique mechanism of inducing β-catenin degradation without the need for phosphorylation by CK1α. Considering these findings, compound 60 holds promise as a potential lead compound for developing Wnt signal inhibitors targeting GSK3β (Ohishi *et al.* 2015).

Additional inquiries are needed to comprehend the mechanisms from which β-Cs and their derivatives inhibit GSK3, as well as to assess their safety and efficacy as potential therapeutics. Nevertheless, the inhibition of GSK3 via β-Cs presents a promising therapeutic approach to manage the diverse diseases linked to dysregulated GSK3 activity.

4.2.9 β-Carbolines as Inhibitors of Miscellaneous Kinases

Figure 4.13 represents several β-C analogs that were active against several kinases such as haspin, aurora, B-Raf, Brk, PDGF, and SK1. Haspin kinase, a type of serine/threonine protein kinase, is significant in mitosis progression and histone H3 phosphorylation. Researchers have recognized it as a propitious target for developing anticancer drugs (Amoussou *et al.* 2018). Depletion of Haspin kinase led to decreased proliferation and reduced activation of the cGAS (cyclic GMP-AMP synthase)-stimulator of interferon genes pathway (Melms *et al.* 2020). Different classes of small molecule inhibitors have been created to investigate the mitotic roles of Haspin kinase. (Wang *et al.* 2020).

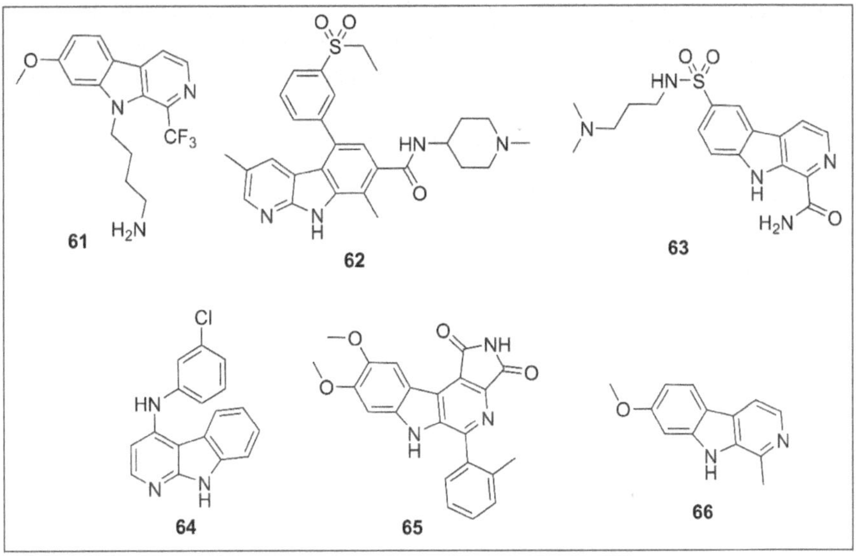

FIGURE 4.13 Various β-carbolines having different kinases inhibitory activities.

A study focused on identifying potential cancer therapeutic targets by inhibiting the haspin kinase through screening 140,000 compounds. β-Cs harmine (1) and harmol (6) were identified as slightly effective inhibitors, and harmine derivatives were designed to increase inhibitory potency further. The incorporation of a tethered primary amine and replacement of the methyl group with a trifluoromethyl group yielded compound 61, displaying considerable in vitro metabolic stability in mouse liver microsomes and moderate selectivity, inhibiting only six kinases in common. These findings suggest that compounds from the harmine series, like 61, could serve as effective molecular tools for exploring the cellular functions of haspin kinase and potentially hold promise for therapeutic applications in cancer treatment (Cuny *et al.* 2012).

Aurora kinases are enzymes that play a critical role in mitosis and are expressed differentially in rapidly dividing cells. Heightened expression of Aurora kinases correlates with polyploidy and cancer pathology (Ahmed *et al.* 2021). Aurora kinases isoforms A, B, and C plays a crucial role in cellular division and are frequently elevated in cancer. Due to their significance, they are attractive targets for treatment. Recently, a new multitargeted inhibitor of Aurora B kinase, 62, has been developed from an innovative azacarboline kinase hinge-binder chemotype. It exhibits potent and time-dependent inhibition specifically targeting Aurora B rather than Aurora A. This inhibition led to repression of histone H3 phosphorylation and induction of polyploidy. It inhibits the proliferation of diverse human cancer cell lines, displaying effective concentration values spanning from 40 to 500 nmol/L. In intact cells, it also hinders the activity of other kinases, such as FLT3 and FGFR2, but not as effectively as Aurora B. In rodent xenografts, it showed strong efficacy against different types of human solid tumors, including ovarian cancer and various leukaemia models. Biomarker studies conducted in vivo revealed that the compound prompted pharmacodynamic responses consistent with Aurora B kinase inhibition, correlating with its presence in cancerous tissue. Based on the promising preclinical findings, compound 62 has advanced to phase I clinical investigations for treating diverse cancer types (Farrell *et al.* 2013).

The B-RAF gene encodes B-Raf kinase, a serine/threonine-protein kinase. The crucial function of BRAF in tumor initiation has been confirmed, and recent structural analyses have demonstrated how regulatory domains participate in controlling the activity of the BRAF kinase domain. Inhibition of BRAF kinase has been studied in the treatment of several diseases (Gunderwala, Cope, and Wang 2022). Two sequences of novel 6-sulfonamide and 1-carboxamide-substituted β-C derivatives were synthesized and evaluated for their inhibitory activities against wild-type B-Raf kinas. These derivatives play a crucial role in the MAPK signal-transduction passage that controls cell differentiation, survival, and proliferation. Most of the tested derivatives demonstrated moderate-to-excellent inhibitory activities. Compound 63, which is 1-carboxamide-6-(N-(3-(dimethylamino)propyl)-sulfamoyl)-β-C, exhibited considerable activity with an IC_{50} value of 1.62 μM, presenting a potential for further exploration as a model compound (Xin *et al.* 2012).

Protein tyrosine kinase 6 (PTK6), also referred to as breast tumor kinase (Brk), is a type of tyrosine kinase enzyme that is exaggerated in a majority of breast tumors and other epithelial tumors. In breast tumor cell lines, Brk is co-expressed with members of the ErbB family and has been associated with tumor progression (Ostrander *et al.* 2007). The study aimed to identify the most promising substituents to improve the inhibitory activity of Brk protein kinase inhibitors among 4-anilino α-Cs. Modifications were introduced to the 2- and 3-position of the aniline residual, and it was discovered that substitution at the 3-position demonstrated greater promise in comparison to solely substituting the 2-position. Compound 64 with a 3-chloro substitution was identified as the most potent, exhibiting a considerable increase in inhibitory activity. Docking studies were conducted, indicating a shared binding mode for 4-anilino α-Cs within the ATP-binding cavity within a Brk homology model. The addition of a 3-hydroxy function capable of forming a hydrogen bond with Ile262 produced the best results. The study also examined the effects of combined substituents in disubstituted derivatives. Docking studies showed that 3-substituted derivatives were most active among the series due to their potential to form hydrogen/halogen bonding. However, compounds with lipophilic substituents also displayed promising nanomolar activities (Mahmoud *et al.* 2014).

Platelet-derived growth factor (PDGF) is a glycoprotein consisting of two subunits that serves as a powerful cell mitogen originating from mesenchyme, such as smooth muscle cells, fibroblasts, and glial cells (Heldin 1992). There are various isoforms of PDGF, and it activates two closely related receptor tyrosine kinases called α- and β-receptors. Although PDGF is crucial for normal development and tissue repair, its abnormal regulation has been linked to several diseases, comprising cancer and fibrotic disorders (Iwayama and Olson 2013).

Researchers developed a new synthetic method for generating imido-β-Cs in three steps from known chemical entities. The newly synthesized β-Cs underwent evaluation for their capability to inhibit tyrosine phosphorylation in Swiss 3T3 fibroblasts upon PDGF (platelet-derived growth factor) stimulation. This assessment revealed a spectrum of compounds exhibiting varying degrees of inhibitory activity. The most potent compound, 2,3-dihydro-8,9-dimethoxy-5-(2-methylphenyl) 1H,6H-pyrrolo[3,4-c] pyrido[3,4-b]indole-1,3-dione 65, demonstrated selective inhibition of PDGF receptor signal transduction and PDGF-regulated cell growth. It showcased IC50 values of 0.4 μM and 2.6 μM for inhibiting purified PDGF receptor kinase and autophosphorylation of PDGF-receptor in intact cells, respectively. Moreover, this analog displayed limited impact on EGF (epidermal growth factor) receptor kinase activity or Src, exhibiting selectivity in inhibiting SCF (Kit/stem cell factor) receptor autophosphorylation or bFGF (basic fibroblast growth factor)-stimulated tyrosine phosphorylation. The authors proposed that β-Cs demonstrated a promising class of tyrosine kinase inhibitors with potential for specifically interfering with PDGF-induced cell growth and PDGF receptor directed signal transduction (Teller *et al.* 2000).

SK1 is an enzyme that belongs to the lipid family and has the ability to transform proapoptotic lipids sphingosine and ceramide into the antiapoptotic lipid SIP (sphingosine-1-phosphate). This process induces the activation of various signaling pathways, resulting in cell proliferation, migration, activation of inflammatory responses, and inhibition of apoptosis (Olivera and Spiegel 2001). SK1 is a main controller of the dynamic ceramide/S1P balance, and it is crucial in pathological cancer genesis, metastasis processes, and progression (Zhang *et al.* 2014). Inhibiting SK1 was explored as a therapeutic technique for various diseases, covering cancer, inflammation, and autoimmune disorders, as it restores the ceramide/S1P balance and promotes apoptosis.

The SK1/sphingosine-1-phosphate pathway is recognized to be linked with cancer progression and the survival of chemotherapy-resistant cells. SK1 is a promising drug target for cancer care, but natural inhibitors for SK1 are limited. In this investigation, researchers analyzed the inhibitory impact of harmaline, a β-C alkaloid, on SK1. The outcomes of fluorescence binding studies and molecular docking exhibit that harmaline has a high binding nature with the substrate-binding pocket of SK1 and suppresses its kinase activity. The IC_{50} value was found to be 6.1328 ± 0.33 μM. The researchers also observed that harmaline had significant cytotoxicity against H1299 cancer cells of lung when compared with the A549 cells. Harmaline (3) was found to stimulate apoptosis process in non-small-cell lung cancerous cells, which may be through the intrinsic passage. The findings propose that harmaline could serve as a basis for crafting potent anticancer agents exhibiting SK1 inhibitory properties (Roy *et al.* 2020).

The studies showed the potential of β-C derivatives for the inhibition of haspin, aurora, B-Raf, Brk, PDGF, and SK1 kinases but exhaustive research is needed to explore the mechanistic study of inhibition and furthermore derivatives are required to be synthesized and for activity investigation towards kinase inhibition.

4.3 CONCLUSION

In conclusion, β-carbolines and their derivatives have shown promise as potential kinase inhibitors in cancer treatment. The unique structural features of β-Cs allow for ATP-mimetic binding with the kinase hinge zone, leading to inhibition of kinase activity. Furthermore, the diverse range of β-C derivatives that can be synthesized and modified makes them attractive candidates for further drug development. Studies have shown that certain naturally occurring β-C derivatives, such as harmine, harmaline, manzamine A, etc., showed immense inhibitory activity against several kinases, and further modification in these compounds can enhance the activity and selectivity.

Despite the promising results, additional research is necessary to thoroughly comprehend the mechanisms underlying β-C's inhibition of kinases and its potential side effects. Moreover, clinical trials are essential to assess the efficacy and safety of β-C-based kinase inhibitors in the treatment of human cancer.

In summary, among cancer treatment the unique structural features and diverse range of β-C derivatives make them an attractive target for drug development. Additional research is necessary to thoroughly grasp their potential as kinase inhibitors and their practical implementation in clinical contexts.

REFERENCES

Aaghaz, S., Sharma, K., Jain, R., and Kamal, A. 2021. β-Carbolines as potential anticancer agents. *European Journal of Medicinal Chemistry* 216:113321.

Abinaya, R., Srinath, S., Soundarya, S., Sridhar, R., Balasubramaniam, K., and Baskar, B. 2022. Recent developments on synthesis strategies, SAR studies and biological activities of β-carboline derivatives–An update. *Journal of Molecular Structure* 1261:132750.

Adayev, T., Wegiel, J., and Hwang, Y.-W. 2011. Harmine is an ATP-competitive inhibitor for dual-specificity tyrosine phosphorylation-regulated kinase 1A (Dyrk1A). *Archives of Biochemistry and Biophysics* 507 (2):212–218.

Ahmed, A., Shamsi, A., Mohammad, T., Hasan, G. M., Islam, A., and Hassan, M. I. 2021. Aurora B kinase: A potential drug target for cancer therapy. *Journal of Cancer Research and Clinical Oncology* 147 (8):2187–2198.

Akabli, T., Toufik, H., Yasri, A., Bih, H., and Lamchouri, F. 2018. Combining ligand-based and structure-based drug design approaches to study the structure-activity relationships of a β-carboline derivative series. *Structural Chemistry* 29:1637–1645.

Amoussou, N. G., Bigot, A., Roussakis, C., and Robert, J.-M. H. 2018. Haspin: A promising target for the design of inhibitors as potent anticancer drugs. *Drug Discovery Today* 23 (2):409–415.

Ayipo, Y. O., Mordi, M. N., Mustapha, M., and Damodaran, T. 2021. Neuropharmacological potentials of β-carboline alkaloids for neuropsychiatric disorders. *European Journal of Pharmacology* 893:173837.

Banoth, K. K., ChandraSekhar, K. V. G., Adinarayana, N., and Murugesan, S. 2020. Recent evolution on synthesis strategies and anti-leishmanial activity of β-carboline derivatives–An update. *Heliyon* 6 (9):e04916.

Barr, F. A., Sillјé, H. H., and Nigg, E. A. 2004. Polo-like kinases and the orchestration of cell division. *Nature Reviews Molecular Cell Biology* 5 (6):429–441.

Bartholome, A., Kampkötter, A., Tanner, S., Sies, H., and Klotz, L.-O. 2010. Epigallocatechin gallate-induced modulation of FoxO signaling in mammalian cells and C. elegans: FoxO stimulation is masked via PI3K/Akt activation by hydrogen peroxide formed in cell culture. *Archives of Biochemistry and Biophysics* 501 (1):58–64.

Becker, W., and Sippl, W. 2011. Activation, regulation, and inhibition of DYRK1A. *The FEBS Journal* 278 (2):246–256.

Bhutani, J., Sheikh, A., and Niazi, A. K. 2013. Akt inhibitors: Mechanism of action and implications for anticancer therapeutics. *Infectious Agents and Cancer* 8 (1):1–4.

Bossemeyer, D. 1995. Protein kinases—structure and function. *FEBS Letters* 369 (1):57–61.

Buschiazzo, A., Ugalde, J. E., Guerin, M. E., Shepard, W., Ugalde, R. A., and Alzari, P. M. 2004. Crystal structure of glycogen synthase: Homologous enzymes catalyze glycogen synthesis and degradation. *The EMBO Journal* 23 (16):3196–3205.

Cao, R., Peng, W., Wang, Z., and Xu, A. 2007. β-Carboline alkaloids: Biochemical and pharmacological functions. *Current Medicinal Chemistry* 14 (4):479–500.

Castro, A. C., Dang, L. C., Soucy, F., *et al.* 2003. Novel IKK inhibitors: β-carbolines. *Bioorganic & Medicinal Chemistry Letters* 13 (14):2419–2422.

Chen, J., Peng, F., Zhang, Y., *et al.* 2017. Synthesis, characterization, cellular uptake and apoptosis-inducing properties of two highly cytotoxic cyclometalated ruthenium (II) β-carboline complexes. *European Journal of Medicinal Chemistry* 140:104–117.

Coulombe, P., and Meloche, S. 2007. Atypical mitogen-activated protein kinases: Structure, regulation and functions. *Biochimica et Biophysica Acta (BBA)-Molecular Cell Research* 1773 (8):1376–1387.

Cui, G., Shu, B., Veeran, S., Yuan, H., Yi, X., and Zhong, G. 2019. Natural β-carboline alkaloids regulate the PI3K/Akt/mTOR pathway and induce autophagy in insect Sf9 cells. *Pesticide Biochemistry and Physiology* 154:67–77.

Cuny, G. D., Ulyanova, N. P., Patnaik, D., *et al.* 2012. Structure–activity relationship study of beta-carboline derivatives as haspin kinase inhibitors. *Bioorganic & Medicinal Chemistry Letters* 22 (5):2015–2019.

Ding, L., Cao, J., Lin, W., *et al.* 2020. The roles of cyclin-dependent kinases in cell-cycle progression and therapeutic strategies in human breast cancer. *International Journal of Molecular Sciences* 21 (6):1960.

Drung, B., Scholz, C., Barbosa, V. A., Nazari, A., Sarragiotto, M. H., and Schmidt, B. 2014. Computational & experimental evaluation of the structure/activity relationship of β-carbolines as DYRK1A inhibitors. *Bioorganic & Medicinal Chemistry Letters* 24 (20):4854–4860.

ElHady, A. K., El-Gamil, D. S., Abadi, A. H., Abdel-Halim, M., and Engel, M. 2022. An overview of cdc2-like kinase 1 (Clk1) inhibitors and their therapeutic indications. *Medicinal Research Reviews* 43(2):343–398.

Embi, N., Rylatt, D. B., and Cohen, P. 1980. Glycogen synthase kinase-3 from rabbit skeletal muscle: Separation from cyclic-AMP-dependent protein kinase and phosphorylase kinase. *European Journal of Biochemistry* 107 (2):519–527.

Farrell, P., Shi, L., Matuszkiewicz, J., *et al.* 2013. Biological characterization of TAK-901, an investigational, novel, multitargeted Aurora B kinase inhibitor TAK-901, a novel multitargeted Aurora B kinase inhibitor. *Molecular Cancer Therapeutics* 12 (4):460–470.

Formagio, A. S. N., Santos, P. R., Zanoli, K., *et al.* 2009. Synthesis and antiviral activity of β-carboline derivatives bearing a substituted carbohydrazide at C-3 against poliovirus and herpes simplex virus (HSV-1). *European Journal of Medicinal Chemistry* 44 (11):4695–4701.

Fousteris, M. A., Papakyriakou, A., Koutsourea, A., *et al.* 2008. Pyrrolo [2, 3-a] carbazoles as potential cyclin dependent kinase 1 (CDK1) Inhibitors. Synthesis, biological evaluation, and binding mode through docking simulations. *Journal of Medicinal Chemistry* 51 (4):1048–1052.

Frost, D., Meechoovet, B., Wang, T., *et al.* 2011. β-carboline compounds, including harmine, inhibit DYRK1A and tau phosphorylation at multiple Alzheimer's disease-related sites. *Plos one* 6 (5):e19264.

García, M. D., Wilson, A. J., Emmerson, D. P., Jenkins, P. R., Mahale, S., and Chaudhuri, B. 2006. Synthesis, crystal structure and biological activity of β-carboline based selective CDK4-cyclin D1 inhibitors. *Organic & Biomolecular Chemistry* 4 (24):4478–4484.

Garvey, D. R., Chhabra, G., Ndiaye, M. A., and Ahmad, N. 2021. Role of Polo-Like Kinase 4 (PLK4) in epithelial cancers and recent progress in its small molecule targeting for cancer management PLK4 in epithelial cancers and small molecule targeting. *Molecular Cancer Therapeutics* 20 (4):632–640.

Ghasemi, J. B., and Davoudian, V. 2014. 3D-QSAR and docking studies of a series of β-carboline derivatives as antitumor agents of PLK1. *Journal of Chemistry* 2014.

Göckler, N., Jofre, G., Papadopoulos, C., Soppa, U., Tejedor, F. J., and Becker, W. 2009. Harmine specifically inhibits protein kinase DYRK1A and interferes with neurite formation. *The FEBS Journal* 276 (21):6324–6337.

Gunderwala, A., Cope, N., and Wang, Z. 2022. Mechanism and inhibition of BRAF kinase. *Current Opinion in Chemical Biology* 71:102205.

Guo, E., Hu, Y., Du, T., *et al.* 2019. Effects of Picrasma quassioides and its active constituents on Alzheimer's disease in vitro and in vivo. *Bioorganic Chemistry* 92:103258.

Hacker, H., and Karin, M. 2006. Regulation and function of IKK and IKK-related kinases. *Science's STKE* 357:re13.

Han, X., Zhang, J., Guo, L., *et al.* 2012. A series of beta-carboline derivatives inhibit the kinase activity of PLKs 7(10): e46546.

He, L., Liao, S. Y., Tan, C. P., *et al.* 2013. Ruthenium–arene–β-Carboline complexes as potent inhibitors of cyclin-dependent kinase 1: Synthesis, characterization and anticancer mechanism studies. *Chemistry–A European Journal* 19 (36):12152–12160.

He, L., Liao, S.-Y., Tan, C.-P., *et al.* 2014. Cyclometalated iridium (III)–β-carboline complexes as potent autophagy-inducing agents. *Chemical Communications* 50 (42):5611–5614.

Heldin, C.-H. 1992. Structural and functional studies on platelet-derived growth factor. *The EMBO Journal* 11 (12):4251–4259.

Huber, K., Brault, L., Fedorov, O., *et al.* 2012. 7, 8-Dichloro-1-oxo-β-carbolines as a versatile scaffold for the development of potent and selective kinase inhibitors with unusual binding modes. *Journal of Medicinal Chemistry* 55 (1):403–413.

Iwayama, T., and Olson, L. E. 2013. Involvement of PDGF in fibrosis and scleroderma: Recent insights from animal models and potential therapeutic opportunities. *Current Rheumatology Reports* 15(2):304.

Jenkins, P. R., Wilson, J., Emmerson, D., *et al.* 2008. Design, synthesis and biological evaluation of new tryptamine and tetrahydro-β-carboline-based selective inhibitors of CDK4. *Bioorganic & Medicinal Chemistry* 16 (16):7728–7739.

Jeyapal, G. P., Krishnasamy, R., Suzuki, C. K., Venkatesh, S., and Chandrasekar, M. 2019. In-silico design and synthesis of N9-substituted β-Carbolines as PLK-1 inhibitors and their in-vitro/in-vivo tumor suppressing evaluation. *Bioorganic Chemistry* 88:102913.

Karin, M. 1999. How NF-κB is activated: The role of the IκB kinase (IKK) complex. *Oncogene* 18 (49):6867–6874.

Kumar, S., Wang, Y.-H., Chen, P.-J., *et al.* 2021. Design and synthesis of β-carboline and combretastatin derivatives as anti-neutrophilic inflammatory agents. *Bioorganic Chemistry* 111:104846.

Li, D., Liu, W., Huang, Y., *et al.* 2022. Facile synthesis of C1-substituted β-carbolines as CDK4 inhibitors for the treatment of cancer. *Bioorganic Chemistry* 121:105659.

Li, Y., Liang, F., Jiang, W., *et al.* 2007. DH334, a β-carboline anti-cancer drug, inhibits the CDK activity of budding yeast. *Cancer Biology & Therapy* 6 (8):1204–1210.

Lin, L.-C., Kuo, T.-T., Chang, H.-Y., Liu, W.-S., Hsia, S.-M., and Huang, T.-C. 2018. Manzamine A exerts anticancer activity against human colorectal cancer cells. *Marine Drugs* 16 (8):252.

Ling, Y., Feng, J., Luo, L., *et al.* 2017. Design and synthesis of C3-substituted β-carboline-based histone deacetylase inhibitors with potent antitumor activities. *ChemMedChem* 12 (9):646–651.

Ling, Y., Guo, J., Yang, Q., *et al.* 2018. Development of novel β-carboline-based hydroxamate derivatives as HDAC inhibitors with antiproliferative and antimetastatic activities in human cancer cells. *European Journal of Medicinal Chemistry* 144:398–409.

Ling, Y., Li, Y., Zhu, R., *et al.* 2019. Hydroxamic acid derivatives of β-carboline/hydroxycinnamic acid hybrids inducing apoptosis and autophagy through the PI3K/Akt/mTOR pathways. *Journal of Natural Products* 82 (6):1442–1450.

Liu, W., Liu, X., Liu, W., *et al.* 2022. Discovery of novel β-carboline derivatives as selective AChE inhibitors with GSK-3β inhibitory property for the treatment of Alzheimer's disease. *European Journal of Medicinal Chemistry* 229:114095.

Llona-Minguez, S., Baiget, J., and Mackay, S. P. 2013. Small-molecule inhibitors of IκB kinase (IKK) and IKK-related kinases. *Pharmaceutical Patent Analyst* 2 (4):481–498.

Luo, B., and Song, X. 2021. A comprehensive overview of β-carbolines and its derivatives as anticancer agents. *European Journal of Medicinal Chemistry* 224:113688.

Mahmoud, K. A., Krug, M., Wersig, T., *et al.* 2014. Discovery of 4-anilino α-carbolines as novel Brk inhibitors. *Bioorganic & Medicinal Chemistry Letters* 24 (8):1948–1951.

Malumbres, M., and Barbacid, M. 2007. Cell cycle kinases in cancer. *Current Opinion in Genetics & Development* 17 (1):60–65.

Marosi, M., Arman, P., Aceto, G., D'ascenzo, M., and Laezza, F. 2022. Glycogen synthase kinase 3: Ion channels, plasticity, and diseases. *International Journal of Molecular Sciences* 23 (8):4413

Martín Moyano, P., Němec, V., and Paruch, K. 2020. Cdc-Like kinases (CLKs): Biology, chemical probes, and therapeutic potential. *International Journal of Molecular Sciences* 21 (20):7549.

Melms, J. C., Vallabhaneni, S., Mills, C. E., *et al.* 2020. Inhibition of haspin kinase promotes cell-intrinsic and extrinsic antitumor activity targeting haspin for cancer therapy. *Cancer Research* 80 (4):798–810.

Nayana, R. S., Bommisetty, S. K., Singh, K., *et al.* 2009. Structural analysis of carboline derivatives as inhibitors of MAPKAP K2 using 3D QSAR and docking studies. *Journal of Chemical Information and Modeling* 49 (1):53–67.

Ohishi, K., Toume, K., Arai, M. A., *et al.* 2015. 9-Hydroxycanthin-6-one, a β-carboline alkaloid from Eurycoma longifolia, is the first Wnt signal inhibitor through activation of glycogen synthase kinase 3β without depending on casein kinase 1α. *Journal of Natural Products* 78 (5):1139–1146.

Olivera, A., and Spiegel, S. 2001. Sphingosine kinase: A mediator of vital cellular functions. *Prostaglandins & Other Lipid Mediators* 64 (1–4):123–134.

Ostrander, J. H., Daniel, A. R., Lofgren, K., Kleer, C. G., and Lange, C. A. 2007. Breast tumor kinase (protein tyrosine kinase 6) regulates heregulin-induced activation of ERK5 and p38 MAP kinases in breast cancer cells. *Cancer Research* 67 (9):4199–4209.

Park, S.-Y., Kim, Y.-H., Kim, Y.-H., Park, G.-T., and Lee, S.-J. 2010. Beta-carboline alkaloids harmaline and harmalol induce melanogenesis through p38 mitogen-activated protein kinase in B16F10 mouse melanoma cells. *BMB Reports* 43 (12):824–829.

Patel, K., Gadewar, M., Tripathi, R., Prasad, S., and Patel, D. K. 2012. A review on medicinal importance, pharmacological activity and bioanalytical aspects of beta-carboline alkaloid "Harmine". *Asian Pacific Journal of Tropical Biomedicine* 2 (8):660–664.

Pearson, G., Robinson, F., Beers Gibson, T., *et al.* 2001. Mitogen-activated protein (MAP) kinase pathways: Regulation and physiological functions. *Endocrine Reviews* 22 (2):153–183.

Poindexter Jr, E., and Carpenter, R. 1962. The isolation of harmane and norharmane from tobacco and cigarette smoke. *Phytochemistry* 1 (3):215–221.

Proud, C. G., Rylatt, D. B., Yeaman, S. J., and Cohen, P. 1977. Amino acid sequences at the two sites on glycogen synthetase phosphorylated by cyclic AMP-dependent protein kinase and their dephosphorylation by protein phosphatase-III. *FEBS Letters* 80 (2):435–442.

Regnier, C. H., Song, H. Y., Gao, X., Goeddel, D. V., Cao, Z., and Rothe, M. 1997. Identification and characterization of an IκB kinase. *Cell* 90 (2):373–383.

Roux, P. P., and Blenis, J. 2004. ERK and p38 MAPK-activated protein kinases: A family of protein kinases with diverse biological functions. *Microbiology and Molecular Biology Reviews* 68 (2):320–344.

Roy, S., Mohammad, T., Gupta, P., *et al.* 2020. Discovery of harmaline as a potent inhibitor of sphingosine kinase-1: A chemopreventive role in lung cancer. *ACS Omega* 5 (34):21550–21560.

Sarkar, S., Pandya, P., and Bhadra, K. 2014. Sequence specific binding of beta carboline alkaloid harmalol with deoxyribonucleotides: Binding heterogeneity, conformational, thermodynamic and cytotoxic aspects. *PLoS One* 9 (9):e108022.

Seifert, A., Allan, L. A., and Clarke, P. R. 2008. DYRK1A phosphorylates caspase 9 at an inhibitory site and is potently inhibited in human cells by harmine. *The FEBS Journal* 275 (24):6268–6280.

Sireesha, R., Sreenivasulu, R., Chandrasekhar, C., *et al.* 2021. Design, synthesis, anti-cancer evaluation and binding mode studies of benzimidazole/benzoxazole linked β-carboline derivatives. *Journal of Molecular Structure* 1226:129351.

Song, Y., Kesuma, D., Wang, J., *et al.* 2004. Specific inhibition of cyclin-dependent kinases and cell proliferation by harmine. *Biochemical and Biophysical Research Communications* 317 (1):128–132.

Song, Y., Wang, J., Teng, S. F., *et al.* 2002. β-Carbolines as specific inhibitors of cyclin-dependent kinases. *Bioorganic & Medicinal Chemistry Letters* 12 (7):1129–1132.

Soni, S., Anand, P., and Padwad, Y. S. 2019. MAPKAPK2: The master regulator of RNA-binding proteins modulates transcript stability and tumor progression. *Journal of Experimental & Clinical Cancer Research* 38:1–18.

Tarpley, M., Oladapo, H. O., Strepay, D., *et al.* 2021. Identification of harmine and β-carboline analogs from a high-throughput screen of an approved drug collection; profiling as differential inhibitors of DYRK1A and monoamine oxidase A and for in vitro and in vivo anti-cancer studies. *European Journal of Pharmaceutical Sciences* 162:105821.

Teller, S., Eluwa, S., Koller, M., *et al.* 2000. Pyrrolo [3, 4-c]-β-carboline-diones as a novel class of inhibitors of the platelet-derived growth factor receptor kinase. *European Journal of Medicinal Chemistry* 35 (4):413–427.

Thatikayala, M., Wadhwa, P., Kaur, P., *et al.* 2022. Beta-carboline as a promising heterocyclic nucleus: Synthetic aspects, pharmacological potential and structure activity relationship. *European Journal of Medicinal Chemistry Reports*:100096.

Trujillo, J. I., Meyers, M. J., Anderson, D. R., *et al.* 2007. Novel tetrahydro-β-carboline-1-carboxylic acids as inhibitors of mitogen activated protein kinase-activated protein kinase 2 (MK-2). *Bioorganic & Medicinal Chemistry Letters* 17 (16):4657–4663.

Uko, N. E., Güner, O. F., Matesic, D. F., and Bowen, J. P. 2020. Akt pathway inhibitors. *Current Topics in Medicinal Chemistry* 20 (10):883–900.

Wang, P., Alvarez-Perez, J.-C., Felsenfeld, D. P., *et al.* 2015. A high-throughput chemical screen reveals that harmine-mediated inhibition of DYRK1A increases human pancreatic beta cell replication. *Nature Medicine* 21 (4):383–388.

Wang, P., Hua, X., Bryner, Y. H., Liu, S., Gitter, C. B., and Dai, J. 2020. Haspin inhibition delays cell cycle progression through interphase in cancer cells. *Journal of Cellular Physiology* 235 (5):4508–4519.

Wang, Y., Jin, Q., Lin, G., *et al.* 2012. Synthesis and structure of the β-carboline derivatives and their binding intensity with cyclin-dependent kinase 2. *Chemical and Pharmaceutical Bulletin* 60 (4):435–441.

Wang, Y., Wang, C., Jiang, C., Zeng, H., and He, X. 2015. Novel mechanism of harmaline on inducing G2/M cell cycle arrest and apoptosis by up-regulating Fas/FasL in SGC-7901 cells. *Scientific Reports* 5 (1):18613.

Woo Yoon, J., Ku Kang, J., Ro Lee, K., *et al.* 2005. β-Carboline alkaloid suppresses NF-κB transcriptional activity through inhibition of IKK signaling pathway. *Journal of Toxicology and Environmental Health, Part A* 68 (23–24):2005–2017.

Wu, J.-P., Wang, J., Abeywardane, A., *et al.* 2007. The discovery of carboline analogs as potent MAPKAP-K2 inhibitors. *Bioorganic & Medicinal Chemistry Letters* 17 (16):4664–4669.

Xin, B., Tang, W., Wang, Y., *et al.* 2012. Design, synthesis and biological evaluation of β-carboline derivatives as novel inhibitors targeting B-Raf kinase. *Bioorganic & Medicinal Chemistry Letters* 22 (14):4783–4786.

Xu, Z., Chang, F.-R., Wang, H.-K., *et al.* 2000. Anti-HIV agents 45(1) and Antitumor Agents 205.(2) two new sesquiterpenes, leitneridanins A and B, and the cytotoxic and anti-HIV principles from Leitneria floridana. *Journal of Natural Products* 63 (12):1712–1715.

Yang, M.-L., Kuo, P.-C., Hwang, T.-L., *et al.* 2011. Synthesis, in vitro anti-inflammatory and cytotoxic evaluation, and mechanism of action studies of 1-benzoyl-β-carboline and 1-benzoyl-3-carboxy-β-carboline derivatives. *Bioorganic & Medicinal Chemistry* 19 (5):1674–1682.

Yi, Y. W., You, K. S., Han, S., *et al.* 2022. Inhibition of IκB kinase is a potential therapeutic strategy to circumvent resistance to epidermal growth factor receptor inhibition in triple-negative breast cancer cells. *Cancers* 14 (21):5215.

Yuan, K., Li, Z., Kuang, W., *et al.* 2022. Targeting dual-specificity tyrosine phosphorylation-regulated kinase 2 with a highly selective inhibitor for the treatment of prostate cancer. *Nature Communications* 13 (1):2903.

Zhang, J., Li, L., Dan, W., *et al.* 2015. Synthesis and antimicrobial activities of 3-methyl-β-carboline derivatives. *Natural Product Communications* 10 (6):1934578X1501000627.

Zhang, J., Li, Y., Guo, L., *et al.* 2009. DH166, a beta-carboline derivative, inhibits the kinase activity of PLK1. *Cancer Biology & Therapy* 8 (24):2374–2383.

Zhang, X.-F., Sun, R.-q., Jia, Y.-f., *et al.* 2016. Synthesis and mechanisms of action of novel harmine derivatives as potential antitumor agents. *Scientific Reports* 6 (1):1–16.

Zhang, X., Wei, C., Liang, H., and Han, L. 2021. Polo-like kinase 4's critical role in cancer development and strategies for Plk4-targeted therapy. *Frontiers in Oncology* 11:587554.

Zhang, Y., Wang, Y., Wan, Z., Liu, S., Cao, Y., and Zeng, Z. 2014. Sphingosine kinase 1 and cancer: A systematic review and meta-analysis. *PloS One* 9 (2):e90362.

5 Sulfur-containing β-Carbolines

Synthesis and Biological Study

Vnira R. Akhmetova, Danil V. Leont'ev, and Nail S. Akhmadiev

5.1 INTRODUCTION

β-Carboline scaffolds, which are secondary metabolites of some plant alkaloids, can rightfully be considered a privileged class of heterocyclic systems, owing to the wide range of their unique biological activities (Stonik and Tolstikov 2008). For example, a group of harmala alkaloids isolated from the plant *Peganum harmala* can induce a variety of psychoactive effects (Figure 5.1). While considering β-carbolines, which are structurally tricyclic pyrido[3,4-b]indoles, it should be noted that, according to Taylor's classification, the indole ring is in the top 100 in terms of the frequency of occurrence in low-molecular-weight drugs included in the FDA Orange Book and is in the 21st position, being markedly ahead of related systems such as 1-benzofuran (90th position) and 1-benzothiophene (92nd position) (Shearer et al. 2022). The combination of two heterocyclic pharmacophores, indole **1** and pyridine, in the β-carboline molecule gives rise to diverse biological activities, evidently, resulting from the simultaneous action on several biomolecular targets. Furthermore, the peripheral pyridine ring can have different degrees of saturation, ranging from aromatic (norharmane **2** and harmine **5**) to partially reduced (dihydro-β-carbolines **4**) and finally to reduced piperidine moiety (tryptoline **3**) (Figure 5.1).

About nine commercialized drugs, possessing different therapeutic action, contain various derivatives of β-carboline heterocycles. For example, vincamine is a vasodilator; vinpocetine improves cerebral circulation; yohimbine and tadalafil are used for the treatment of erectile dysfunction; abecarnil is an anxiolytic agent; reserpine, rescinnamine, and deserpidine are antihypertensive agents; cipargamin is an experimental antimalarial agent; and ajmaline and prajmaline are antiarrhythmic drugs (Abinaya et al. 2022). Numerous investigations of the versatile biological activities of β-carboline derivatives according to the latest trends of drug design and DOS strategy open up new potential applications of these compounds.

DOI: 10.1201/9781351058032-5

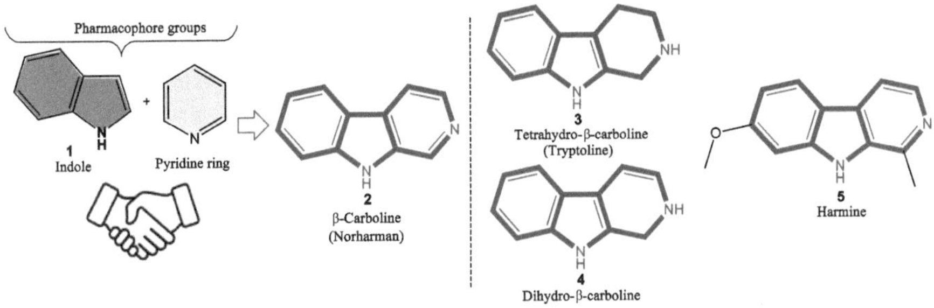

FIGURE 5.1 Structures of indole **1**, β-carboline **2**, tetrahydro-β-carboline **3**, dihydro-β-carboline **4** and Harmine **5**.

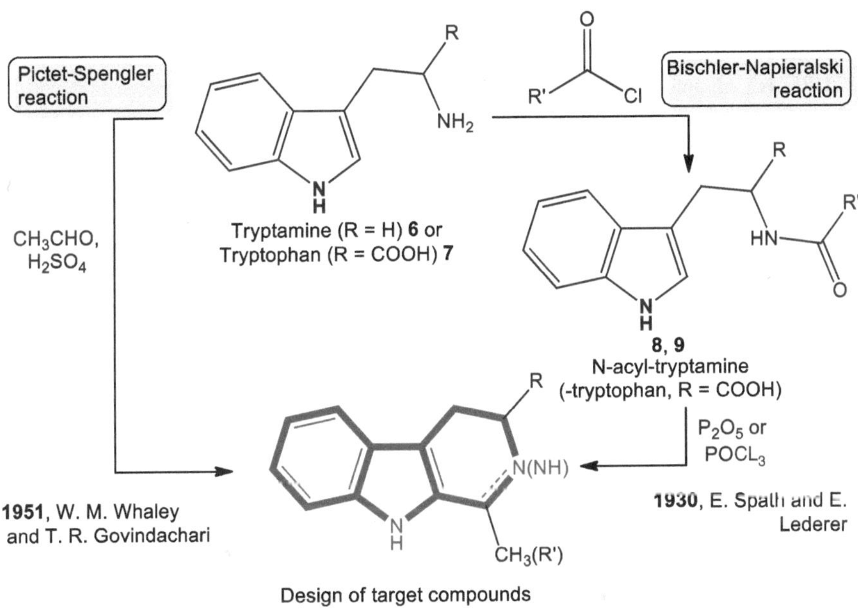

SCHEME 5.1 Main synthetic pathways for the formation of the β-carboline skeleton.

The traditional synthetic routes to the β-carboline skeleton are based on modified versions of two name reactions, the Pictet–Spengler and Bischler–Napieralski reactions (Scheme 5.1). In both cases, tryptamine **6** or tryptophan **7** is used as the starting molecule (Whaley and Govindachari 1951; Brian 1996; Li 2006; Spath and Lederer 1930). Therefore, while describing synthetic examples, we often omit the steps of synthesis of the starting β-carbolines, but focus on the molecular modification of the heterocyclic core by introduction of the sulfur atoms to the periphery of the β-carboline scaffold. Most often, the description includes the medical and biological properties of the resulting sulfur derivatives.

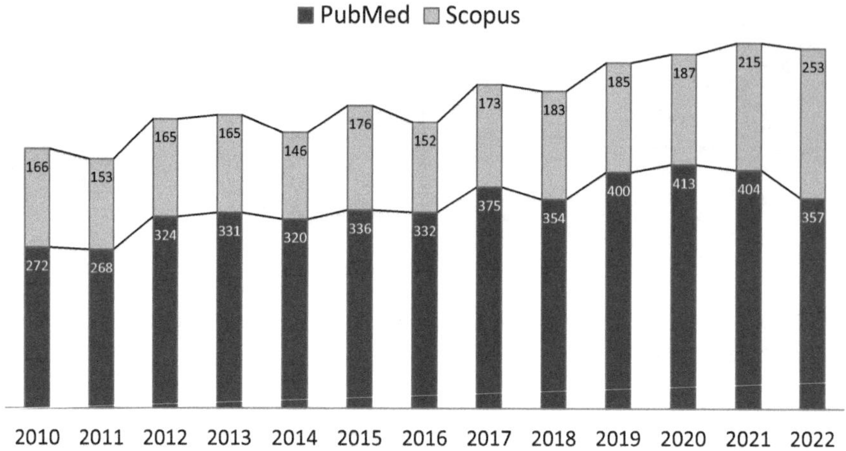

FIGURE 5.2 Number of publications with the keyword "β-carboline" between 2010 and 2022 pointed in the PubMed® and Scopus® databases.

A number of reviews and book chapters published in the last five years systematize the synthetic possibilities for the design of natural products based on β-carbolines (Chundawat 2020), including the most recent advances in the synthesis of β-carboline alkaloids with improved procedures for their synthesis (Devi et al. 2018; Dai et al. 2018) and their pharmacological potential (Szabó et al. 2021; Ayipo et al. 2021c; Thatikayala et al. 2022); the presence of β-carbolines in foodstuffs and prevention of neurodegenerative diseases (Piechowska et al. 2019); combination of the β-carboline core with various pharmacophores and screening of the products for anticancer properties (Soni et al. 2021; Sanctis et al. 2022; Luo and Song 2021; Aaghaz et al. 2021; Sahoo et. Al. 2019); metal complexes with β-carboline-based ligands (Ayipo et al. 2021b); and also studies of the fungicidal (Dai et al. 2022), antimicrobial (Almeida et al. 2021), antimalarial (Kamboj et al. 2021), and antileishmanial (Banoth et al. 2020) activities. There are also publications that systematize the syntheses of the 1,2,3,4-tetrahydro-β-carboline core with various pharmacological properties (Wang et al. 2021a; Maity et al. 2019), including their use as effective pesticides (Song et al. 2021). A recent review addresses the structure–activity relationships of β-carbolines in SAR analysis (Abinaya et al. 2022). Numerous β-carboline alkaloids were proposed as potential agents for the treatment of nervous and mental disorders (Ayipo et al. 2021a), in particular Alzheimer's disease (Beato et al. 2021).

The growth of research interest in the synthesis and unique biological properties of the β-carboline core is confirmed by the increasing number of relevant publications in reference databases such as PubMed® (2023) and Scopus® (2023) (Figure 5.2).

However, to our knowledge, no systematic reviews on the sulfur-containing β-carboline derivatives are currently available from the scientific literature. This chapter describes the advances made in the last two decades in the synthesis of sulfur-containing β-carboline scaffolds and elucidation of their biological activities.

The literature search was mainly performed using the Scopus®, Pub Med®, and SciFinder® databases (2022).

5.2 SULFUR-CONTAINING β-CARBOLINE: SYNTHESIS AND DRUG DESIGN

Sulfur-containing compounds represent an important class of functional molecules, including active pharmaceutical intermediates (APIs) and materials. The most common groups of organosulfur compounds are classified, depending on the sulfur oxidation state and type of bonding, into thioethers, disulfides, trisulfides, sulfoxides, sulfones, sulfonates, sulfates, sulfonamides, sulfamates, thioesters, thioamides, isothiocyanates, thiazoles, and thiazolines (Wang et al. 2020). All of these compounds can serve as building blocks for the preparation of pharmacologically active products. The sulfur atoms present in biological molecules as sulfide or thiol groups (allicin and amino acids – methionine, cysteine, and taurine) exhibit antibacterial and antioxidant properties, while the S-CH_3 thiomethyl groups incorporated in coenzymes participate in the methyl transport and fixation (e.g., S-adenosylmethionine) (Parcell 2002). Recently, high potential of sulfanyl azoles for the design of molecules with multiple biological activity was demonstrated (Akhmetova et al. 2022).

In continuation of the studies into the chemistry of sulfur-containing compounds, we addressed a biologically active class of carboline heterocycles that rank high in the medical and biological properties.

The S-derivatives of β-carbolines are most often formed by replacement of hydrogen atoms by the S atom in the C-(1,3) and N-(2) positions of the pyridine ring (routes a, b, and c) or in the C-(6,7) and N(9) positions of the indole moiety (routes d, e, f). The preferable way is to introduce the sulfur atoms *via* Spacers, which are represented by methylene or ethylene units, or aromatic rings (Figure 5.3). It

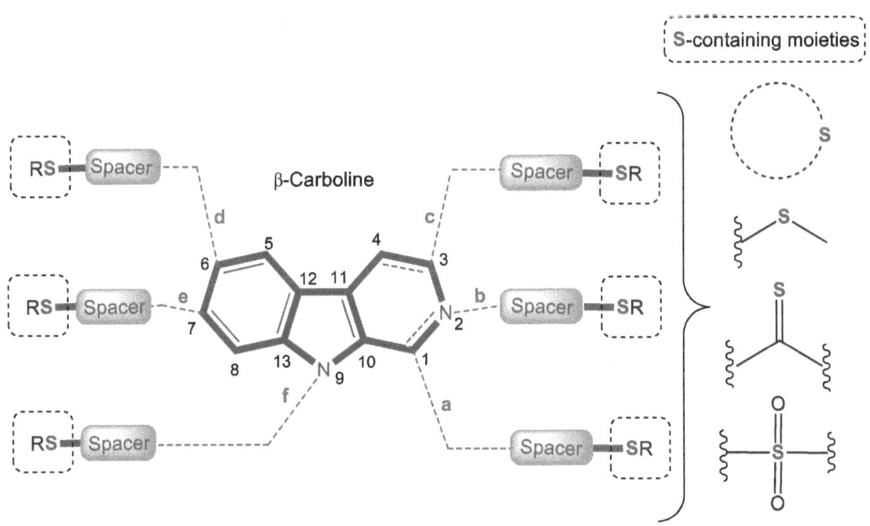

FIGURE 5.3 The design of β-carboline S-derivatives.

is noteworthy that studies on thiomethylation or thiolation of the β-carboline core are virtually missing from the literature. Most studies are devoted to the design of S-containing β-carbolines *via* bulky Spacers (Figure 5.3). The atom numbering in the carboline core corresponded to the International Union of Pure and Applied Chemistry rules (Abramovitch and Spenser 1964). The material in the review is arranged in terms of the methods used for functionalization of β-carboline, depending on the position of S-containing substituent in the ring. In some cases, quite a few examples of β-carboline synthesis that are beyond the scope of this chapter are deliberately omitted.

An important circumstance is that a number of sulfur β-carboline derivatives were isolated from natural compounds (metabolites present in marine organisms); their structures were proven and the strategies for their synthesis were developed. Most of the presented compounds exhibit versatile biological activities (Figure 5.4). For example, Gesashidine A has an antibacterial activity against *Micrococcus luteus* (MIC = 16.6 μg/mL) (Suntornchashwej et al. 2005), while Hyrtimomines D, E, J show an inhibitory activity against *Candida albicans* (IC_{50}, 4 and 8 lg/mL; IC_{50}, 2 μg/mL) and *Cryptococcus neoformans* (IC_{50} = 4 and 8 lg/mL, respectively) (Tanaka et al. 2013; Tanaka et al. 2014). Meanwhile, the compounds Hyrtimomine H and K exhibit no antimicrobial activities (Tanaka et al. 2014). Other Dragmacidonamine A, B derivatives were isolated from the sponge *Dragmacidon*; only Dragmacidonamine A had a moderate cytotoxic activity against the L5178Y mouse lymphoma cells (Ebada et al. 2008; Pedpradab et al. 2004). Mention should be made of *Lurbinectedin*, which was used to treat metastatic small cell lung cancer and was found to be cytotoxic against other types of cancer (He et al. 2019; Li et al. 2022a). The Orthoscuticellines E metabolite showed a moderate antiplasmodial activity against *Plasmodium falciparum* (Prinsep et al. 1991; Kleks et al. 2020). Eudistomidin J, isolated from the marine tunicate *Eudistoma glaucus*, showed cytotoxic activity *in vitro* against P388 (IC_{50} = 0.043 μg/mL) and L1210 (IC_{50} = 0.047 μg/mL) mouse leukemia cells and against human epidermoid carcinoma KB cells (IC_{50} = 0.063 μg/mL) (Suzuki et al. 2011; Rashid et al. 2001). Stolonine C taurinamide derivatives, isolated from the marine tunicate *Cnemidocarpa stolonifera*, exhibited a moderate cytotoxicity against the PC3 prostate cancer cells when present in concentration of 20 μM (Tran et al. 2015). Shishijimicin A, which was obtained from the ascidian *Didemnum proliferum*, deserves special attention. Testing of this compound as an innovative antitumor antibiotic revealed anticancer properties against the P388 leukemia cells (IC_{50} = 0.48 pg/mL) (Nicolaou et al. 2015; Oku et al. 2003). Didemnoline A-D metabolites, differing in the N-substitution of the β-carboline core, were isolated from the tunicate *Didemnum* and identified (Palanisamy et al. 2017). Ishigadine A, a new indole alkaloid with a moderate cytotoxic activity against the L1210 mouse leukemia cells, was isolated from the Okinawa sea sponge *Hyrtios* sp. (Takahashi et al. 2018).

5.2.1 Synthesis of Hybrid β-Carbolines Modified in the C-1 Position by Thione, Sulfone Groups and Benzothiophene Fragment

In order to prepare molecules with potential fungicidal properties, targeted synthesis of β-carboline derivatives **12** containing a thione group was carried out (Wang et al.

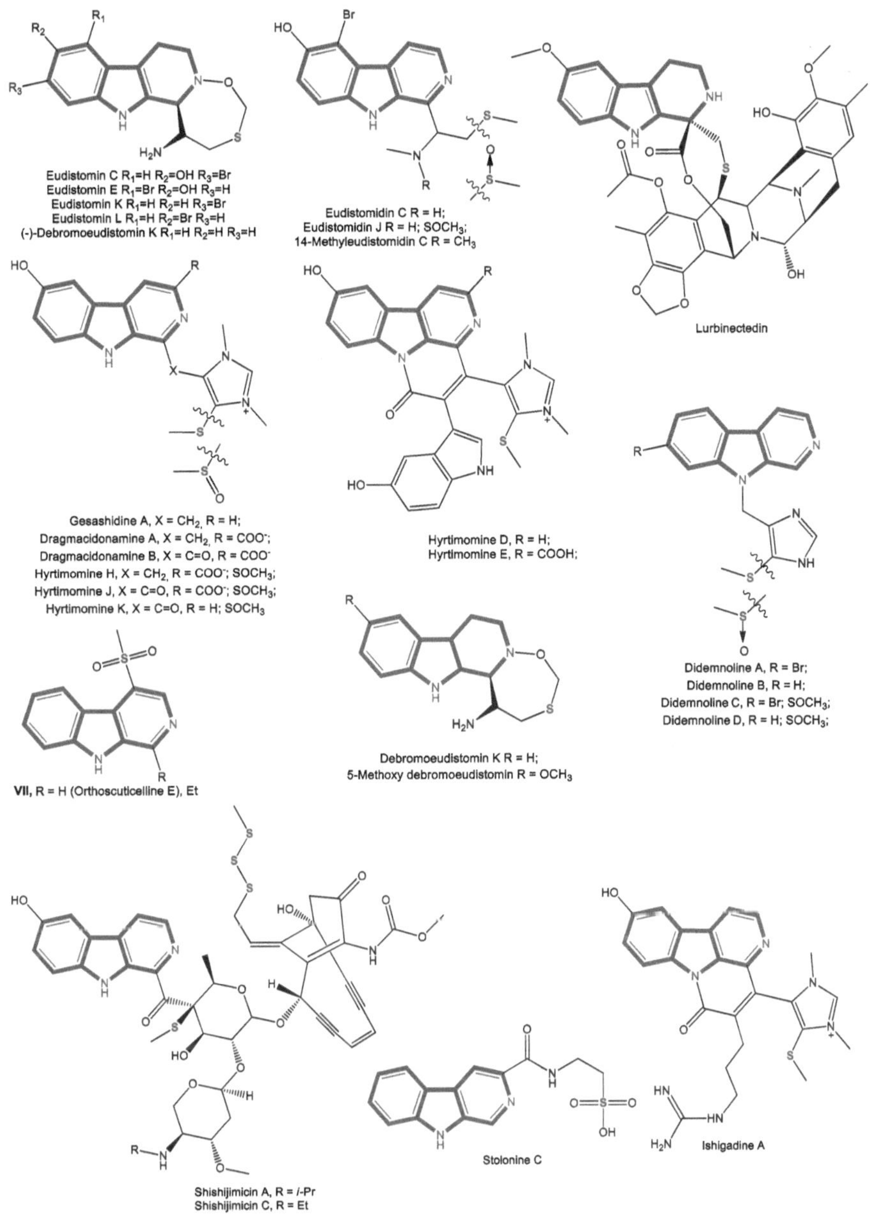

FIGURE 5.4 Sulfur-containing β-carboline moieties in the composition of natural metabolites.

2021b) (Scheme 5.2). The series of products was tested for the antiviral (against the tobacco mosaic virus) and fungicidal activities (*Fusarium oxysporium*, *Cercospora arachidicola Hori*, *Physalospora piricola*, *Rhizoctonia cerealis*, *Alternaria solani*, *Pyricularia grisea*, *Phytophthora capsici*, *Sclerotinia sclerotiorum*). Among this series of compounds, the highest activity against the tobacco mosaic virus exceeding

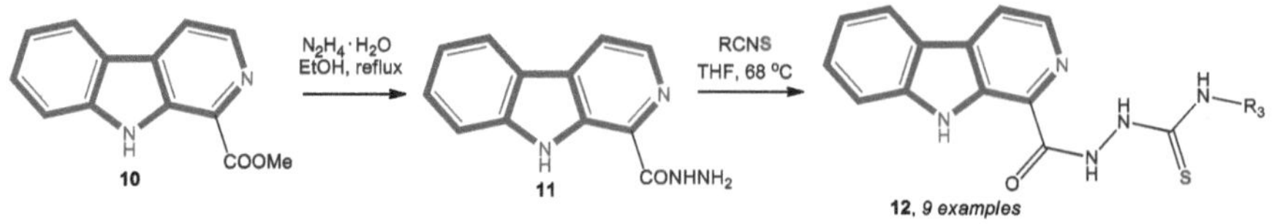

SCHEME 5.2 Design of hybrid molecules **12** based on β-carboline scaffold with thiohydrazide substituent in C-1 position.

the activity of ribavirin (protective effect = 40 ± 1%) was found for the compounds containing benzyl and *para*-bromophenyl substituents at the terminal nitrogen atom of the thiourea moiety, tested in 500 μg·mL^{-1}concentration (protective effects of 58 ± 3% and 60 ± 2%, respectively). Compounds containing butyl, *para*-methoxyphenyl, and benzyl substituents present in concentrations of 50 μg·mL^{-1} showed >97% selectivity of the fungicidal action against the phytopathogenic fungi *Physalospora piricola* and *Rhizoctonia cerealis*.

Recently, a regioselective method was proposed for the synthesis of mono-*ortho*-sulfonamide β-carboline derivatives **15** by the ruthenium-catalyzed reaction of aryl-substituted β-carbolines **13** with sulfonyl azides **14** (Scheme 5.3). This reaction can take place both with electron-withdrawing and with electron-donating substituents in the benzene ring of the C-1-aryl substituent. The synthetic pathway involves the formation of intermediate ruthenium complex **A** of N-ligand **13** and azide **14**. The elimination of N_2 affords the target *ortho*-sulfonamides **15**. The primary screening of the antiproliferative action against the cancer cells – HCT-116 (human colorectal carcinoma) and A549 (human lung carcinoma) – showed that the compounds provide a high percentage of inhibition of HCT-116 cells, being more active than harmine (Bora et al. 2021).

In 2022, Singh's research team developed a facile and efficient synthesis of β-carboline-linked 2-acylbenzothiophenes **18** in the KI–DMSO system (Singh et al. 2022). This approach is implemented by the crotonic condensation of 1-acetyl-β-carboline **15** with 2-nitrobenzaldehyde **16**, which gives the intermediate 2-nitrochalcone-β-carboline **17**. The thiophene ring is formed *via* the reaction of chalcone **17** with elemental sulfur in the presence of an equimolar amount of KI (Scheme 5.4).

5.2.2 Design of N(2)-substituted β-Carbolines with S-function

The effect of N(2)-modification of the tetrahydro-β-carboline scaffold on the antileishmanial activity was considered in relation to *Leishmania donovani* promastigotes and axenic amastigotes (Manda et al. 2014). The target N-alkylated β-carbolines **20** and **22** were prepared by the condensation of tetrahydro-β-carboline **3** with thienylalkyl(acyl) halides **19** and **21** in the presence of bases (Scheme 5.5). Among the series of thienyl(thiazole)-substituted tetrahydro-β-carbolines, the highest activity against promastigotes (IC_{50} = 9.1 μM) was exhibited by the compound containing a methyl-5-chlorothienyl moiety; however, this compound showed no activity against axenic amastigotes. The same derivative **20** had an antitrypanosomal

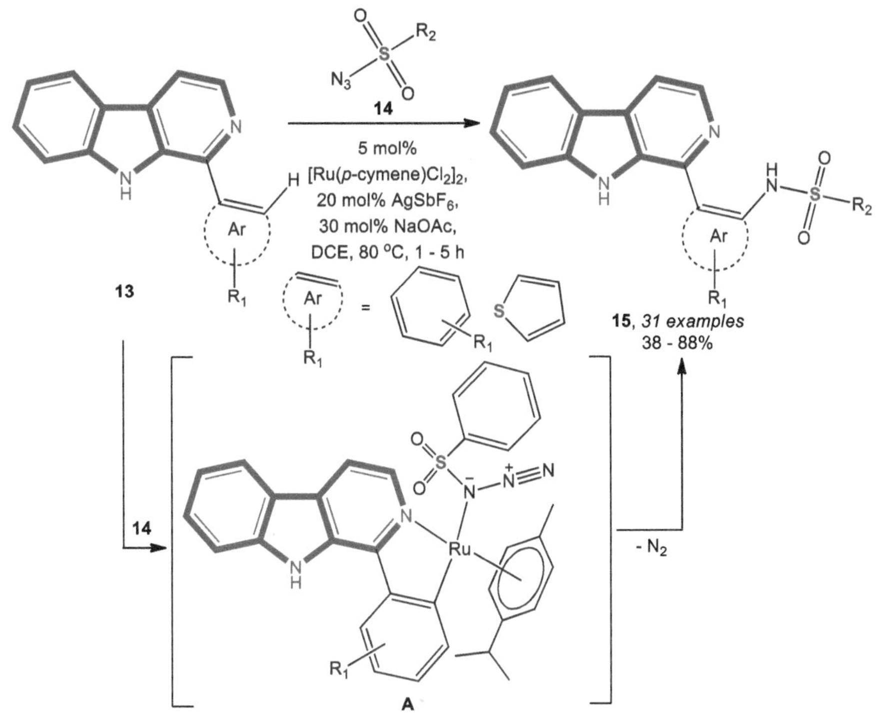

SCHEME 5.3 Ruthenium(II)-catalysed *ortho*-sulfonamidation of 1-aryl-β-carbolines.

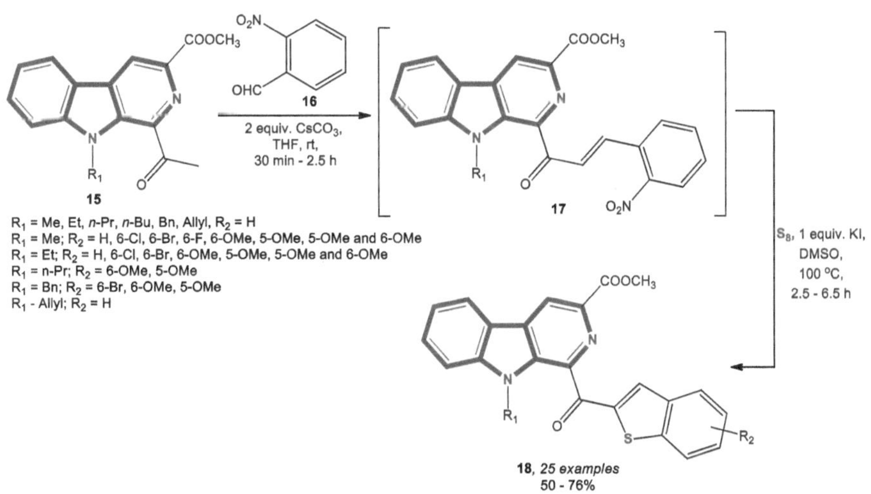

SCHEME 5.4 KI promoted synthesis of 2-acylbenzothiophene-β-carbolines **18** based on the condensation between 1-acetyl-β-carbolines **15**, 2-nitrobenzaldehyde **16** and elemental sulfur.

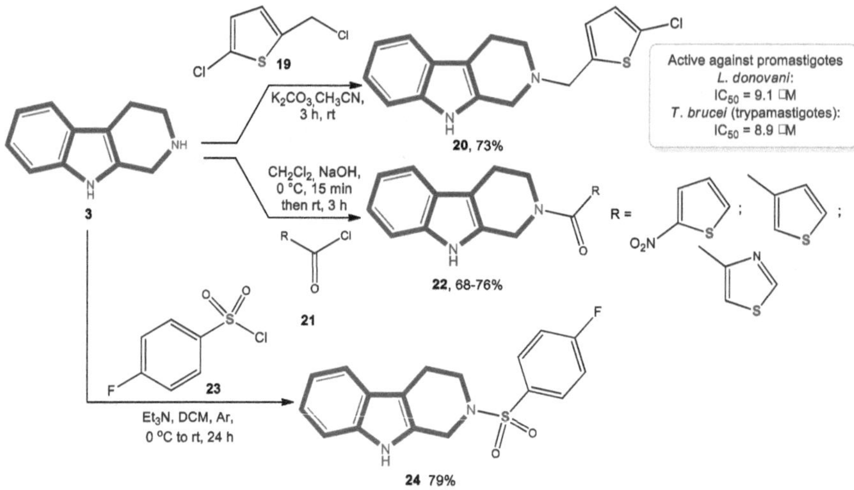

SCHEME 5.5 Synthesis of N(2)-substituted tetrahydro-β-carbolines.

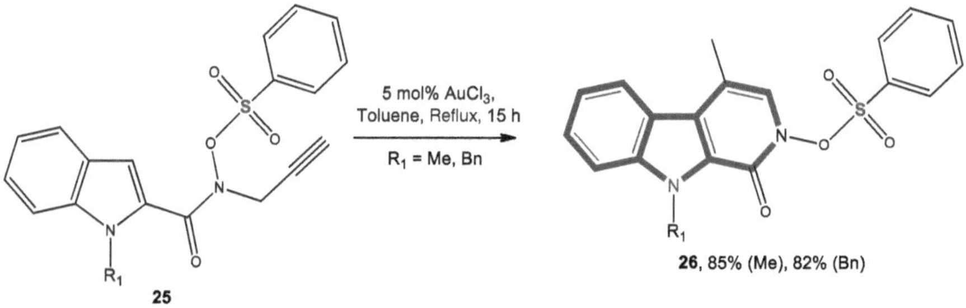

SCHEME 5.6 Gold(III)-catalyzed cycloisomerization of indole-substituted N-propargylamides **25** in the design of 2,9-dihydro-β-carboline-1-ones **26.**

activity against *Trypanosoma brucei* (IC_{50} = 8.9 μM). There is a single example of the synthesis of tetrahydro-β-carbolinesulfonamide **24**; the synthesis includes the reaction of tetrahydro-β-carboline **3** with 4-fluorobenzenesulfonyl chloride **23** in the presence of triethylamine (Ren et al. 2022). The authors note that this functionalization increases the solubility and polarity of the target molecule **24** (Scheme 5.5).

A convenient method for the design of substituted 2,9-dihydro-β-carbolin-1-ones **26** was developed by Padwa and co-workers (2010). The essence of the method is the intramolecular cyclization of indole-substituted N-propargylamides **25** in the presence of gold(III) chloride (Scheme 5.6).

5.2.3 C-3 Functionalization of β-Carbolines with Sulfanyl-, Thione-, Sulfone Groups *via* Spacers

Analysis of the data published over the past few years showed a considerable interest in the design of molecules containing sulfur atoms of different valence: divalent sulfur in thioethers and thiones; hexavalent sulfur in sulfates and sulfonic acids, their salts, and amides. A recent review addresses S-containing scaffolds (five-, six-, and seven-membered heterocycles) as important structural elements for the drug design related to sulfur-containing molecules with cytotoxic properties (Laxmikeshav et al. 2022). According to this concept, β-carboline hybrid molecules were obtained as multipurpose antitumor agents.

5.2.3.1 Sulfanyl Derivatives of β-Carbolines

In order to block the C-1 position of β-carbolines, the C-1 atom was substituted with the 3-hydrophenyl pharmacophore moiety, which is a structural unit of the monastrol drug (Abdelsalam et al. 2018). First, key hydrazide **27** was obtained in four preparative steps from tryptophan **7** and 3-hydroxybenzaldehyde. The presence of the active functional group in the side chain enabled subsequent transformations to prepare a series of sulfur-containing thiosemicarbazide **28** and benzosulfohydrazide **29** derivatives and various heterocyclic systems (thiazolidinone **30** and thiazoline **31**, mono- and bis-1,3,4-oxadiazole-2-thiones **28, 30, 31**). The range of sulfur-containing β-carbolines was investigated for anticancer activities against MCF-7 (breast cancer) and A-549 (lung cancer) cells using colorimetric MTT assay with vinblastine as the reference compound. Screening of the cytotoxic activity *in vitro* revealed two groups of most active 3-hydroxyphenyl-9*H*-β-carbolines in terms of the IC_{50} values: these are compounds containing a thiosemicarbazide moiety at C-5 and 4-bromophenyl or tolyl substituent at nitrogen and the S-methyl sulfanyloxadiazole derivative (Scheme 5.7). Enzyme inhibition assay in relation to topo-I and KSP/Eg5 ATPase demonstrated high inhibitory activity of 4-bromophenyl-substituted thiosemicarbazide derivative

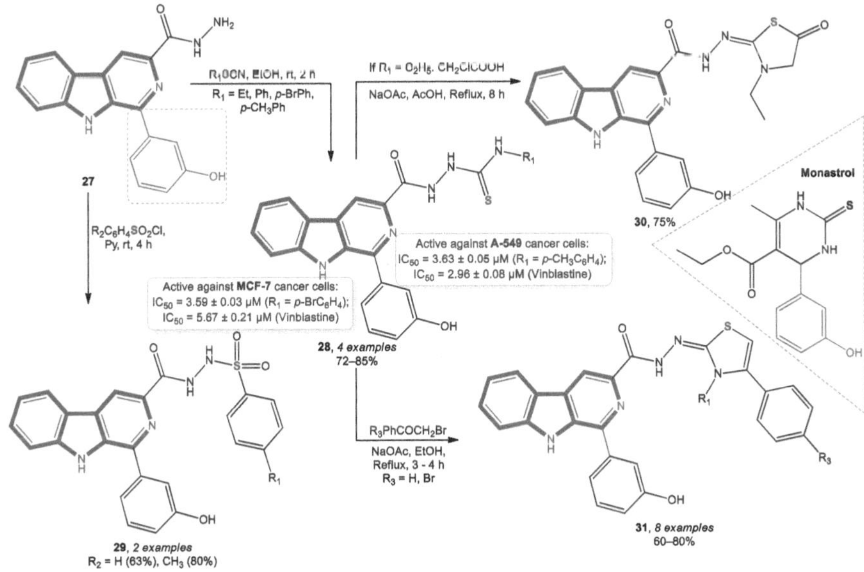

SCHEME 5.7 C-3-Functionalization of 3-hydroxyphenyl-β-carbolines.

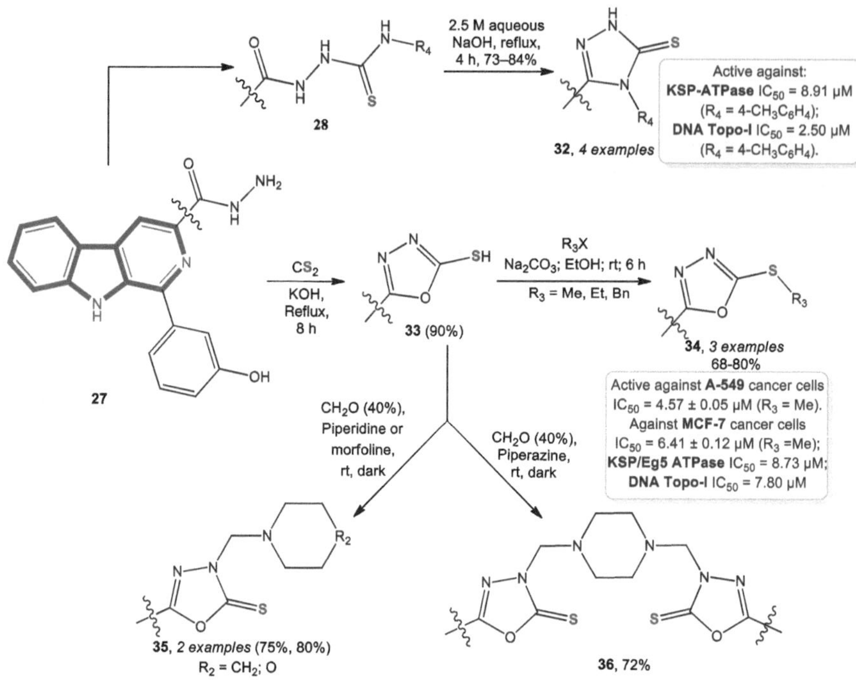

SCHEME 5.8 Diversification of sulfur-containing β-carbolines as multi-target anticancer agents.

28 against both enzymes (IC_{50} = 5.90 μM for topo-I and IC_{50} = 13.37 μM for KSP-ATPase) exceeding that of camptothecin (IC_{50} = 6.90 μM for topo-I) and monastrol (IC_{50} = 16.62 μM for KSP-ATPase), while S-methyl sulfanyloxadiazole derivative **29** showed twice higher inhibitory activity against KSP/Eg5-ATPase (IC_{50} = 8.73 μM) compared with the monastrol activity.

Notably, the bioisosteric replacement of the 1,3,4-oxadiazole moiety by the 1,2,4-triazole-3-thione moiety resulted in a higher anticancer activity in the single-dose NCI *in vitro* assay (Abdelsalam et al. 2018). Experiments for studying the mechanism of cytotoxicity in relation to MDA-MB-435 melanoma cells showed that lead compound **32** containing a tolyl substituent in position N-4 causes Pre-G1 apoptosis and arrests the cell cycle in the G2/M phase by dual inhibition of both topo-I and KSP-ATPase, with its inhibitory activity being twice as high as those of camptothecin or monastrol used as reference compounds (Scheme 5.8).

Later (Barbosa et al. 2016), β-carboline-4-thiazolidinones were tested for the anticancer activity *in vitro* against human cancer cell lines and for the antiviral activity against the herpes simplex virus type 1 (HSV-1). It was found that the nature of the 4-thiazolidinone moiety and substitutions in the β-carboline C-3 and C-1 positions determine the biological activity. The presence of 4-dimethylaminophenyl group in the C-1 position and N-(2-aryl-4-thiazolidinone)-carboxamide moiety in the C-3 position of β-carboline **37** endows the molecule with enhanced anti-HSV-1

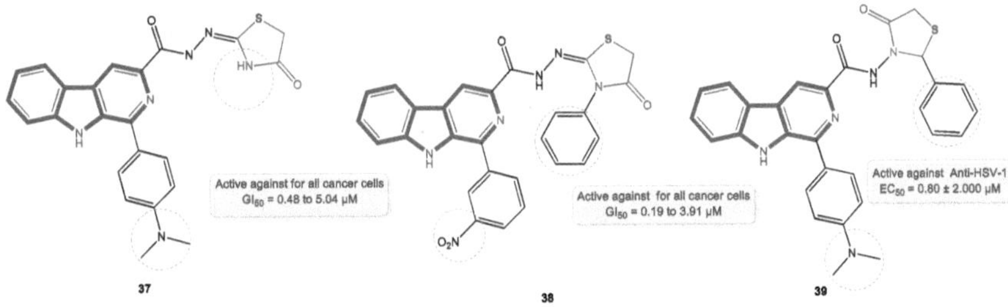

FIGURE 5.5 Structures of 4-thiazolidinone-β-carbolines **37-39** exhibiting anticancer and antiviral activity.

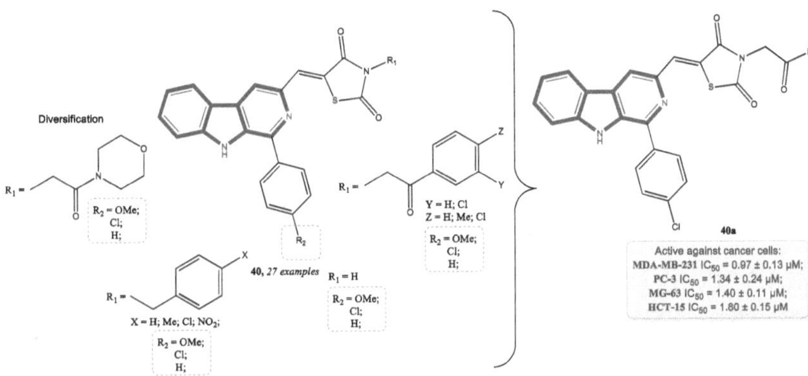

FIGURE 5.6 C-3 Functionalization of the β-carboline skeleton by diversification in the thiazolidinedione moiety.

activity (EC_{50} =0.80 ± 2.000 μM). Meanwhile, compound **39**, which has the same C-1-substituent, but N'-(2-ylidene-4-thiazolidinone)-3-carbohydrazide moiety at C-3, demonstrates anticancer activity *in vitro* with GI_{50} below 5 μM against any of the tested cell lines (U251, MCF-7, NCI/ADR-RES, 786-0, NCI-H460, OVCAR-3, and HT-29) (Figure 5.5).

In 2018, Tokala et al. obtained a series of novel hybrid molecules **40** containing a thiazolidinedione moiety in the C-3 position of β-carboline; in other words, the sulfur atom in these compounds is linked *via* a vinyl spacer (Figure 5.6). Cytotoxic activity assays *in vitro* using PC-3 (prostate cancer cells), A549 (lung cancer cells), MG-63 (osteosarcoma cells), HCT-15 (colon cancer cells), MDA-MB-231 (breast cancer cells), A431 (skin cancer cells), PANC-1 (pancreatic cancer cells), and L-132 normal human lung epithelial cells demonstrated that the molecule containing a *para*-chlorophenyl substituent in the carboline C-1 position and a morpholinethiazolidinedione moiety in the C-3 position has the highest activity against the MDA-MB-231 breast cancer cells (IC_{50} = 0.97±0.13 mM). The half-maximal inhibitory concentration of doxorubicin used as the reference is IC_{50} = 1.18 ± 0.06 mM. Although this compound has high cytotoxicity against cancer cells, it is also toxic (IC_{50} = 3.33 ± 0.93 mM) against the conditionally normal L-132 cells, which attests to non-specific action of this

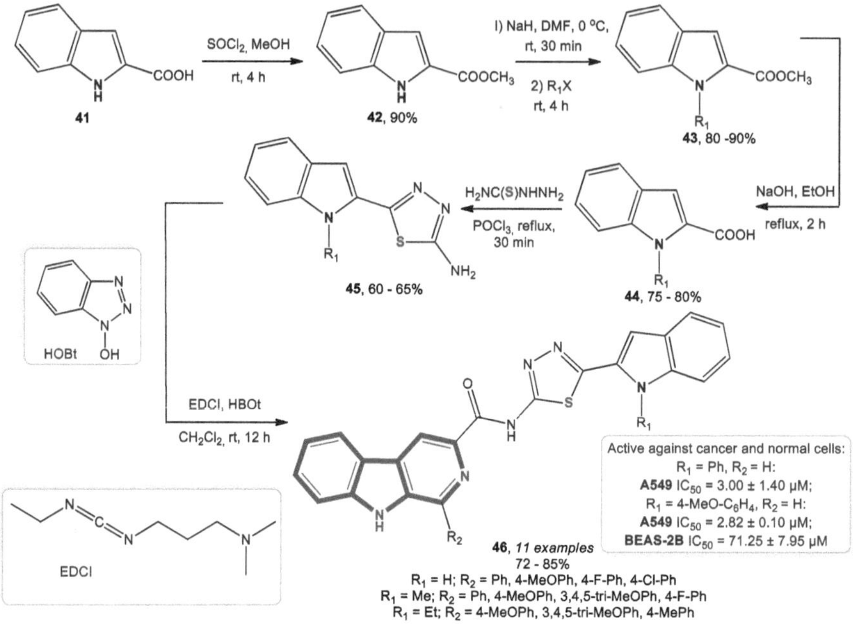

SCHEME 5.9 Design of potential antitumor agents based on β-carboline-indoles **46** linked by a thiadiazolo-carboxamide spacer.

compound. Meanwhile, a recent review noted positive dynamics for the antitumor activity of hybrid molecules upon the introduction of the thiazolidinedione moiety (Tilekar et al. 2022).

N. Shankaraiah's research team (Tokala et al. 2020) developed a synthetic strategy towards hybrid molecules based on C-3-substituted phenyl-β-carbolines and indoles linked by a thiadiazole-carboxamide spacer. The sequential synthesis of hybrid molecules **46** starting from 1*H*-indole-2-carboxylic acid **41** is depicted in Scheme 5.9. Screening of the cytotoxic potential *in vitro* for this series of compounds demonstrated that the compound containing a *para*-methoxyphenyl moiety in the C-1 position of β-carboline has a relatively high activity (IC_{50}< 5 μM) against all of the tested cancer cell lines (A549, MDA-MB-231, BT-474, HCT-116, and THP-1), with the highest cytotoxic potential being found in relation to the lung cancer (A549) cells: IC_{50} = 2.82 ± 0.10 μM, the selectivity index (SI) is 25.26 (IC_{50} = 71.25 ± 7.95 μM (BEAS-2B)). Additional experiments dealing with the mechanism of cytotoxic action showed that thiadiazole indole derivatives of β-carbolines arrest the G0/G1 phase of the cell cycle in A549 cells in a dose-dependent manner.

Tokala et al. proposed the use of β-carboline acid **47** (Scheme 5.10) for subsequent transformations (Tokala et al. 2021). The synthetic strategy towards the target compounds implies the preparation of potentially cytotoxic hybrid molecules by linking the β-carboline core and benzothiazole by a carboxamide spacer. The series of compounds obtained in this way was screened *in vitro* for the cytotoxic activity by measuring the half-maximal inhibitory concentrations for attached cancer cells such

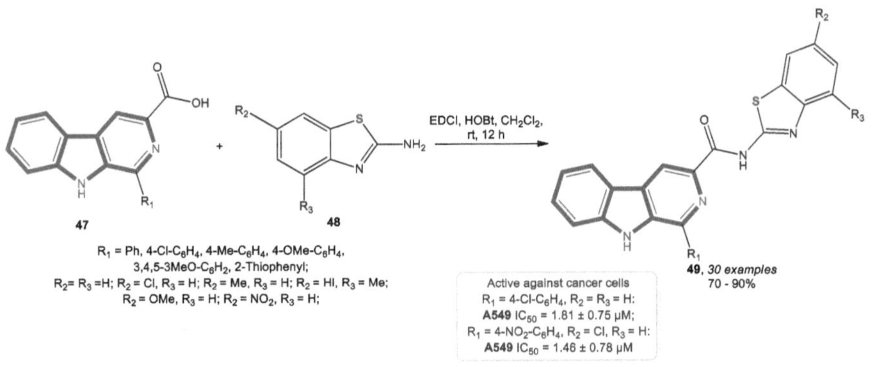

SCHEME 5.10 Synthesis of hybrid molecules based on β-carboline-benzothiazoles with a carboxamide linker.

as A549 and NCI-H460 (lungs), HCT-116 (colon), and BT-474 (breast) and by determining the affinity for the suspension leukemia cell lines, including MOLT-4 (acute lymphoblastic leukemia), THP-1 (acute monocytic leukemia), and HL-60 (acute myeloblastic leukemia). The primary screening of the cytotoxic activity demonstrated that two compounds containing *para*-nitro- or *para*-chloro-phenyl substituent in the C-1 position of β-carboline have the most pronounced inhibitory effect against A549 lung cancer cells (IC_{50} = 1.46± 0.78μM and 1.81± 0.75 μM, respectively). It is noteworthy that these values are close to the results for the cytostatic doxorubicin used as the reference (IC_{50} = 1.92 ± 0.86 μM); also, these compounds have a high cytospecificity with respect to BEAS-2B normal human lung epithelial cells: IC_{50} = 72.21± 1.89 μM and 58.41 ± 1.61 μM. The selectivity index varies from 49.4 ± 7.2 to 32.2 ± 5.2 (the compound is selective when SI ≥ 10), respectively.

5.2.3.2 Thiocarbamide and Thione Derivatives of β-Carbolines

In order to form thione groups in 3-amino-β-carbolines **50**, Zhong and co-workers performed functionalization with thioisonitriles **51**. The preparation of the target products is illustrated in Scheme 5.11. The fungicidal activity of the resulting β-carboline derivatives containing the pharmacophore acylthiourea group was assessed *in vitro* in relation to *Rhizoctonia solani*, which causes damping off diseases of rice (Zhang et al. 2018). It was found that the bioisosteric replacement of the sulfur atom by oxygen enhances the fungicidal activity. The fungicidal activity of the whole series of functionalized β-carbolines (EC_{50} = 0.131–1.227 mmol L^{-1}) is higher than the activity of the initial harmine (EC_{50} = 2.453 mmol L^{-1}). Among arylthiourea β-carbolines, compound **52** (R_2 = R_3) with 3,4,5-trimethoxyphenyl substituents has the highest activity characterized by EC_{50} = 0.293 mmol L^{-1} (validamycin A, EC_{50} = 0.397 mmol L^{-1}).

An original approach to C-3-functionalization of β-carboline to thioethers *via* thione groups was proposed by Sarragiotto and co-workers (Formagio et al. 2008). They developed the synthesis of β-carbolines containing a phenyl group at C-1 as well as 1,3,4-oxadiazole- and 1,2,4-triazole substituents at the carboline C-3 atom

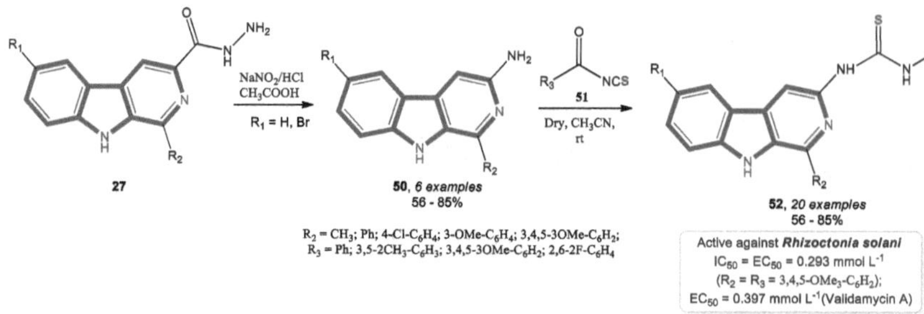

SCHEME 5.11 Design of potential fungicidal agents based on β-carbolinaryl thiourea.

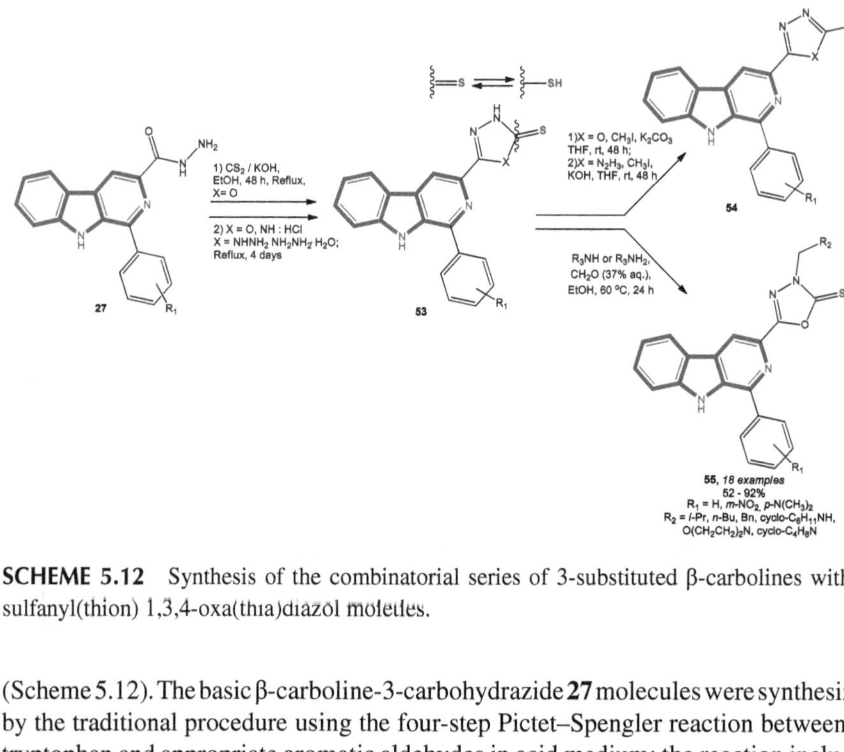

SCHEME 5.12 Synthesis of the combinatorial series of 3-substituted β-carbolines with 2-sulfanyl(thion) 1,3,4-oxa(thia)diazol moieties.

(Scheme 5.12). The basic β-carboline-3-carbohydrazide **27** molecules were synthesized by the traditional procedure using the four-step Pictet–Spengler reaction between L-tryptophan and appropriate aromatic aldehydes in acid medium; the reaction included the oxidation of β-carboline-3-carboxylates with sulfur in refluxing xylene followed by refluxing of the products with hydrazine hydrate in ethanol. The authors also noted that 3-(2-thioxo-1,3,4-oxadiazol-5-yl) β-carbolines and 3-(5-thioxo-1,2,4-triazol-3-yl) β-carbolines **53** occur in solution as the thione–thiol tautomers. The cyclization of β-carboline-3-carbohydrazides **27** to β-carboline 1,3,4-oxadiazole **53** proceeds in 24 h (Zhang et al. 2017). The same study also addressed the structure–activity relationships for the obtained combinatorial series of compounds by measuring the cytotoxic activity *in vitro* against Sf9 cells and the inhibition of the growth of *Spodoptera litura*. The highest activity against Sf9 cells was found for the compound containing a 4-F-Ph substituent in the C-3 position of carboline: IC_{50} = 16.25 μM (for harmine, IC_{50} = 140.68 μM; for camptothecin, IC_{50} = 18.95 μM). This compound also provides significant inhibition of the growth of *S. litura* larvae with a decrease in the weight of the larvae and chrysalis.

In continuation of these studies (Savariz et al. 2010), the same research team carried out the aminomethylation of the previously obtained β-carboline **53** with various amines and formaldehyde (Scheme 5.12). Evidently, because of the higher nucleophilicity, the reaction involved only the amino group of the 2-thioxo-1,3,4-oxadiazole moiety and gave Mannich bases **55** in more than 52% yield. The obtained combinatorial series of substituted β-carbolines (~ 44 compounds (Formagio et al. 2008; Savariz et al. 2010) were subjected to the primary screening *in vitro* to reveal the anticancer activity against a broad range of cancer cells such as melanoma (UACC-62) and breast (MCF7), resistant ovarian (NCI/ADR), kidney (786-0), lung (NCI-460), prostate (PCO-3), ovarian (OVCAR), and colorectal (HT-29) cancer cells. The highest activity was revealed for 1-(N,N-dimethylaminophenyl)-3-(5-thioxo-1,2,4-triazol-3-yl) β-carboline **55**, which is especially effective against lung cancer (GI_{50} = 0.06 μM) and kidney and ovarian cancer cell lines. Furthermore, the introduction of the methylamine group into the 1,3,4-oxadiazole moiety enhances the anticancer activity in substituted β-carbolines.

It is known that imidazolidine-2,4-dione scaffold (hydantoin) and its sulfur-containing analogue, thiohydantoin, are considered to be privileged heterocyclic scaffolds. These moieties can be encountered in drug precursors (e.g., enzalutamide) or in compounds that exhibit wide ranges of pharmacological activities (Cho et al. 2019).

The annulated thiohydantoin structures with tetrahydro-β-carboline scaffold **57** were synthesized by the reaction of tetrahydro-β-carboline ester **56** with isothiocyanates **51** in the presence of excess triethylamine (Scheme 5.13) (Huang et al. 2018). Evaluation of the series of products in relation to the tobacco mosaic virus *in vitro* demonstrated that most of the compounds have moderate activities, and the lead compound is that containing the cyclohexyl substituent in the thiohydantoin moiety. The activity of this compound against the virus is higher than the activity of the reference compound ribavirin (Scheme 5.13) or β-carboline ester **56** (inhibition rate = 22 ± 1 % at 100 mg/mL); this attests to the synergistic effect of the combination of the thiohydantoin motif with the β-carboline scaffold.

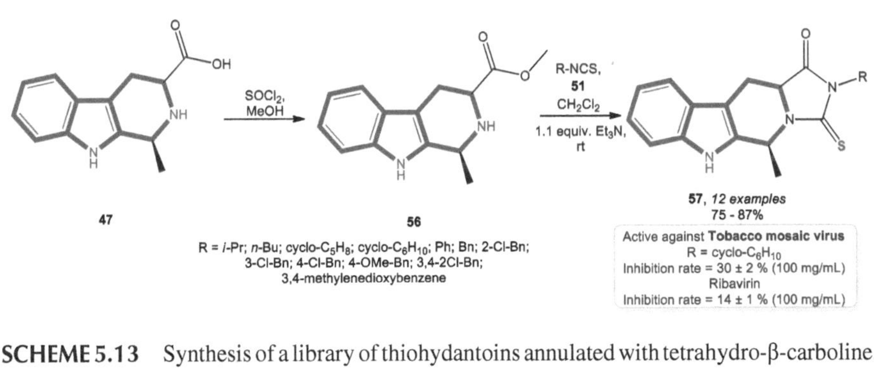

SCHEME 5.13 Synthesis of a library of thiohydantoins annulated with tetrahydro-β-carboline.

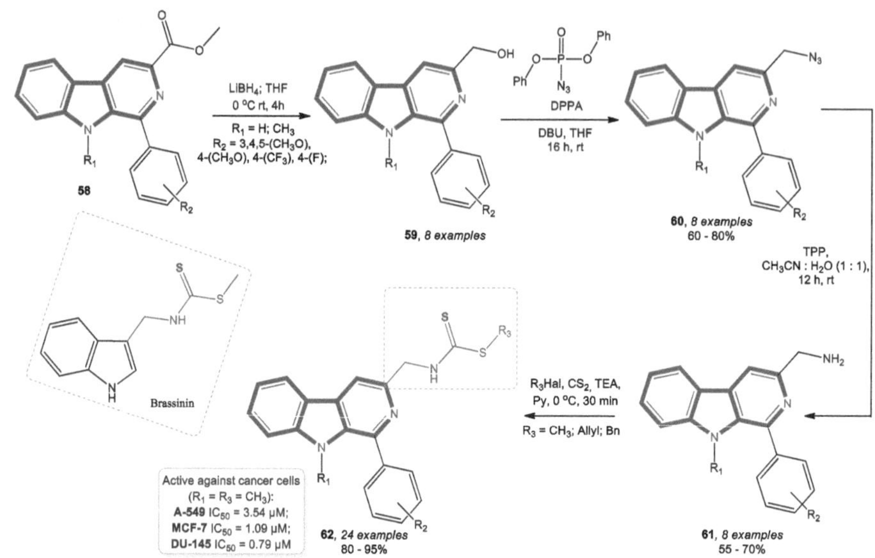

SCHEME 5.14 Construction of dithiocarbamate derivatives of β-carboline **62**.

A multistep synthesis (seven steps starting from L-tryptophan) of molecules combining two pharmacophore moieties, β-carboline and dithiocarbamate in the C-3 position of the pyridine ring, was proposed (Scheme 5.14) (Kamal et al. 2015). As the base molecule, the authors used the natural compound brassinin, which comprises an indole ring and a dithiocarbamate moiety and is known to have chemopreventive effect (cancer immunosuppression) (Chripkova et al. 2014). According to cytotoxicity assays of β-carbolines containing methylamine (**61**) and dithiocarbamate (**62**) moieties against A549, MCF-7, and DU-145 cancer cells, the molecules containing dithiocarbamate together with β-carboline **62** are much more active than their precursors. Derivative **62** containing methyl groups in the N(9) position of carboline and in the S-thiocarbamate moiety was the lead compound, exhibiting a cytotoxic activity comparable with that of doxorubicin reference (IC_{50} =1.58 μM (A549), 0.96 μM (MCF-7), 1.38 μM (DU-145)).

5.2.3.3 Design of Sulfone Derivatives of β-Carbolines

The molecules containing an 1-acylindole moiety in the C-1 position together with the amide group in the C-3 position in the tricyclic pyrido[3,4-b]indole ring are of no less interest. These compounds are structural analogues of the family of indole alkaloids, particularly, pityriacitrin B (Xu et al. 2019). A weak antiproliferative activity of pityriacitrin alkylamides against breast cancer (MCF-7, MDA-231) and prostate cancer (PC3) cell lines was reported (Zhang et al. 2011). Meanwhile, pityriacitrin was used as the basic scaffold (Xu et al. 2019) and functionalized to amides by treatment with various aromatic and aliphatic amines. The products and the starting molecules (19 compounds) were tested *in vitro* for the cytotoxic activity against several cell lines: SGC-7901 (gastric cancer), A875 (human melanoma),

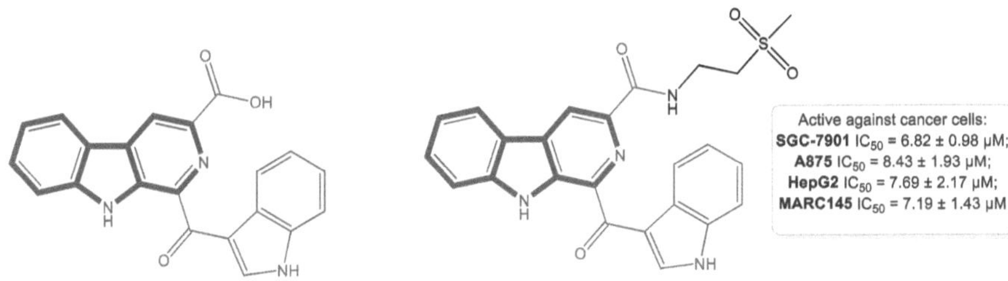

Pityriacitrin B

63

FIGURE 5.7 Pityriacitrin B and its S- containing derivative **63**.

HepG2 (human hepatocellular carcinoma), and MARC145 (subclone of the MA-104 African green monkey kidney cells). Modification of pityriacitrin B by introducing a sulfone group into the C-3 position was found to be very promising for the development of new drugs. It is known that sulfonic group is present in the molecules of a great number of clinical pharmacological substances (Feng et al. 2016). The β-carboline derivative with the C3-methylsulfonyl group is the lead compound with the highest activity (Figure 5.7). The activity of compound **63** against all four cell lines is markedly higher than the activity of the natural alkaloid pityriacitrin B with IC_{50} = 67.69 ± 11.79 μM (SGC-7901), 48.75 ± 6.25 μM (A875), 51.82 ± 5.99 μM (HepG2), and 41.88 ± 9.32 μM (MARC145).

A recent review (Dai et al. 2022) noted high antifungal efficiency of natural and synthetic β-carbolines functionalized by quaternization in the N-2 position. In addition, carboline carboxamide **63** (Figure 5.7) also showed a good result in antifungal activity assays *in vitro* against phytopathogenic fungal strains such as *Phytophthora capsici*, *Sclerotinia sclerotiorum*, *Botrytis cinereal*, and *Rhizoctonia solani* (Huang et al. 2022). The inhibition percentages for a concentration of 50 μg/mL^{-1} were 48.18 (*P. Capsici*), 22.73 (*S. Sclerotiorum*), 39.64 (*B. Cinereal*), and 23.09 (*R. Solani*), respectively.

Tran et al. reported for the first time *Stolonine C*, a taurine conjugate with β-carboline-3-carboxylic acid (Tran et al. 2015). This sulfur-containing β-carboline was isolated from the marine tunicate *Cnemidocarpa stolonifera* (Australia) and identified by spectral methods. A synthetic approach to *Stolonine C* consisting of three key steps was proposed (Scheme 5.15). In the first step, l-tryptophan **7** is mixed with a 37% aqueous solution of formaldehyde to give methyl (3S)-1,2,3,4-tetrahydro-β-carboline-3-carboxylate **56**. Oxidation of the product with manganese(IV) oxide followed by hydrolysis with an aqueous solution of NaOH and MeOH (1:4) affords β-carboline-3-carboxylic acid **64**. Then amidation of the acid with taurine **65** in the presence of EDCI/HOBt in DMF at room temperature for 16 h furnishes the target compound in 40% yield. Cytotoxicity screening against PC3 prostate cancer cells (in 20 μM concentration) showed 26% inhibition of cell growth. In addition, *Stolonine C* increases the size of cells by inducing elongation of the mitochondrial structure, which results in PC3 cell death *via* the apoptotic pathway.

One more example of binding of an arylsulfonyl moiety to β-carboline-3-carboxylic acid **64** is the piperazine spacer (Scheme 5.16) (Manasa et al. 2020). The series of

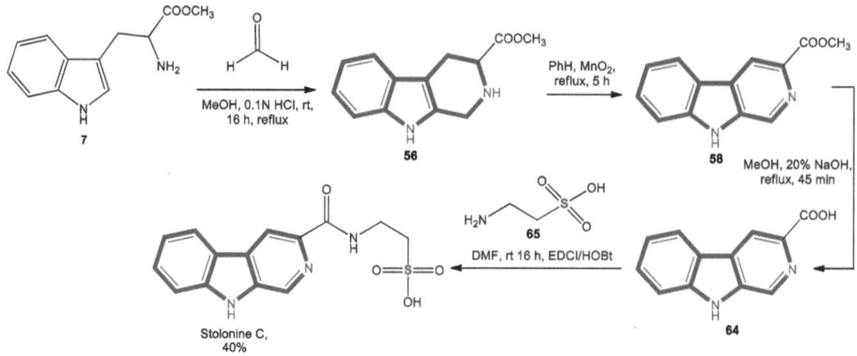

SCHEME 5.15 Directed C-3 functionalization of 3-carboxyl-β-carboline **64** by the sulfonyl group.

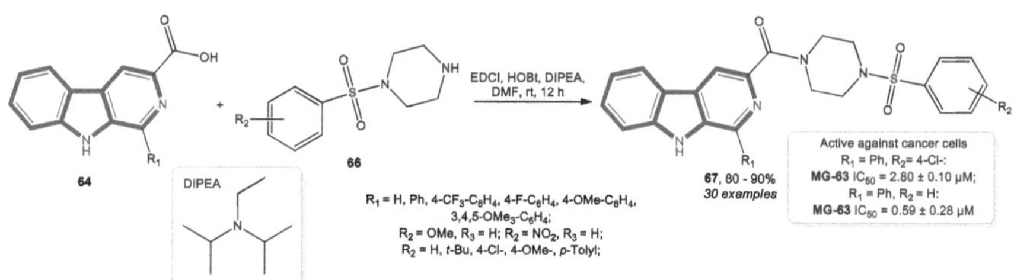

SCHEME 5.16 Pharmacophore hybridization of β-carboline derivatives **67** linked aryl sulfonyl piperazine.

obtained compounds were evaluated *in vitro* for the cytotoxic effect against six human cancer cell lines: HT-29 (colorectal adenocarcinoma), MDA-MB-231 (breast cancer), MG-63 (osteosarcoma), U87MG (brain cancer), PC3 (prostate cancer), and HeLa (cervical cancer), and against normal African green monkey kidney cell line (Vero). Screening of the cytotoxic activity demonstrated that an important structural detail is the phenyl group in the C-1 position of β-carboline (**67**). The highest cytotoxicity (IC_{50} = 0.59 ± 0.28 μM and 2.80 ± 0.10 μM) against the MG-63 cell line was found for compounds **67** that contains phenyl or *para*-chlorophenyl substituent in the sulfonyl moiety. Additional experimental studies of the putative mechanism of the cytotoxic action showed that these compounds arrest the cell cycle in the sub-G1 phase, thus inducing the apoptosis, and that they are able to inhibit the enzymatic activity of topoisomerase II.

The targeted synthesis of hybrid molecules **73** based on β-carbolines containing a furoxan pharmacophore was performed by the reaction of β-carboline acids **64** with benzenesulfonylfuroxans **72** in the presence of tandem promoters, 4-dimethylaminopyridine (DMAP) and 1-ethyl-3-(3-dimethylaminopropyl) carbodiimide hydrochloride (EDCI), or by acylation in the presence of EDCI and 1-hydroxybenzotriazole (HOBT), depending on the used furoxan substrate **72** (Scheme 5.17). A study of the antiproliferative activity of the combinatorial series

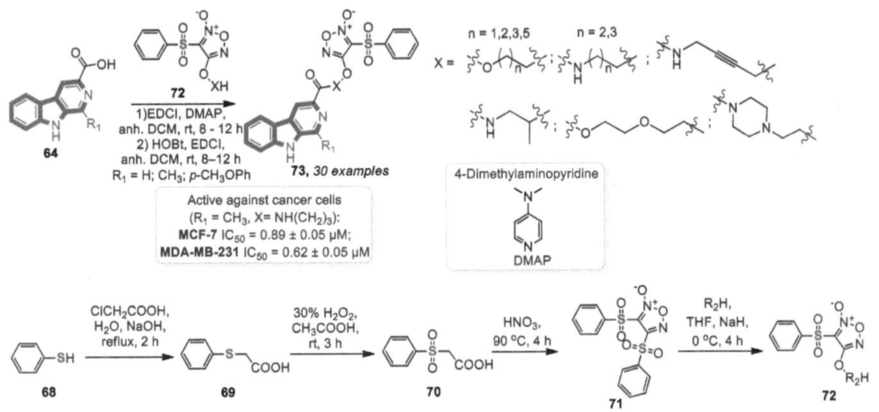

SCHEME 5.17 Synthesis of a combinatorial series of hybrid molecules based on β-carboline-(phenylsulfonyl)furoxan.

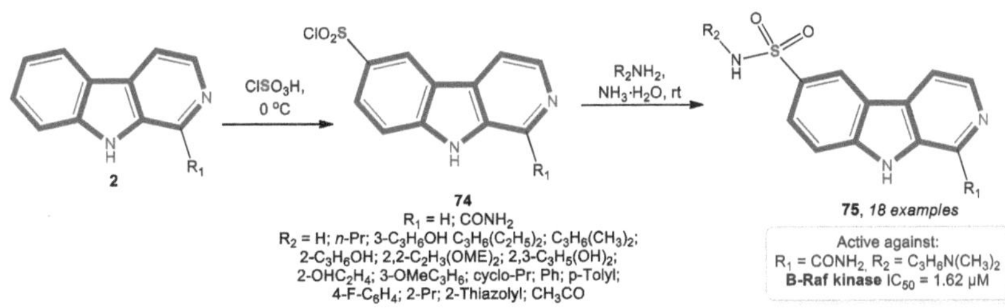

SCHEME 5.18 Design of 6-sulfonamide-substituted β-carbolines as B-Raf kinase inhibitors.

of the obtained (30 examples) and starting compounds demonstrated that the highest activity is inherent in the molecule with a methylene group in the C-1 position of carboline in which the carboline scaffold is linked to a benzenesulfonylfuroxan moiety *via* an aminopropyl spacer. The activity of the lead compound exceeds the activity of harmine(IC_{50} =16.94 ± 0.47 μM (MCF-7) and IC_{50} = 21.91 ± 1.25μM (MDA-MB-231)) and doxorubicin(IC_{50} = 3.09 ± 0.16 μM andIC_{50} = 2.86 ± 0.18μM, respectively) used as the positive control (Hu et al. 2021).

5.2.4 C-6 Sulfur-containing β-Carbolines

An innovative trend in the therapy of the malignant metastatic melanoma is to develop cellular signaling pathway inhibitors targeting B-Raf kinase (Menzies and Long 2014). Potential B-Raf kinase inhibitors of a new type were synthesized under mild conditions from β-carboline derivatives **75** containing a 6-sulfonamide moiety in the C-6 position (Xin et al. 2012). The two-step synthetic pathway is presented in Scheme 5.18. Determination of the IC_{50} values for B-Raf kinase inhibition identified 1-carboxamide-6-(N-(3-(dimethylamino)propyl)sulfamoyl)-β-carboline as the lead compound with the high activity characterized by IC_{50} = 1.62 μM.

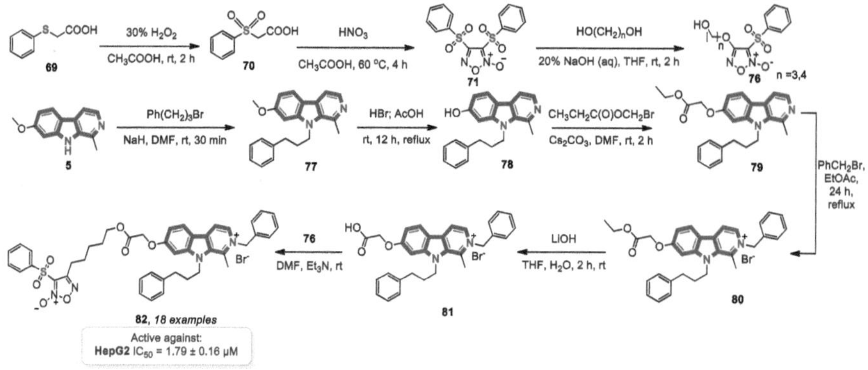

SCHEME 5.19 Synthesis of hybrid molecules **82** in the form of ammonium salts.

5.2.5 Diversification into C-7 Position of β-Carbolines

In 2022, the synthesis of an anticancer agent based on molecular hybridization of β-carboline was reported (Li et al. 2022b). The idea of the synthesis was to prepare a hybrid molecule combining an NO-donor furoxan ring connected to the C-7 position of the β-carboline core through an ester linker (Scheme 5.19). Compound **82** prepared in this way, precursor acid **81**, and harmine **5** were evaluated for the antiproliferative activity against five human cancer cell lines (HepG2, human hepatocarcinoma cells; BGC-823, human gastric cancer cells; HCT-116, human colorectal carcinoma cells; MCF-7, human breast cancer cells; and A549, human lung cancer cells) and against L02 normal human liver cells. The highest cytotoxic activity was identified for target compound **82** against HepG2 cells (IC_{50} = 1.79 ± 0.16 μM). However, this compound also has a moderate cytotoxicity against the L02 cells (IC_{50} = 10.04 ± 1.17μM). It was found that the putative mechanism of antitumor action is determined by the dose-dependent induction of HepG2 apoptosis. The lead compound arrests the cell cycle of the tumor cells in the G2/M phase and suppresses the cell proliferation *in vitro*.

5.2.6 S-Functionalization at the N(9)-position of β-Carbolines

The intermediate cyclic block synthons based on sultam **84** were prepared in satisfactory yield by successive sulfonylation of the indole moiety of the β-carboline scaffold at the nitrogen atom by treatment with pyridine-2-sulfonyl chloride in the presence of KOH or NaH and the subsequent intramolecular radical cyclization in the presence of tris(trimethylsilyl)silane (TTMSS) and azobisisobutyronitrile (AIBN) (Scheme 5.20). Tricyclic compound **84** is of interest as a precursor in the synthesis of ethers **85**. The reaction with alcohols takes place in the presence of excess cesium carbonate as the SNAr substitution accompanied by sultam ring opening in **84** (Recio et al. 2019).

Xu and co-workers proposed an original synthesis of 3-phenyl-β-carbolines **94** based on iodine-mediated electrophilic cyclization of 2-alkynyl-1-methylene azide indoles **91** (Liu et al. 2012). The authors noted that the introduction of the sulfonyl

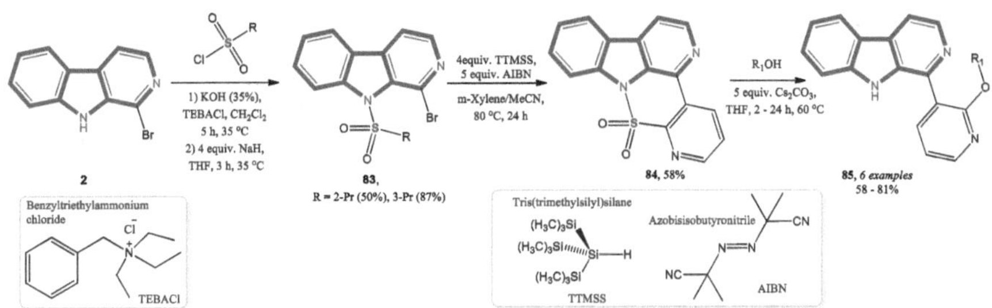

SCHEME 5.20 Synthesis of sultam **84** as an effective precursor for the synthesis of esters **85**.

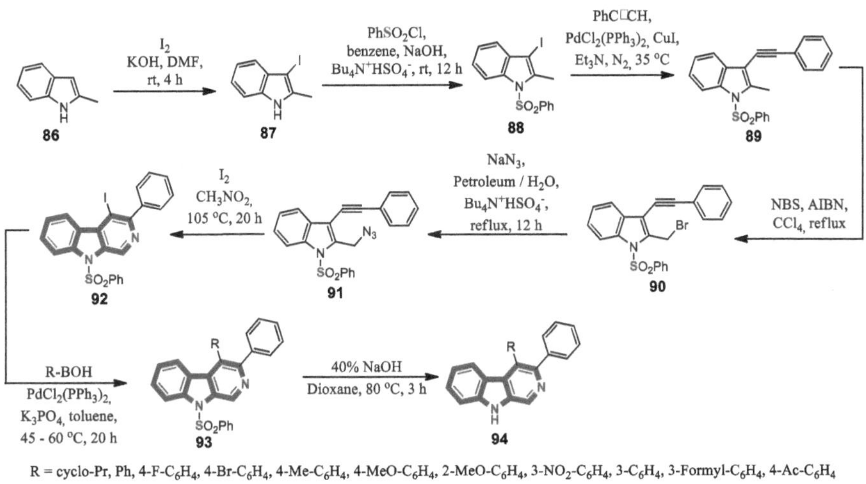

SCHEME 5.21 Synthesis of N-sulfonyl 3-phenyl-β-carbolines **934** based on 3-iodo-2-methyl-1-(phenylsulfonyl)-1*H*-indole **88**

group in the second step of the synthesis is required to stabilize 3-iodoindole-2-methylindole **88** (Scheme 5.21). The primary screening of the inhibitory activity of the final products of synthesis against proliferation of three cell lines (HeLa, MCF-7, and A549) showed that the activity is sensitive to substituents in the benzene ring at C-4 of the carboline system. Indeed, compounds with electron-donating groups in the benzene ring are more active against these three cell lines than compounds with electron-withdrawing substituents (Liu et al. 2012).

5.3 SYNTHESIS OF TRANSITION METAL COMPLEXES INVOLVING β-CARBOLINES AS N-LIGANDS

A number of sulfur derivatives containing the pharmacologically active β-carboline scaffold exhibit pronounced antitumor properties (Kumar et al. 2017). The use of chemotherapy is still a priority and an efficient way of treatment of numerous malignant neoplasms, which increases the life expectancy of cancer patients (Akhmetova et al.

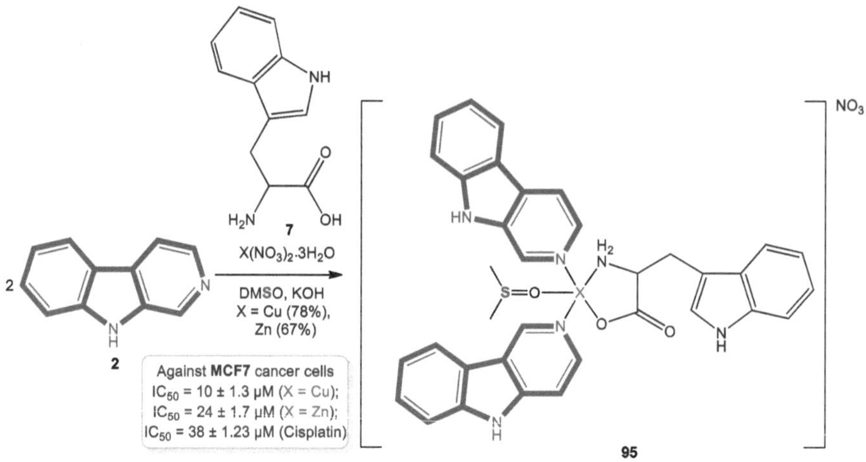

SCHEME 5.22 Synthesis of ternary complexes [X(tryp)(β-carboline)$_2$(DMSO)]NO_3, where X = Zn or Cu.

2020). Meanwhile, the increasing occurrence of resistance to known drugs and severe side effects stimulate the search for new innovative compounds based on β-carboline scaffold (Aaghaz et al. 2021; Tzouras et al. 2022). A particular trend is the design of effective multimechanistic anticancer agents based on metal complexes (Ayipo et al. 2021b).

Sulfur atoms can be introduced into β-carbolines *via* metal spacers, that is, *via* the formation of metal complexes. DMSO molecule, which readily forms complexes (solvates) with transition metals, often serves as the carrier of the sulfur atom (Akhmadiev et al. 2018).

Khan and co-workers (Khan et al. 2020) proposed the synthesis of new copper(II) and zinc(II) complexes **95** with norharmane **2** and tryptophan **7** as ligands (Scheme 5.22). The resulting complexes were solvates containing one DMSO molecule, which was confirmed by theoretical DFT calculations (B3LYP). The primary screening of the anticancer potential of the synthesized complexes was performed *in vitro* using two cancer cell lines, MCF7 (human breast cancer) and HepG2 (human hepatocellular carcinoma), and non-oncogenic HEK293 cell line (human embryonic kidney cells). According to MTT assay, the highest cytotoxic activity against MCF7 neoplastic cells is inherent in the copper(II) complex (IC_{50} = 10 ± 1.3 μM), which is much more active than the zinc(II) complex (IC_{50} = 24 ± 1.7 μM) or than cisplatin used as the reference (IC_{50} = 38 ± 1.23 μM). The subsequent investigation of the mechanism of action of the copper(II) complex confirmed that the cell cycle is arrested in the G2/M phase *via* activation of apoptotic pathways. The toxicity of copper(II) complex **95** against conditionally normal HEK293 cells is low (IC_{50}> 100 μM), which attests to the selective cytotoxic action against cancer cell lines.

The same group of authors carried out the subsequent transformation of the initial tryptophan ligand into imine on treatment with salicylaldehyde; this enhanced the cytotoxicity of an analogous copper complex against MCF7 cells (IC_{50} = 7.8 ± 0.4 μM) (Alharbi et al. 2021).

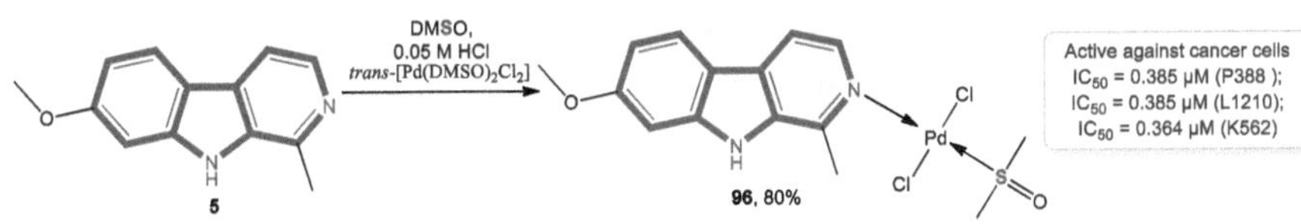

SCHEME 5.23 Synthesis of *trans*-palladium(II) complex **96** with DMSO and β-carboline ligands having antitumor activity.

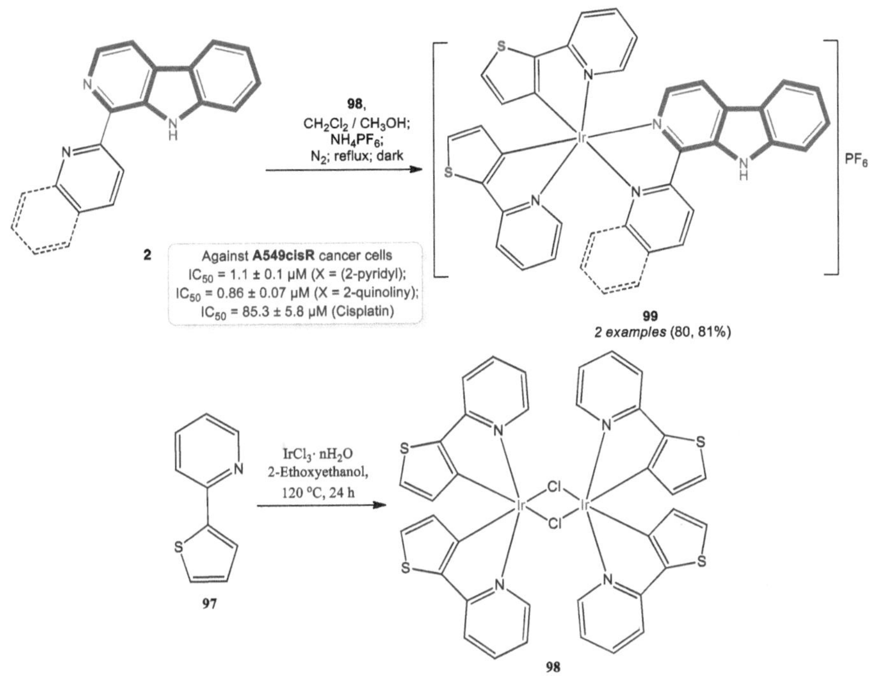

SCHEME 5.24 Synthesis of Ir(III)-β-carboline complexes **99** by ligand exchange of cyclometallated complex **98**.

An example of the synthesis of a complex with coordinated DMSO molecule (Scheme 5.23), *trans*-[Pd(DMSO)(harmine)Cl_2] **96**, was reported (Al-Allaf and Rashan 1998). Screening of the cytotoxic activity demonstrated a considerable efficiency of *trans*-palladium complex **96** *in vitro* against three cell lines; compound **96** is more active than cisplatin used as the reference, which demonstrates IC_{50} of 0.500 μM (P388), 0.833 μM (L1210), and 20.0 μM (K562).

Another approach to the synthesis of sulfur-containingβ-carboline complexes is based on the use of thienyl derivatives of the ligands (Scheme 5.24). Owing to their unique properties, photoluminescent cyclometallated iridium(III) complexes attract increasing research attention not only in the field of materials science, but also in medicinal chemistry as alternatives to platinum-based drugs (Caporale and Massi

2018; Lo and Zhang 2012). For example, He and co-workers (He et al. 2014) reported the synthesis of two cyclometallated Ir(III)-β-carboline complexes **99** involving 2-(2-thienyl)pyridine as an auxiliary ligand and β-carbolines as functional ligands. The mononuclear heteroligand complex **99** was obtained by ligand exchange of binuclear complex **98** synthesized from 2-(2-thienyl)pyridine **97** with $IrCl_3 \cdot nH_2O$. The anticancer activity was assessed *in vitro* using several human cancer cell lines (A549, A549cisR, HepG2, HeLa) and L02 human liver cells. It was found that, unlike the starting β-carboline ligands, Ir(III)-β-carboline complex **99** has a higher antiproliferative potential against all cancer cell lines. Its anticancer efficiency against the A549cisR cisplatin-resistant cell line is 100 times as high as that of cisplatin. The authors note that this effect is due to the mechanism of action of this compound differing from that of cisplatin, namely, the induction of ROS-mediated caspase-independent cell death *via* the autophagic pathway (inhibition of mTOR signaling) in the absence of apoptosis. Mention should also be made of the selectivity of action of Ir(III) complexes, which is manifested as a lower (by a factor of ~4) cytotoxicity against L02 cells in comparison with the HepG2 human cancer cells.

Recently, the synthesis monocationic bicyclic metallated photosensitizers based on Ir(III) complexes **101** with β-carboline ligands was proposed (Sanz-Villafruela et al. 2021); the photocatalytic properties of the products were studied (Scheme 5.25). The Ir(III) complexes were obtained by the reaction between the intermediate *rac*-[Ir(μ-Cl)$(ppy)_2]_2$ **101** and β-carboline ligand **100** with a thiazole substituent in the C-3 position. The reaction was carried out at the boiling point of the MeOH/CH_2Cl_2 solvent mixture (2:1) and was accompanied by the dichlorine bridge cleavage. The resulting Ir(III) complexes showed a high catalytic efficiency in the chemoselective and C-3-regioselective oxidative thiocyanation of indolines **103** to yield the corresponding 3-thiocyanatoindoles **104**.

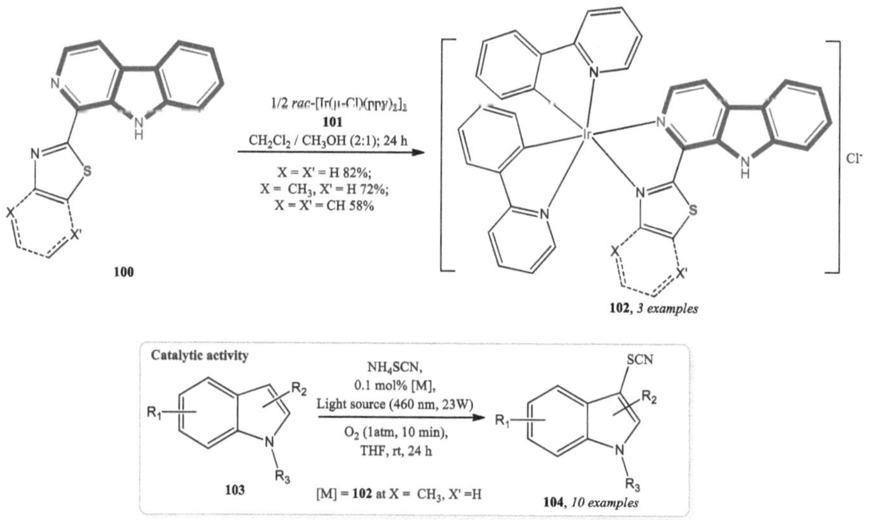

SCHEME 5.25 Design of Ir(III) complexes with β-carboline ligands. Photocatalytic thiocyanation of 1*H*-indole in the presence of Ir(III) complex **102**.

It is noteworthy that in the presented examples, β-carbolines act as monodentate ligands that form complexes *via* the pyridine nitrogen atom (Ayipo et al. 2021b).

5.5 CONCLUSION

The interest of the scientific community in β-carboline alkaloids is due to their occurrence in natural products and the possibility of synthesizing diverse structures, which explains the wide range of their pharmacological properties. In recent years, the development of the chemistry of sulfur-containing β-carboline compounds has tended to focus on the design of new hybrid molecules resorting to both traditional and innovative synthetic pathways. Depending on the reagents, functionalization involves C-1, C-3, C-6, C-7, N-2, and N-9 positions of the β-carboline scaffold. Note that there are no examples of S-functionalization into the C-4, C-5, or C-8 positions. The obtained synthetic products containing structurally diverse sulfur atoms (sulfide, thione, or sulfoxide sulfur) are of interest as compounds possessing a substantial pharmacological potential. As a result, depending on the position of sulfur atoms relative to the β-carboline scaffold, the modified molecules exhibit pronounced anticancer, antiviral (tobacco mosaic virus and HSV-1), fungicidal (*Rhizoctonia cerealis* and *Rhizoctonia solani*), and antitrypanosomal activities. The synergistic effect of hybrid molecules is noticeable when β-carbolines are S-functionalized *via* N-2 metal spacers; this may serve for the preparation of promising cytostatic agents for tumor cells (e.g., A549cisR) that are resistant to a number of anticancer agents. In addition, the catalytic activity of the Ir-N-β-Cr complex in the oxidative thiocyanation of indolines is noteworthy. We believe that the multiple biological activity of sulfur-containing β-carbolines may serve as an important starting point for the development of multifunctional drugs for the therapy of comorbid diseases.

ACKNOWLEDGMENTS

This study was carried out under the research plans of the IPC UFRC RAS, state No. assignments FMRS-2022-0074, FMRS-2022-0079 and of the Russian Ministry of Education and Science (Federal target program No. 2021-0291-FP5-0001).

REFERENCES

Aaghaz, S., K. Sharma, R. Jain, and A. Kamal. 2021. β-Carbolines as potential anticancer agents. *Eur. J. Med. Chem.* 216:113321.

Abdelsalam, M. A., O. M. AboulWafa, E.-S. A. M. Badawey, M. S. El-Shoukrofy, M. M. El-Miligy, N. Gouda, and M. M. Elaasser. 2018. Design, synthesis, anticancer screening, docking studies and in silico ADME prediction of some β-carboline derivatives. *Future Med. Chem.* 10:1159–75.

Abdelsalam, M. A., O. M. AboulWafa, E.-S. A. Badawey, M. S. El-Shoukrofy, M. M. El-Miligy, and N. Gouda. 2018. Design and synthesis of some β-carboline derivatives as multi-target anticancer agents. *Future Med. Chem.* 10:2791–814.

Abinaya, R., S. Srinath, S. Soundarya, R. Sridhar, K. K. Balasubramanian, and B. Baskar. 2022. Recent developments on synthesis strategies, SAR studies and biological activities of β-carboline derivatives – an update. *J. Mol. Struct.* 1261:132750.

Abramovitch, R. A., and I. D. Spenser. 1964. The Carbolines. *Adv. Heterocycl. Chem.* 3:79–207.

Akhmadiev, N. S., E. S. Mescheryakova, R. A. Khisamutdinov, A. N. Lobov, M. F. Abdullin, A. G. Ibragimov, R. V. Kunakova, and V. R. Akhmetova. 2018. Synthesis, structure and catalytic activity of novel five-membered Pd(II) and Pt(II) metallaheterocycles based on 1,2-bis(3,5-dimethylisoxazol-4-yl-methylsulfanyl)ethane. *J. Organomet. Chem.* 872:54–62.

Akhmetova, V. R., N. S. Akhmadiev, and A. G. Ibragimov. 2022. Sulfur-containing pyrazoles, pyrazolines and indazoles. In *N-Heterocycles. Synthesis and Biological Evaluation*, ed. K. L. Ameta, R. Kant, A. Penoni, A. Maspero, L. Scapinello. Springer, Singapore, pp. 275–312.

Akhmetova, V. R., N. S. Akhmadiev, M. F. Abdullin, L. U. Dzhemileva, and V. A. D'yakonov. 2020. Synthesis of new N,N′-Pd(Pt) complexes based on sulfanyl pyrazoles, and investigation of their *in vitro* anticancer activity. *RSC Adv.* 10:15116–23.

Al-Allaf, T. A. K., and L. J. Rashan. 1998. Synthesis and cytotoxic evaluation of the first trans-palladium(II) complex with naturally occurring alkaloid harmine. *Eur. J. Med. Chem.* 33:817–20.

Alharbi, W., I. Hassan, R. A. Khan, S. Parveen, K. H. Alharbi, I. I. Bin Sharfan, I. M. Alhazza, H. Ebaid, and A. Alsalme. 2021. Bioactive tryptophan-based copper complex with auxiliary β-carboline spectaclepotential on human breast cancer cells: In vitro and in vivostudies. *Molecules* 26:1606.

Almeida, M. C., D. I. S. P Resende, P. M da Costa, M. M. M Pinto, and E. Sousa. 2021. Tryptophan derived natural marine alkaloids and synthetic derivatives as promising antimicrobial agents. *Eur. J. Med. Chem.* 209:112945.

Ayipo, Y. O., M. N. Mordi, M. Mustapha, and T. Damodaran. 2021a. Neuropharmacological potentials of β-carboline alkaloids for neuropsychiatric disorders. *Eur. J. Pharmacol.* 893:173837.

Ayipo, Y. O., W. A. Osunniran, and M. N. Mordi. 2021b. Metal complexes of β-carboline: Advances in anticancer therapeutics. *Coord. Chem. Rev.* 432:213746.

Ayipo, Y. O., S. N. Yahaya, H. F. Babamale, I. Ahmad, H. Patel, and M. N. Mordi. 2021c. ß-Carboline alkaloids induce structural plasticity and inhibition of SARS-CoV-2 nsp3 macrodomain more potently than remdesivir metabolite GS-441524: Computational approach. *Turk. J. Biol.* 45:503–517.

Banoth, K. K., Faheem, K. V. G. ChandraSekhar, N. Adinarayana, and S. Murugesan. 2020. Recent evolution on synthesis strategies and anti-leishmanial activity of β-carboline derivatives – An update. *Heliyon* 6:04916.

Barbosa, V. A., P. Baréa, R. S. Mazia, T. Ueda-Nakamura, W. Ferreira da Costa, M. A. Foglio, A. L. T. Goes Ruiz, J. Ernesto de Carvalho, D. B. Vendramini-Costa, C. V. Nakamura, and M. H. Sarragiotto. 2016. Synthesis and evaluation of novel hybrids β-carboline-4-thiazolidinones as potential antitumor and antiviral agents. *Eur. J. Med. Chem.* 124:1093–104.

Beato, A., A. Gori, B. Boucherle, M. Peuchmaur, and R. Haudecoeur. 2021. β-Carboline as a privileged scaffold for multitarget strategies in Alzheimer's disease therapy. *J. Med. Chem.* 64:1392–422.

Bora, D., S. R. Dannarm, S. E. John, S. Sana, R. Sonti, and N. Shankaraiah. 2021. Regioselective *ortho*-sulfonamidation: Exploration of intrinsic directing property of β-carbolines and their photophysical studies. *Asian J. Org. Chem.* 10:3384–89.

Brian, E. L. 1996. Synthesis of β-carbolines. A review. *Org. Prep. Proced. Int.* 28:1–64.

Caporale, C., and M. Massi.2018. Cyclometalated iridium(III) complexes for life science. *Coord. Chem. Rev.* 363:71–91.

Cho, S. H., S.-H. Kim, and D. Shin. 2019. Recent applications of hydantoin and thiohydantoin in medicinal chemistry. *Eur. J. Med. Chem.* 164:517–45.

Chripkova, M., D. Drutovic, M. Pilatova, J. Mikes, M. Budovska, J. Vaskova, M. Broggini, L. Mirossay, and J. Mojzis. 2014. Brassinin and its derivatives as potential anticancer agents. *Toxicol. in Vitro* 28(5):909–15.

Chundawat,T. S. 2020. Design and strategic synthesis of some β-carboline-based novel natural products of biological importance. In *Organic Synthesis – A Nascent Relook*, ed. B. P. Nandeshwarappa 248. IntechOpen.

Dai, J., W. Dan, U. Schneider, and J. Wang. 2018. β-Carboline alkaloid monomers and dimers: Occurrence, structural diversity, and biological activities. *Eur. J. Med. Chem.* 157:622–56.

Dai, J.-K., W.-J. Dan, and J.-B. Wan. 2022. Natural and synthetic β-carboline as a privileged antifungal scaffolds. *Eur. J. Med. Chem.* 229:114057.

De Sanctis, R., F. Jacobs, C. Benvenuti, M. Gaudio, R. Franceschini, R. Tancredi, P. Pedrazzoli, A. Santoro, and A. Zambelli. 2022. From seaside to bedside: Current evidence and future perspectives in the treatment of breast cancer using marine compounds. *Front. Pharmacol.* 13:909566.

Devi, N., S. Kumar, S. K. Pandey, and V. Singh. 2018. 1(3)-Formyl-β-carbolines: Potential aldo-X precursors for the synthesis of β-carboline-based molecular architectures. *Asian J. Org. Chem.* 7:6–36.

Ebada, S. S., R. A. Edrada, W. Lin, and P. Proksch. 2008. Methods for isolation, purification and structural elucidation of bioactive secondary metabolites from marine invertebrates. *Nat. Protoc.* 3:1820–31.

Feng, M., B. Tang, S. H. Liang, and X. Jiang. 2016. Sulfur containing scaffolds in drugs: Synthesis and application in medicinal chemistry. *Curr. Top. Med. Chem.* 16:1200–16.

Formagio, A. S. N., L. T. D. Tonin, M. A. Foglio, C. Madjarof, J. E. de Carvalho, W. F. da Costa, F. P. Cardoso, and M. H. Sarragiotto. 2008. Synthesis and antitumoral activity of novel 3-(2-substituted-1,3,4-oxadiazol-5-yl) and 3-(5-substituted-1,2,4-triazol-3-yl) β-carboline derivatives. *Bioorg. Med. Chem.* 16:9660–67.

He, L., S.-Y. Liao, C.-P. Tan, Y.-Y. Lu, C.-X. Xu, L.-N. Ji, and Z.-W. Mao. 2014. Cyclometalated iridium(III)–β-carboline complexes as potent autophagy-inducing agents. *Chem. Commun.* 50:5611–14.

He, W., Z. Zhang, and D. Ma. 2019. A Scalable total synthesis of the antitumor agents Et-743 and Lurbinectedin. *Angew. Chem. Int. Ed.Engl.* 58:3972–75.

Hu, X., X. Gao, G. Gao, Y. Wang, H. Cao, D. Li, and H. Hua. 2021. Discovery of β-carboline-(phenylsulfonyl)furoxan hybrids as potential anti-breast cancer agents. *Bioorg. Med. Chem. Lett.* 40:127952.

Huang, D., Z. Zhang, Y. Li, F. Liu, W. Huang, Y. Min, K. Wang, J. Yang, C. Cao, Y. Gong, and S. Ke. 2022. Carboline derivatives based on natural pityriacitrin as potential antifungal agents. *Phytochem. Lett.* 48:100–5.

Huang, Y., Z. Guo, H. Song, Y. Liu, L. Wang, and Q. Wang. 2018. Design, synthesis, and biological activity of β-carboline analogues containing Hydantoin, Thiohydantoin, and urea moieties. *J. Agric. Food Chem.* 66:8253–61.

Kamal, A., M. Sathish, V. L. Nayak, V. Srinivasulu, B. Kavitha, Y. Tangella, D. Thummuri, C. Bagul, N. Shankaraiah, and N. Nagesh. 2015. Design and synthesis of dithiocarbamate linked β-carboline derivatives: DNA topoisomerase II inhibition with DNA binding and apoptosis inducing ability. *Bioorg. Med. Chem.* 23:5511–26.

Kamboj, A., B. Sihag, D. S. Brar, A. Kaur, and D. B. Salunke. 2021. Structure activity relationship in β-carboline derived anti-malarial agents. *Eur. J. Med. Chem.* 221:113536.

Khan, R. A., M. R. Khan, M. Usman, F. Sayeed, H. A. Alghamdie, S. Alrumman, W. Alharbi, N. N. Farshorig, M. M. Al-Oqail, M. R. Siddiqui, M. A. Khanjer, and A. Alsalme. 2020. β-Carboline copper complex as a potential mitochondrial-targeted anticancer chemotherapeutic agent: Favorable attenuation of human breast cancer MCF7 cells via apoptosis. *Saudi J. Biol. Sci.* 27:2164–73.

Kleks, G., S. Duffy, L. Lucantoni, V. M. Avery, and A. R. Carroll. 2020. Orthoscuticellines A–E, β-carboline alkaloids from the bryozoan *Orthoscuticella ventricosa* collected in Australia. *J. Nat. Prod.* 83:422–8.

Kumar, S., A. Singh, K. Kumar, and V. Kumar. 2017. Recent insights into synthetic β-carbolines with anti-cancer activities. *Eur. J. Med. Chem.* 142:48–73.

Laxmikeshav, K., P. Kumari, and N. Shankaraiah. 2022. Expedition of sulfur-containing heterocyclicderivatives as cytotoxic agents in medicinalchemistry: A decade update. *Med. Res. Rev.* 42:513–75.

Li, J. J. 2006. *Nominal Reactions. Mechanisms of Organic Reactions.* M. Binom. Knowledge Laboratory, pp. 456.

Li, D., J. Meng, and X. Jiang. 2022a. Total synthesis of Trabectedin, Lurbinectedin, and Renieramycin T. *CCS Chem.* 1–7.

Li, Z., Y. Apizi, C. Zhang, Z. Wang, H. He, X. Li, Y. Zhu, J. Yang, L. Xiao, and M. Wang. 2022b. Synthesis of harmine-nitric oxide donor derivatives as potential antitumor agents. *Bioorg. Med. Chem. Lett.* 65:128698.

Liu, L., Y. Y. Xu, Z. Q. Yang, J. N. Xiang, and G. Y. Xu. 2012. Synthesis and cytotoxic activity of 3-phenyl-4-substituted-β-carbolines. *Chin. Chem. Lett.* 23:1230–32.

Lo, K. K.-W., and K. Y. Zhang. 2012. Iridium(III) complexes as therapeutic and bioimaging reagents for cellular applications. *RSC Adv.* 2:12069–83.

Luo, B., and X. Song. 2021. A comprehensive overview of β-carbolines and its derivatives as anticancer agents. *Eur. J. Med. Chem.* 224:113688.

Maity, P., D. Adhikari, and A. K. Jana. 2019. An overview on synthetic entries to tetrahydro-β-carbolines. *Tetrahedron* 75:965–1028.

Manasa, K. L., S. Thatikonda, D. K. Sigalapalli, A. Sagar, G. Kiranmai, A. M. Kalle, M. Alvala, C. Godugu, N. Nagesh, and B. N. Babu. 2020. Design and synthesis of β-carboline linked aryl sulfonyl piperazine derivatives: DNA topoisomerase II inhibition with DNA binding and apoptosis inducing ability. *Bioorg. Chem.* 101:103983.

Manda, S., S. I. Khan, S. K. Jain, S. Mohammed, B. L. Tekwani, I. A. Khan, R. A. Vishwakarma, and S. B. Bharate. 2014. Synthesis, antileishmanial and antitrypanosomal activities of N-substituted tetrahydro-β-carbolines. *Bioorg. Med. Chem. Lett.* 24:3247–50.

Menzies, A. M., and G. V. Long. 2014. Systemic treatment for BRAF-mutant melanoma: Where do we go next? *Lancet Oncol.* 15:e371–e81.

Nicolaou, K. C., Z. Lu, R. Li, J. R. Woods, and T.-ik. Sohn. 2015. Total synthesis of Shishijimicin A. *J. Am. Chem. Soc.*137:8716–19.

Oku, N., S. Matsunaga, and N. Fusetani. 2003. Shishijimicins A-C, novel enediyne antitumor antibiotics from the Ascidian *Didemnum proliferum*. *J. Am. Chem. Soc.*125:2044–45.

Pacell, S. 2002. Sulfur in human nutrition and applications in medicine. *Altern. Med. Rev.* 7:22–44.

Palanisamy, S. K., N. M. Rajendran, and A. Marino. 2017. Natural products diversity of Marine Ascidians (Tunicates; Ascidiacea) and Successful drugs in clinical development. *Nat. Prod. Bioprospect.* 7:1–111.

Pedpradab, S., R. Edrada, R. Ebel, V. Wray, and P. Proksch.2004. New β-carboline alkaloids from the Andaman sea sponge *Dragmacidon* sp. *J. Nat. Prod.* 67:2113–6.

Piechowska, P., R. Zawirska-Wojtasiak, and S. Mildner-Szkudlarz. 2019. Bioactive β-carbolines in food: A review. *Nutrients* 11:814.

Prinsep, M. R., J. W. Blunt, and M. H. G. Munro. 1991. New cytotoxic β-carboline alkaloids from the marine bryozoan, Cribricellina cribraria. *J. Nat. Prod.* 54:1068–76.

PubMed – medical and biological abstracts service, https://pubmed.ncbi.nlm.nih.gov, (accessed January 9, 2023)

Rashid, M. A., K. R. Gustafson, and M. R. Boyd. 2001. New cytotoxic N-methylated β-carboline alkaloids from the Marine Ascidian *Eudistomagilboverde*. *J. Nat. Prod.* 64:1454–56.

Recio, J., A. Perez-Redondo, J. Alvarez-Builla, and C. Burgos. 2019. Access to 2-substituted 1-pyridin-3-yl-β-carboline derivatives by intramolecular radical cyclization-ring opening-SNAr substitution. *Org. Chem. Front.* 6:3300–04.

Ren, R., X. Wang, D. A. Leas, C. Haberli, M. Cal, Y. Dong, M. Kaiser, J. Keiser, and J. L. Vennerstrom. 2022. Antischistosomal tetrahydro-γ-carboline sulfonamides. *Bioorg. Med. Chem. Lett.* 59:128546.

Sahoo, C. R., S. K. Paidesetty, and R. N. Padhy. 2019. Norharmane as a potential chemical entity for development of anticancer drugs. *Eur. J. Med. Chem.* 162:752–64.

Sanz-Villafruela, J., C. Martínez-Alonso, I. Echevarría, M. Vaquero, A. Carbayo, Fidalgo, J., Rodríguez, A. M., Cuevas-Vicario, J. V., Lima J. C., A. J. Moro, B. R. Manzano, F. A. Jalón, and G. Espino 2021. One-pot photocatalytic transformation of indolines into 3-thiocyanate indoles with new Ir(III) photosensitizers bearing β-carbolines. *Inorg. Chem. Front.* 8:1253–70.

Savariz, F. C., A. S. N. Formagio, V. A. Barbosa, M. A. Foglio, J. E. de Carvalho, M. C. T. Duarte, B. P. Dias Filhoc, and M. H. Sarragiotto. 2010. Synthesis, antitumor and antimicrobial activity of novel 1-substituted phenyl-3-[3-alkylamino(methyl)-2-thioxo-1,3,4-oxadiazol-5-yl] β-carboline derivatives. *J. Braz. Chem. Soc.* 21:288–98.

SciFinder – chemical abstracts service, https://scifinder.cas.org, (accessed December 20, 2022)

Scopus – bibliographic and abstract database service, www.scopus.com, (accessed January 10, 2023)

Shearer, J., J. L. Castro, A. D. G. Lawson, M. MacCoss, and R. D. Taylor. 2022. Rings in clinical trials and drugs: Present and future. *J. Med. Chem.* 65:8699–712.

Singh, M., R. Jamra, A. K. Paul, C. C. Malakar, and V. Singh. 2022.KI-assisted sulfur activation/insertion/denitration strategy towards dual C–S bond formation for one-pot synthesis of β-carboline-tethered 2-acylbenzothiophenes. *Asian J. Org. Chem.* 11:e2021006.

Song, H., J. Zhang, Y. Li, and Q. Wang. 2021. Pesticide activities evaluation of β-carboline, dihydro-β-carboline, tetrahydro-β-carboline alkaloids, and their derivatives. In *Recent Highlights in the Discovery and Optimization of Crop Protection Products*, ed. P. Maienfisch and S. Mangelinckx, Elsevier Inc.Cambridge. pp. 547–62.

Soni, J. P., Y. Yeole, and Y. Shankaraiah. 2021. β-Carboline-based molecular hybrids as anticancer agents: A brief sketch. *RSC Med. Chem.* 12:730–50.

Spath, E., and E. Lederer. 1930. Synthese der Harmala-alkaloide: Harmalin, Harmin und Harman. *Chem. Ber.* 63:120.

Stonik, V. A., and G. A. Tolstikov. 2008. Natural compounds and the creation of domestic drugs. *Bull. Russ. Acad. Sci.* 78:675–84 (Russian).

Suntornchashwej, S., N. Chaichit, M. Isobe, and K. Suwanborirux. 2005. Hectochlorin and morpholine derivatives from the thai sea hare *Bursatella leachii*. *J. Nat. Prod.* 68:1109–10.

Suzuki, T., T. Kubota, and J. Kobayashi. 2011. Eudistomidins H–K, new β-carboline alkaloids from the Okinawan marine tunicate *Eudistoma glaucus.Bioorg. Med. Chem. Lett.* 21:4220–3.

Szabó, T., B. Volk, and M. Milen. 2021. Recent advances in the synthesis of β-carboline alkaloids. *Molecules* 26(3):663.

Takahashi, H., S.-ichiro Kurimoto, J. Kobayashi, and T. Kubota. 2018. Ishigadine A, a new canthin-6-one alkaloid from an Okinawan marine sponge Hyrtios sp. *Tetrahedron Lett.* 59:4500–02.

Tanaka, N., R. Momose, Y. Takahashi, T. Kubota, A. Takahashi-Nakaguchi, T. Gonoi, J. Fromont, and J. Kobayashi. 2013. Hyrtimomines D and E, bisindole alkaloids from a marine sponge *Hyrtios* sp. *Tetrahedron Lett.* 54:4038–40.

Tanaka, N., R. Momose, A. Takahashi-Nakaguchi, T. Gonoi, J. Fromont, and J. Kobayashi. 2014. Hyrtimomines, indole alkaloids from Okinawan marine sponger *Hyrtios* spp. *Tetrahedron* 70:832–7.

Thatikayala, M., P. Wadhwa, P. Kaur, P. K. Singh, A. Yadav, M. Kaushik, and S. K. Sahu. 2022. Beta-carboline as a promising heterocyclic nucleus: Synthetic aspects, pharmacological potential and structure activity relationship. *Eur. J. Med. Chem. Rep.* 6:100096.

Tilekar, K., O. Shelke, N. Upadhyay, A. Lavecchia, and C. S. Ramaa. 2022. Current status and future prospects of molecular hybrids with thiazolidinedione (TZD) scaffold in anticancer drug discovery. *J. Mol. Struct.* 1250:131767.

Tokala, R., S. Sana, U. J. Lakshmi, P. Sankarana, D. K. Sigalapalli, N. Gadewal, J. Kode, and N. Shankaraiah. 2020. Design and synthesis of thiadiazolo-carboxamide bridged β-carboline-indole hybrids: DNA intercalative topo-IIα inhibition with promising antiproliferative activity. *Bioorg. Chem.* 105:104357.

Tokala, R., S. Mahajan, G. Kiranmai, D. K. Sigalapalli, S. Sana, S. E. John, N. Nagesh, and N. Shankaraiah, 2021. Development of β-carboline-benzothiazole hybrids *via* carboxamide formation as cytotoxic agents: DNA intercalative topoisomerase IIα inhibition and apoptosis induction. *Bioorg. Chem.* 106:104481.

Tokala, R., S. Thatikonda, S. Sana, P. Regur, C. Godugu, and N. Shankaraiah. 2018. Synthesis and *in vitro* cytotoxicity evaluation of β-carboline-linked 2,4-thiazolidinedione hybrids: Potential DNA intercalation and apoptosis-inducing studies. *New J. Chem.* 42:16226.

Tran, T. D., N. B. Pham, M. Ekins, J. N. A. Hooper, and R. J. Quinn. 2015. Isolation and total synthesis of Stolonines A–C, unique taurine amides from the Australian Marine Tunicate *Cnemidocarpa stolonifera.Mar. Drugs* 13:4556–75.

Tzouras, N. V., T. Scattolin, A. Gobbo, S. Bhandary, F. Rizzolio, E. Cavarzerani, V. Canzonieri, K. Van Hecke, G. C., Vougioukalakis, C. S. J. Cazin, and S. P. Nolan. 2022. A Green synthesis of carbene-metal-amides (CMAs) and carboline-derived CMAs with potent *in vitro* and *ex vivo* anticancer activity. *ChemMedChem* 17:e202200135.

Verniest, G., D. England, N. De Kimpe, and A. Padwa. 2010. Synthesis of substituted β-carbolines *via* gold(III)-catalyzed cycloisomerization of N-propargylamides. *Tetrahedron* 66:1496–502.

Wang, N., P. Saidhareddy, and X. Jiang, 2020. Construction of sulfur-containing moieties in the total synthesis of natural products. *Nat. Prod. Rep.* 37:246–75.

Wang, J., F. Gong, T. Liang, Z. Xie, Y. Yang, C. Cao, J. Gao, T. Lu, and X. Chen. 2021a. A review of synthetic bioactive tetrahydro-β-carbolines: A medicinal chemistry perspective. *Eur. J. Med. Chem.* 225:113815.

Wang, T.-N., S. Yang, S.-Y. Shi, W.-Y. Yuan, J.-X. Chen, Z.-Y. Duan, A.-D. Lu, Z.-W. Wang, and Q.-M. Wang. 2021b. Pityriacitrin marine alkaloids as novel antiviral and anti-phytopathogenic-fungus agents. *Pest Manag. Sci.* 77:4691–700.

Whaley, W. M., and T. R. Govindachari. 1951. The Pictet-Spengler synthesis of tetrahydroisoquinolines and related compounds. *Org. React.* 6:151.

Xin, B., W. Tang, Y. Wang, G. Lin, H. Liu, Y. Jiao, Y. Zhu, H. Yuan, Y. Chen, and T. Lu. 2012. Design, synthesis and biological evaluation of β-carboline derivatives as novel inhibitors targeting B-Raf kinase. *Bioorg. Med. Chem. Lett.* 22:4783–86.

Xu, T., L. Shi, Y. Zhang, K. Wang, Z. Yang, and S. Ke, 2019. Synthesis and biological evaluation of marine alkaloid-oriented β-carboline analogues. *Eur. J. Med. Chem.* 168:293–300.

Zhang, P., X. Sun, B. Xu, K. Bijian, S. Wan, G. Li, M. Alaoui-Jamali, and T. Jiang. 2011. Total synthesis and bioactivity of the marine alkaloid pityriacitrin and some of its derivatives. *Eur. J. Med. Chem.* 46:6089–97.

Zhang, Z.-J., Y. Zeng, Z.-Y. Jiang, B.-S. Shu, V. Sethuraman, and G.-H. Zhong. 2018. Design, synthesis, fungicidal property and QSAR studies of novel β-carbolines containing urea, benzoylthiourea and benzoylurea for the control of rice sheath blight. *Pest Manag. Sci.* 74:1736–46.

Zhang, Z.-J., J.-J. Zhang, Z.-Y. Jiang, and G.-H. Zhong, 2017. Design, synthesis and bioactivity evaluation of novel β-carboline 1,3,4-oxadiazole derivatives. *Molecules* 22:1811.

6 Antimalarial Activity and β-Carboline Derivative

A Synthetic Approach

Jagannath S. Godse, Santosh B. Gaikwad, Sunil U. Tekale, Sanjay B. Ubale, and Rajendra P. Pawar

6.1 INTRODUCTIONS OF β-CARBOLINES

Natural and synthesized indole containing heterocyclic compounds with varying degrees of aromaticity (ThanikachalamP. Vet al. 2019 and Mahmoud.Eet al.2022) make up the family of β-carbolines. This class of alkaloids, found from the seeds of Peganum, has a planar tricyclic pyrido indole ring structure. Harmala belongs to the family of flowering plants Zygophillaceae. Historically, this herb has been used to alleviate gastrointestinal issues and disorders such as malaria and malignancies in South Africa (Cravo. P. et al. 2015). There are different β-carboline alkaloids found in the wild. no less than eight orders of plants. These alkaloids can be found in animals as well as plants. Bacteria, insects, food, and cigarettes are all common places to find them. Fumes, human organs and fluids, saltwater worms and crustaceans, sponges, and marine ascidians, among others, contain more than 6,000 compounds with the β-carboline moiety as their central structure. One hundred naturally occurring and synthesized alkaloids and scaffolds with diverse biological functions include those against inflammation, sedation, seizures, parasites, tumours, microbes, viruses, etc.

Recent years have seen a surge in interest in these chemicals due to their potential biological applications. Inhibitors of acetylcholinesterase and butyrylcholinesterase enzymes as well as N-methyl-D-aspartate receptors have been discovered, with the effectiveness of these compounds depending on the location of the attached substituent. β-carboline alkaloids are promising new treatments for several neurological conditions. The tetrazole derivative of 2,3,4,9-tetrahydro-β-carboline is also a candidate antileishmanial chemotype (Purohit. P. and Pandey. A. Ket al. 2017). Several pharmaceuticals on the market, including tadalafil, vinpocetine, brovincamine, abecarnil, cipargamin, reserpine, and lurbinectedin, share this distinctive architecture. Our research centers on the antimalarial properties of β-carbolines. Malaria, caused by hematoprotozoan parasites, is rapidly becoming the world's leading cause of preventable death, killing an estimated one to two million people annually, the vast

DOI: 10.1201/9781351058032-6

majority of whom are children (Kim.K.Set al. 2019). According to the WHO, there have been 219 million (2017), 228 million (2018), 232 million (2019), 245 million (2020), and 247 million (2021) malaria cases worldwide in the last five years (WHO, 2017, 2018, 2019, 2020 and 2021).

There are five different plasmodium species that cause malaria in vertebrates, but only *P. falciparum* and *P. vivax* are considered serious threats to human health. The malaria problem is dynamic, robust, and multifaceted due in large part to the parasitesinnate capacity to develop treatment resistance. The aryl amino alcohol scaffolds like quinine, quinidine, chloroquine, etc., antifolate scaffolds like pyrimethamine chlorproguanil, proguanil, trimethoprim, and Artemisinin scaffolds like artemisinin, dihydroartemisinin, artemether, and artesunate are the three main classes of antimalarials currently on the market. Antimalarial multidrug resistance has emerged as a global concern to treat malaria, even though aartemisinin-based combination therapy and chloroquine are the most extensively used medications for this tropical disease. Finding novel medications with a heterocyclic group based on nitrogen in their structure can halt the development of resistance from scratch. Several β-carboline alkaloids, both synthetic and natural, have been shown to have potent antiplasmodial activity(Chu, X.M. and Wang, Cet al. 2019).

Malaria is a devastating disease that claims the lives of many people every year. Plasmodium parasites, protozoan organisms, are the primary instigators of this disease (Siqueira-Neto J. L and Debnath A.et al. 2018). The female *Anopheles spp*. mosquito harbors these protozoa in her salivary glands, and they are transferred to people by mosquito bites. The Plasmodium falciparum, a Plasmodium vivax, Plasmodium ovale, and Plasmodium malaria are the four most common types of malaria and the most common species of Plasmodium that infect humans and cause malaria. This illness is characterized by high fevers that come and go, chills, and severe headaches. Brain damage and coma are among the potentially fatal outcomes of cerebral malaria. While anyone can contract this virus, infants and young children as well as pregnant women are especially at risk.

Malaria not only takes lives, but also causes harm to local labor forces and the economy. In numerous African countries, efforts to contain the epidemic have been undermined by the diversion of monies intended to ensure the population's food supply. Malaria epidemics are indirectly responsible for widespread hunger and poverty. The extensive use of dichlorodiphenyltrichloroethane (DDT) for mosquito control has been a malaria eradication initiative that is not medically oriented. While it led to a reduction in infections, particularly in North America, the environmental hazards associated with it outweighed its benefits. As a result, governments worldwide banned DDT use, leading to an increase in infectious disease.

The development of a safe and long-lasting vaccination has been studied for almost 50 years, but no real progress has been made. Inoculation against *P. falciparum* became possible during the process of developing the RTS, S/AS01 vaccine (Mosquirix™), although its efficacy is limited at best (Dimala, C. Aet al. 2018; Bell, G. J.et al. 2021). Artemether, artesunate, and arteether are examples of derivatives of artemisinin, and are available as therapeutic options utilizing amodiaquine, piperaquine, and lumefantrine in conjunction with one another. Dihydroartemisinin is an active metabolite that is effective against all strains of *P. falciparum* that are resistant to many pharmacological agents. These treatments aid in recovery without causing serious adverse effects (Chaorattanakawee S et al. 2023).

However, the "Hypnozoite Dilemma" and the "Artemisinin Resistance" could impede our progress in accomplishing total elimination of this illness, despite the combo therapy's promise to aid the global campaign against malaria. Conributing to the issue is the fact that parasite clearance periods have lengthened among patients. Some hypotheses attribute the slowdown to the parasite's early-stage downregulation of metabolism to survive artemisinin's antimalarial effects. The presence of dormant hypnozoites, which continue to be hidden within the body even after the symptoms have been alleviated, is the second major impediment to the development of a permanent treatment for malaria. Because these cells might revive unexpectedly and cause symptoms to emerge for the first time, developing effective malaria treatments is extremely difficult. Primaquine is an effective treatment (Ajima, U et al. 2019), but it has drawbacks for those with glucose-6-phosphate dehydrogenase (Ashley, E. A et al. 2014) deficiency and should be avoided in areas where malaria is common. Tafenoquine, another recently licenced medicine, likewise produces a single serving of radical therapy of P. vivax malaria (Luzzatto, L et al. 2016; Lacerda, M. V et al. 2019), but it is ineffective and causes hemolysis in people who have problems with glucose-6-phosphate dehydrogenase levels.

A tricyclic nucleus of β-carbolines has been identified as the active ingredient in a category of highly effective natural products that fight malaria isolated originating from the roots of the shrub-tree Eurycoma longifolia plants native to Malaysia and Burma. In traditional medicine, this substance is employed for the treatment of persistent fever as well as tertian malaria. Considering these findings, Kuo et al. (2003) extracted and studied β-carbolines from these roots, proving that the β-carbolines played a crucial part in the plant's antimalarial abilities. β-Carbolines, which are tricyclic compounds containing a pyrido[3,4-b] indole ring, may be classified into three distinct scaffolds based on the level of unsaturation in the pyridine ring. Compounds with an aromatic pyridine ring that is completely unsaturated are referred to as β-carbolines. (Figure 6.1). However, those compounds that possess a ring that is only partially saturated are referred to as 3,4-dihydro β-carbolines. On the contrary, 1,2,3,4-tetrahydro β-carbolines are a type of compound that have molecules with a fully saturated pyridine ring. There was also an investigation into the antiplasmodial properties of N-methyl quaternary salts of β-carbolines and tetrahydro β-carbolines.

The development of medications to treat malaria requires a deep comprehension of the relationship between the structure and activity of β-carbolines so that their therapeutic qualities can be utilized. In the following sections we investigate the structure–activity relationship in synthesized β-carbolines.

6.2 A SYSTEMATIC SYNTHETIC APPROACH OF β-CARBOLINES

An extensive number of studies have been carried out on β-carbolines-based natural products over the past two decades (Zhang, M; Sun, D et al. 2015; Riyazahmed, K. S. et al. 2021). Several natural substances and pharmacological drugs have the β-carboline as a central component. Plants, fish, insects, and mammals including

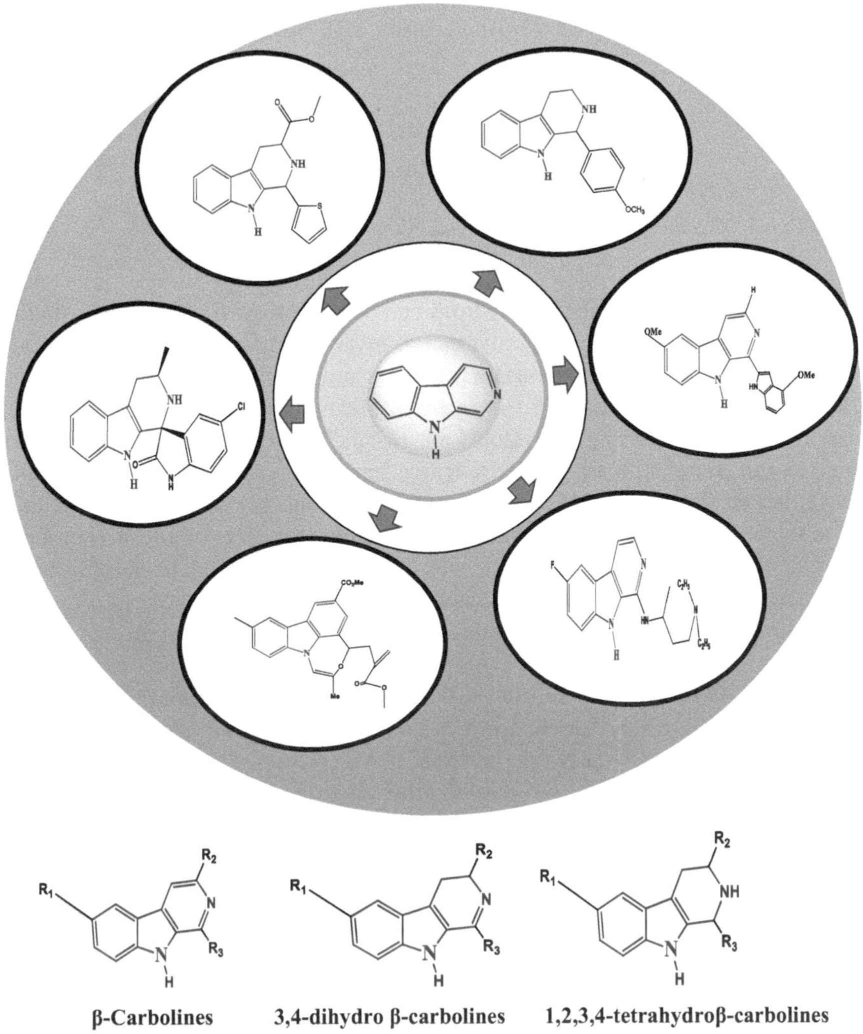

FIGURE 6.1 Structure of β-carbolines.

humans all have this basic unit present in their tissues and fluids as alkaloids or hormones. Antimalarial, antineoplastic, antiviral, and anxiolytic effects in addition to cGMP inhibitors and topoisomerase-II inhibitors are assigned to various β-carboline-based substances of synthesized or derived from nature.

Researchers have substituted or fused synthetic approaches to β-carbolines. (Saha, B et al. 2007; Dalpozzo, R et al. 2016). Since its discovery in 1911, scientists have relied on the Pictet–Spengler reaction referred as a key step. The Pictet–Spengler response provides two alternate strategies for building tetrahydro-β-carbolines cores: either by installing the various substitutions that undergo sequential reactions during the process of cyclization to produce novel tetrahydro-β-carbolines derivatives,

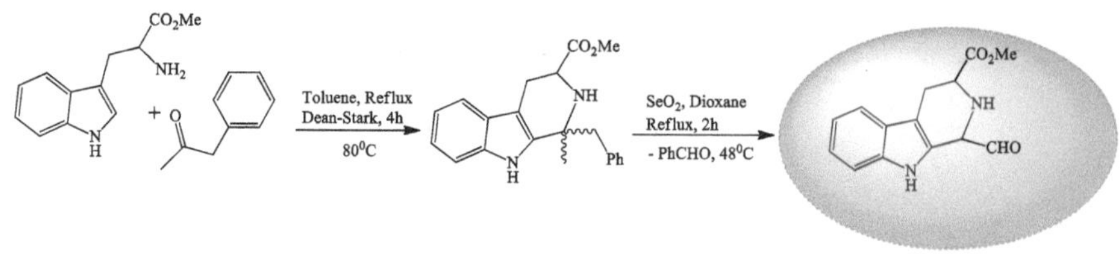

SCHEME 6.1 Synthesis of 1-formyl-9H-β-carbolines.

or by initially synthesizing the tetrahydro-β-carbolines core with suitable substitution. Then, the needed β-carboline derivative can be made by oxidizing these tetrahydro-β-carbolines. However, other methods for producing novel β-carbolines are required due to the great importance attached to this heterocyclic moiety. Synthetically engineered methods of manufacturing substituted or fused β-carbolines could include, for example, producing a β-carboline core with a functional group positioned appropriately. It is a desirable protocol to produce substituted and 1-9 annulated β-carbolines due to the proximity containing an electrophilic site in the form of a formyl group to the indole NH, which is a nucleophilic site. The N-2 could also be used for intramolecular cyclization, yielding 1–2 annulated β-carbolines.

In their research on selenium dioxide-mediated oxidation of differently substituted tetrahydro-β-carbolines (THBCs), Gatta and Misitiwere were the first to report the production of 1-formyl-9H-β-carboline. When they reacted a diastereomeric combination of 1-methyl,1-phenyl tetrahydro-β-carbolines with selenium dioxide in dioxane, they got 1-formyl-9H-β-carboline instead of the predicted 1-methyl,1-phenyl-3-(methoxycarbonyl)-1,4-dihydro-4-oxo-β-carboline (Scheme 6.1) (Chundawat, T S et al. 2019). Initially, the benzylic functional group underwent oxidation to yield benzaldehyde. Subsequently, the C-ring underwent aromatization. Lastly, the C-1-methyl group was oxidized to generate the formyl group.

By oxidizing 1-methyl-3-methoxycarbonyl-β-carboline with selenium dioxide in dioxane, Gatta and coworkers later described an improved method for producing methyl 1-formyl-9-H-pyrido [3,4-b] indole-3-carboxylate.. The synthesis of canthin-6-one using 1-formyl-9H-β-carboline was also reported by other researchers (SinghV, Batra S et al. 2012; Devi, N.; Kumar, S.et al. 2018). To produce pyrimido-[3,4,5-lm], they widened the synthetic utility of pyrido-[3,4-b]-indole derivatives used in the production of several carboline moiety derivatives.

Synthesis of natural alkaloids containing 4,8-dioxygenated β-carboline has been reported (Scheme 6.2) (Chundawat, T. S et al. 2019; Suzuki, H.et al. 1999). The synthesis includes two steps: (i) keeping the tosyl group from oxidizing the phenolic group in an enhanced Fischer indolization to afford a 7-oxygenated indole, and (ii) constructing a 4-methoxy-β-carboline skeleton via the C-3 selective cyclization of the indole's a C-2 substituent. Then, a modified Reissert–Henze reaction is used to transform 4-methoxy-carboline into a 1-nitrilederivative involving diethyl phosphoryl cyanide and N-oxide.

Takasu et al. synthesized some naturally occurring compounds like Kumujancine, 4-methoxy vinyl β-carboline, Creatine, and their respective salts. The Pictet–Spengler

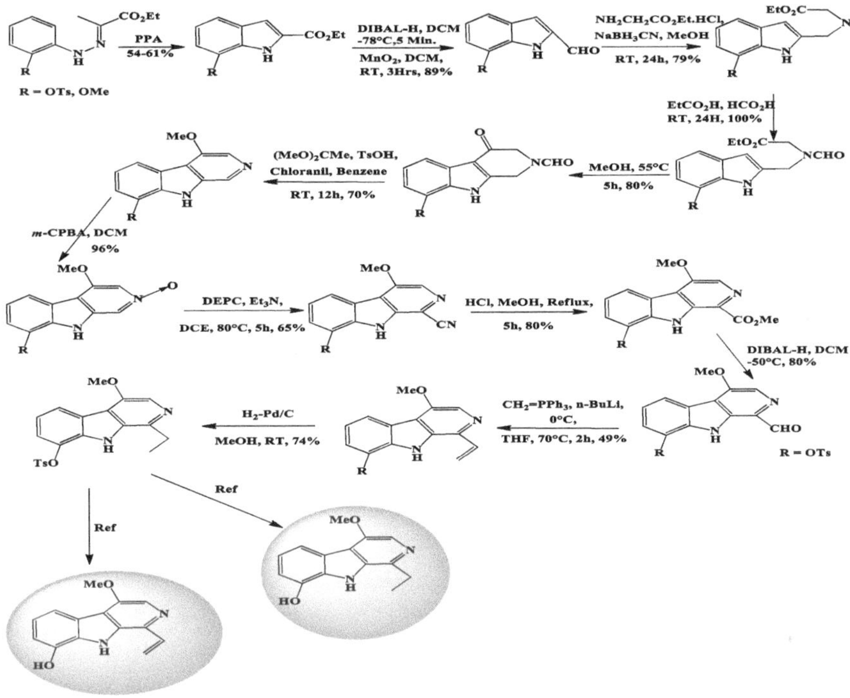

SCHEME 6.2 Synthesis of 4,8-dioxygenated β-carbolines.

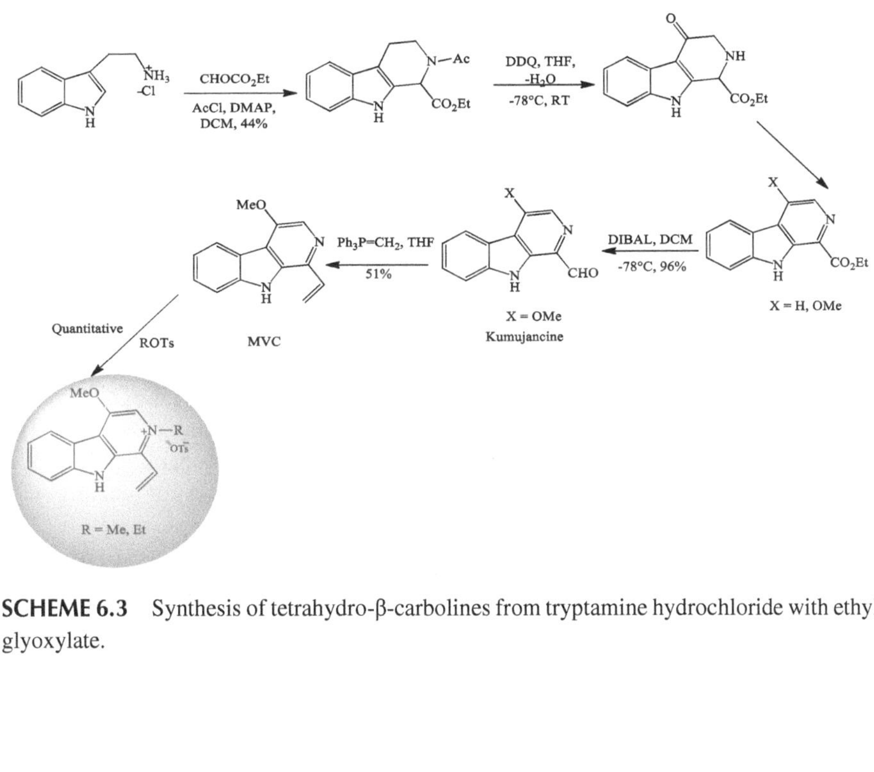

SCHEME 6.3 Synthesis of tetrahydro-β-carbolines from tryptamine hydrochloride with ethyl glyoxylate.

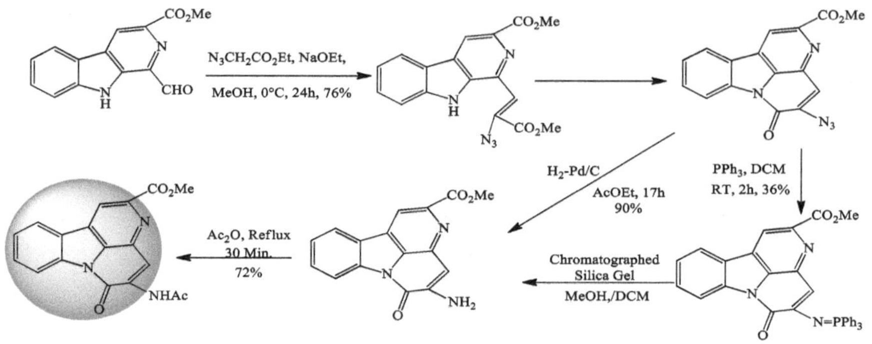

SCHEME 6.4 Cis and trans isomer of tetrahydro-β-carboline.

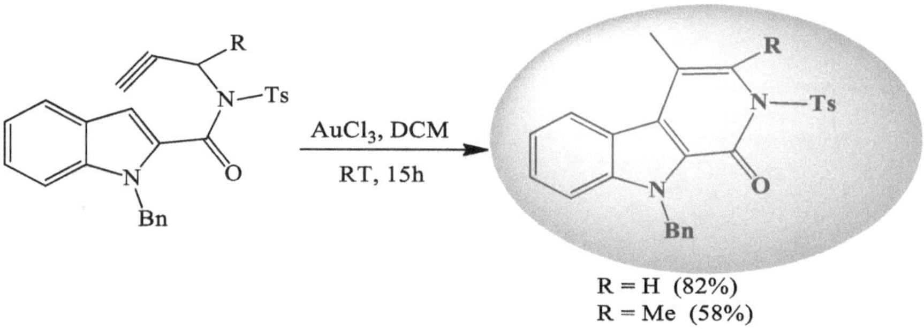

SCHEME 6.5 Synthesis of N-tosyl-β-carbolinones.

reaction was employed to react tryptamine hydrochloride with ethyl glyoxylate in ethanol, and subsequently acylated with acetyl chloride, and the resulting tetrahydro-β-carbolines were produced in yields of 44% (Scheme 6.3) (Devi, N et al. 2018; Takasu, K et al. 2004).

As a result of the condensation of 1-formyl-9H-β-carboline with ethyl azidoacetate, Condie and Bergman observed an intramolecular cyclization that occurred as a result of the attack of the nitrogen of the indole subunit on the ester functionality (Scheme 6.4). Catalytic reduction of the resultant 5-azido-canthin-6-one yielded the 5-aminocanthin-6-one (Condie G. C. et al. 2004; Devi N et al. 2017).

From indole-substituted N-propargylamides, N-tosyl-β-carbolinones were prepared via catalyzed 6-exo-dig cyclization by $AuCl_3$ (Scheme 6.5). The reaction was conducted in dichloromethane (CH_2Cl_2) under low-temperature conditions. The required starting materials were obtained by acylating N-tosylpropargylamines utilizing N-benzyl-1H-2-indolecarbonyl chloride generated in situ. After reacting with $POCl_3$ or PCl_4, the resulting β-carbolines, which were either 3-substituted or 3,4-disubstituted, were transformed into several valuable chemicals (Milen, M.; Ábrányi-Balogh, P et al. 2016).

Liang et al. showed an alternative gold-catalyzed method for constructing a β-carboline scaffold.From (C-2)-alkynyl aziridine indoles, the synthesis of

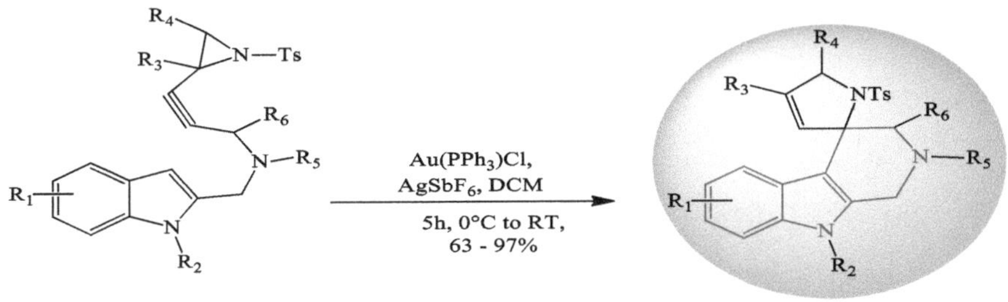

SCHEME 6.6 Synthesis of spiro-tetrahydro-β-carbolines.

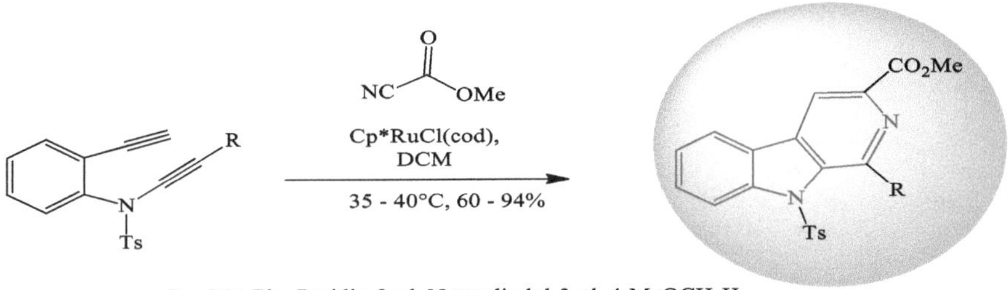

SCHEME 6.7 Synthesis of β-carbolines.

spiro-tetrahydro-β-carboline derivatives involved a Friedel-Crafts-type intramolecular reaction followed by hydroamination of the resulting product (Scheme 6.6). This method's benefits include operability simplicity, benign reaction conditions and compatibility with functional groups (Wang, Jet al. 2021; Milen, Met al. 2016).

Witulski et al. discovered a quick ruthenium-catalyzed technique for constructing the β-carboline skeleton (Scheme 6.7). Starting with readily accessible 2-iodoaniline, the necessary synthesis of yne-ynamides occurred in five stages. Corresponding β-carbolines were obtained through the catalytic reaction of Cp*RuCl(cod) involving a [2+2+2] process The reaction involved the cycloaddition of yne-ynamides with methyl cyanoformate. The marine alkaloid eudistomin U was successfully synthesized using this approach of complete synthesis (Doerksen, R. Set al. 2021; Nissen, F et al. 2011).

The formation of the proper β-carbolinium ions from 4-(2-azidophenyl) pyridinium derivatives was reported in 2012 by Driver's group using rhodium (II) as the catalyst (Scheme 6.8). After the ions were reduced with $NaBH_4$, tryptolines were produced in high yields. The analgesic oxindole alkaloid racemic horsfiline and its chlorinated equivalent were produced using four flasks to showcase the practicality of this synthesis method (Pumphrey, A. Let al. 2014; Pumphrey, A. Let al. 2012).

Palladium-catalyzed one-pot desilylation/Sonogashira/[2+2+2] aryl iodide with terminal alkyne cyclization yielded 3,4-annulated β-carboline (Scheme 6.9) (Varelas, J. Get al. 2015). Three rings formed in a single flask during this reaction, and an 80 percent yield was achieved. In a later paper, six distinct annulated pyrido[3,4-b]

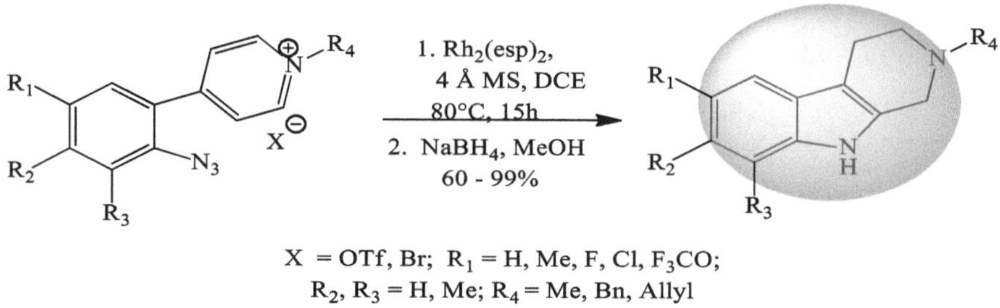

SCHEME 6.8 Synthesis of β-carbolinium ions from 4-(2-azidophenyl) pyridinium.

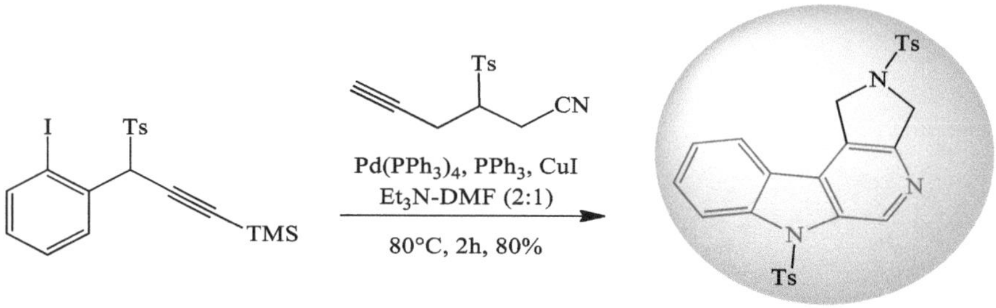

SCHEME 6.9 Synthesis of 3,4-annulated β-carbolines.

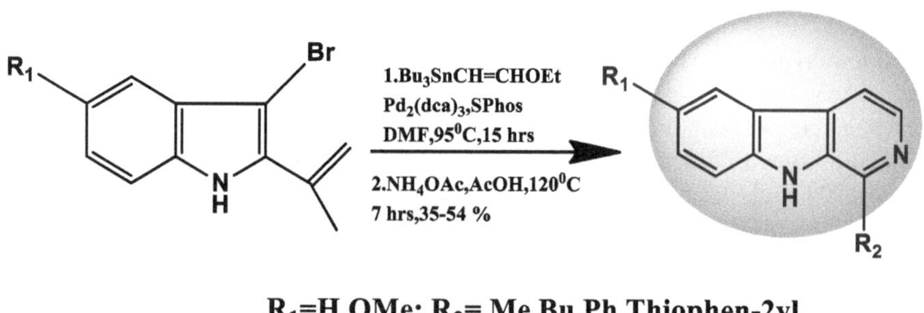

SCHEME 6.10 Synthesis of β-carbolines from 2-acyl-3-bromoindole.

indoles were described using this protocol's expanded version. The $Rh(cod)_2BF_4$ complex, which is a cationic rhodium(I) complex, c.

The 2-acyl-3-bromoindoles were converted into aromatic β-carbolines via Stille cross-coupling with tributyl[(Z)-2-ethoxyvinyl]Stan-nane serving as the fundamental unit of the C2 building (Scheme 6.10) (Kamlah, A.; Lirk, F.; Bracher, F et al. 2016). Moderate yields of β-carbolines were formed after a Pd-catalyzed first step and a ring closure using NH_4OAc in glacial acetic acid.

The process of intra-molecular cyclization and reduction of indoles was achieved using Arhodium (II) as a catalyst in a one-pot reaction, which allowed for the

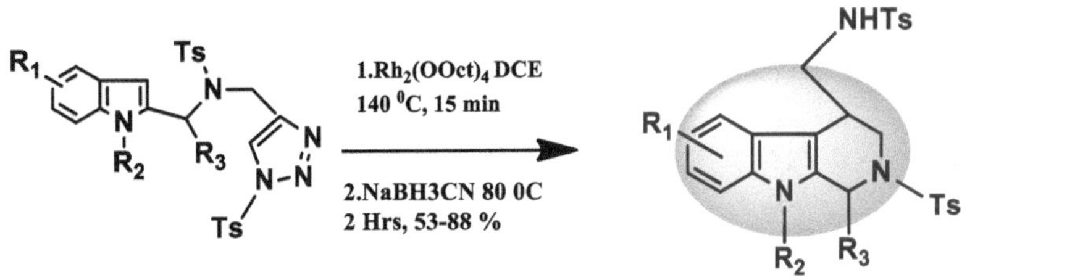

R_1=H,Me,Hal,OMe OBn : R_2= Me,Bn,Ts, PNB,Allyl : R_3: H,Me

SCHEME 6.11 Synthesis of 4-functionalized tetrahydro-β-carbolines.

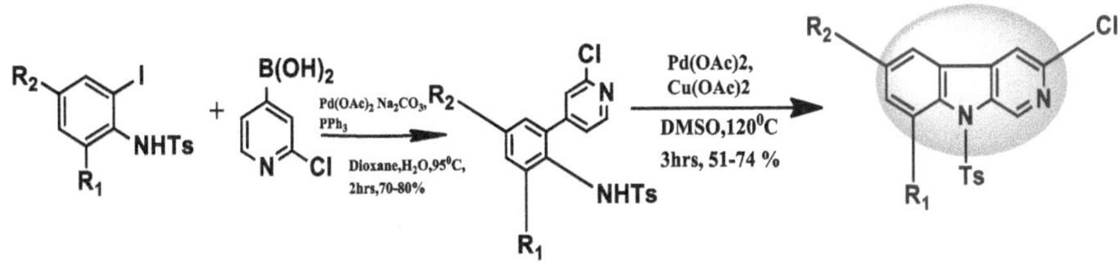

SCHEME 6.12 Synthesis of β-carbolines from N-tosylated 2-iodoanilines with (2-chloropyridin-4-yl) boronic acid.

production of a range of 4-functionalized tetrahydro-β-carbolines (Scheme 6.11) (Shang, H et al. 2016). For the production of tetrahydro-β-carbolines with high functionalization, this three-step technique is ideal.

To get beta β-carbolines, Ray et al. came up with a two-step process. By combining N-tosylated 2-iodoanilines with (2-chloropyridin-4-yl) boronic acid, a reaction can be initiated and precursors were produced (Scheme 6.12) (Dhara, S et al. 2014; Dhiman, S et al. 2016). Pd-mediated C-H/N-H activation led to moderate yields of the desired compounds after ring closure. The readily available chloro-β-carbolines have a wide range of potential applications in cross-coupling processes.

One-pot catalysis using a triple orthogonal metal relay was developed to synthesize 1,3-di- and 1,3,4-tri-substituted β-carbolines using one-pot bimetallic relay catalysis. Additionally, a one-pot bimetallic relay catalysis was created to synthesize 1,3-disubstituted 4-hydroxy-β-carbolines. The widely available 3-(2-aminophenyl)-5-hexenyn-3-ols were used as inputs in both cases (Scheme 6.13) (Dhiman, S et al. 2016)

When palladium or platinum cannot be used, elemental sulphur can be used as an alternative oxidant to convert tetrahydro-β-carbolines to the fully aromatic framework. By heating **a** with sulphur in xylenes under reflux condition (Scheme 6.14) (Devi, N et al. 2018 and Arshad, A. S. M et al. 2021) it was still possible to synthesize aromatic esters **b** in his synthesis of eudistomins **a**.

Using 2,3-dichloro-5,6-dicyano-1,4-benzoquinone (DDQ) to oxidize tetrahydro-β-carbolines is a highly effective synthetic approach for producing 4-alkoxy-β-carbolines (Scheme 6.15) (Huang, Y. Q.et al. 2018; Raghavendra, G. Met al. 2015).

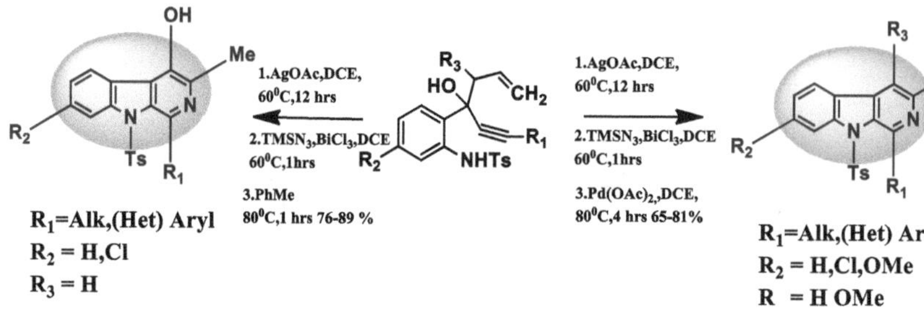

SCHEME 6.13 Synthesize 1,3-di- and 1,3,4-tri-substituted β-carbolines.

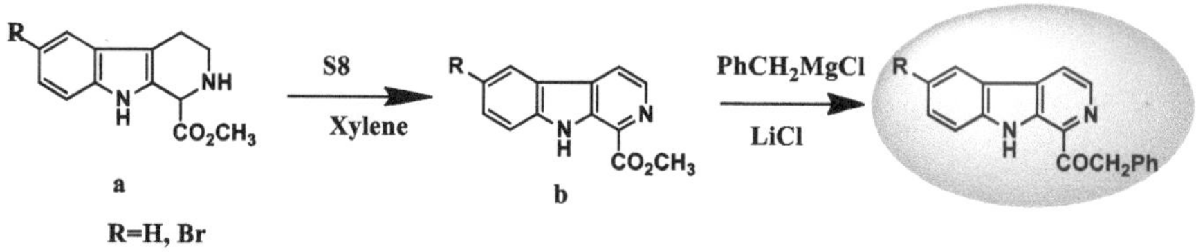

SCHEME 6.14 Oxidation of tetrahydro-β-carbolines using sulphur.

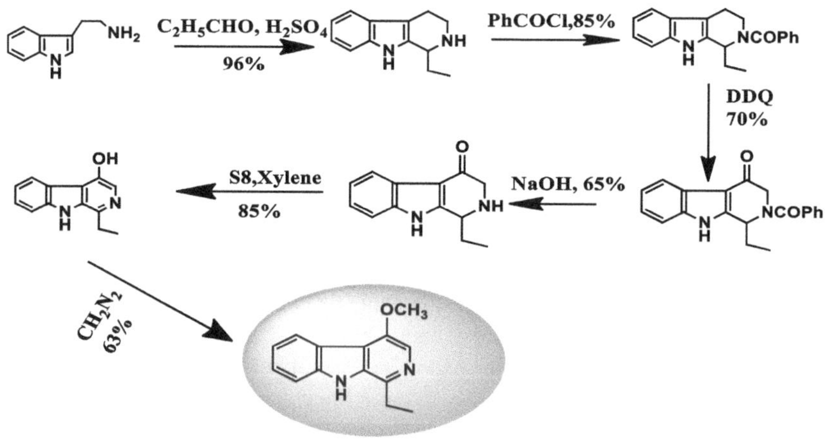

SCHEME 6.15 Synthesis of 4-alkoxy-β-carbolines.

Claisen rearrangements involving allyl alcohol and p-toluenesulfonic acid have been utilized to synthesize several β-carbolines, with a 200°C/30-minute heating time yielding 84% of the target compound (Scheme 6.16) (ShaikhA et al. 2018) Alternatively, 4-amino-β-carboline can be synthesized via the Fischer indole synthesis process by employing hydrazine as a reactant; this is thought to take place via the production of a hydrazone (subsequent to isomerization, there is a loss of ammonia).

The past decade in the field of medicinal chemistry has demonstrated a range of synthetic approaches for the production of the β-carboline molecule and its

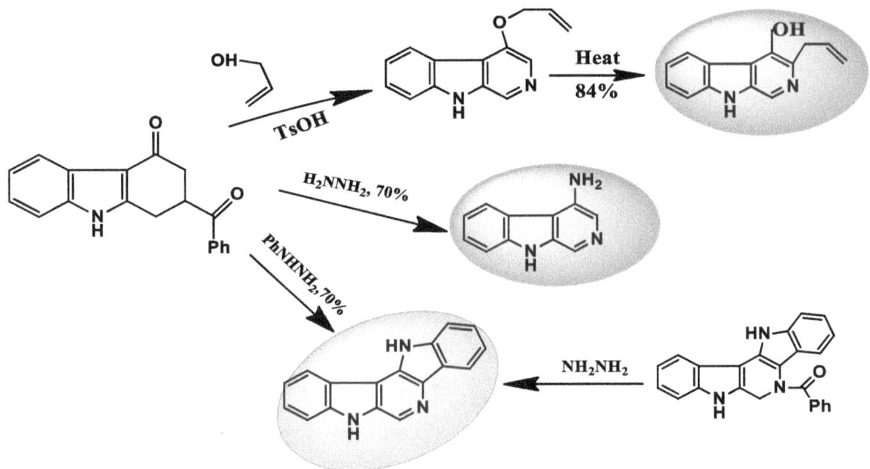

SCHEME 6.16 Synthesis of β-carbolines from allyl alcohol and p-toluenesulfonic acid.

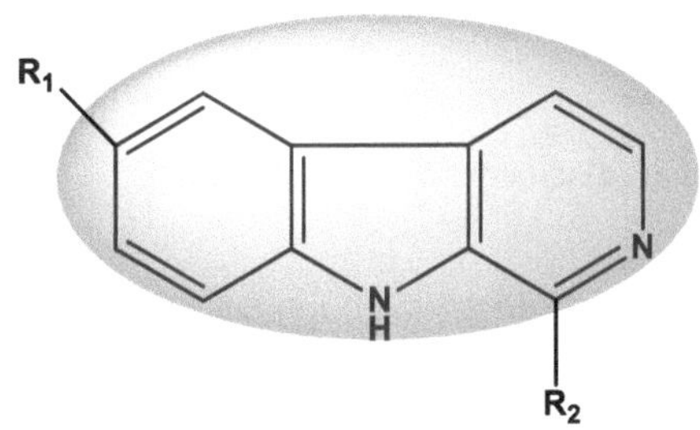

FIGURE 6.2 Antimalarial activities of 1-amino-β-carbolines compounds.

derivatives. These methods employ different reaction conditions. This chapter, however, showed the wide applicability of this prototype involves the design and synthesis of a novel β-carboline compound analogues and antimalarial activity. We anticipate greater benefits for the pharmaceutical industry and think this substrate has promising applications with respect to medicinal chemistry.

6.3 ANTIMALARIAL ACTIVITY OF β-CARBOLINE DERIVATIVES

A group of 1-amino-β-carbolines with C-6 halogen substitutions were synthesized by Thompson et al. 6-substituted β-carboline compounds were shown to have more inhibitory action than unsubstituted 1-amino-β-carbolines. C-6 substitution with -Cl or -F were found to be helpful, apart from 1,4′-bipiperidine derivative. Furthermore, substituent R1 is F or Cl, while R2 is N,N-Diethyl-N'-methyl-propane-1,3-diamine or N,N-Diethyl-N'-methyl-butane-1,3-diamine demonstrated strong inhibitory action against the P. falciparum parasite (K1 strain) (Thompson, M. Jet al. 2012).

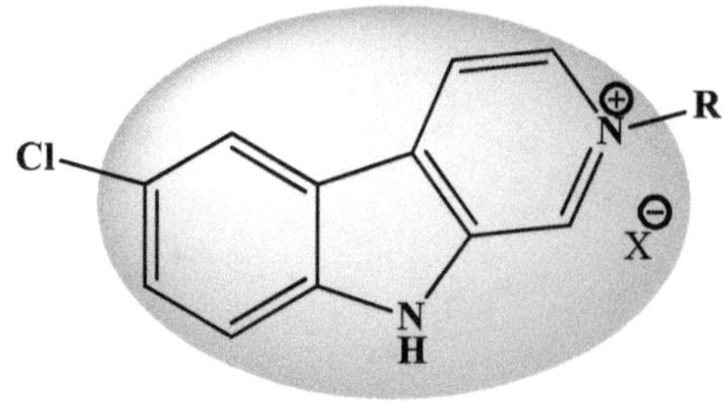

FIGURE 6.3 Nostocarboline and its derivatives.

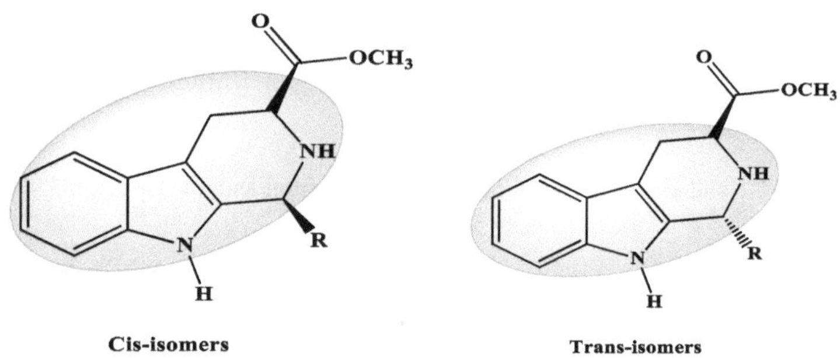

FIGURE 6.4 Stereoisomers of tetrahydro β-carbolines.

The production of these quaternary carbolineum molecules involved chlorination at the C6 position and methylation at the N2 position of norharmane, resulting in the formation of compounds known as "nostocarbolines." In these carbolines, R is the -CH3 group, which is highly active against *Plasmodium* (IC50 = 0.194 M) but inert against other parasites. Furthermore, changing the methyl group to a larger alkyl, aromatic, or allyl group resulted in a decrease in antimalarial activity but a considerable increase in cytotoxicity (Becher, P. G. et al. 2005).

Piperonyl (1a/2a), pyridyl (1b/2b), thiophene (1c/2c), and n-hexyl (1d/2d) substitutions were recently performed to investigate the C1 location of Tetrahydro-β-carboline (Figure 6.4). The methyl ester derivative of each molecule's cis- and trans-isomers was synthesized.

The methyl ester derivative of each molecule's cis- and trans-isomers was synthesized by using the Pictet–Spengler reagent for producing L-tryptophan.

The potential in vitro inhibitory activity of synthetic compounds against P. berghei was also evaluated. With an IC50 of less than 5 g/mL, all the compounds (1a-d and 2a-d) showed significant inhibitory action. Piperonyl (2a) and pyridyl (2b) trans isomers were the more active forms compared to the comparable cis isomers. Compounds 2a and 2b were identified because of this study. Compounds 2a and 2b were identified because this study opted to test for antimalarial activity in a rodent model of malaria. The Piperonyl (2a) Compoundpyridyl (2b) showed superior in vivo action in terms of parasite elimination, and parasite growth was inhibited more effectively in in vitro tests (Gorki, V et al. 2018; Singh, R et al. 2020).

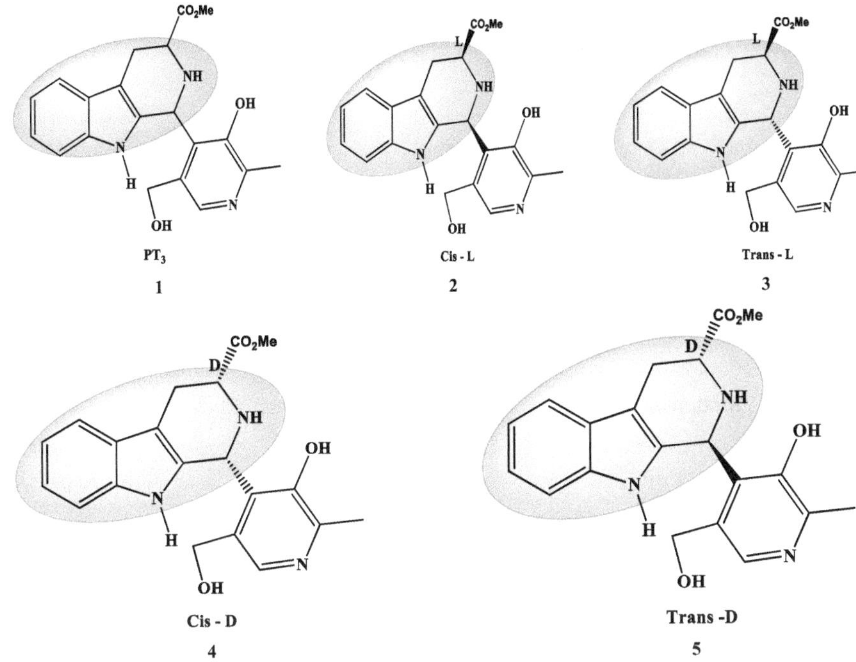

FIGURE 6.5 Derivatives of pyridoxal β- carbolines.

Brokamp et al. also found that the trans isomer of the pyridoxal β-carboline derivative (PT_3, 1) was more active than the cis isomer against *P. falciparum* (IC_{50} = 8 ±1 μM for (-)- trans-D (Figure 6.5) and 22±3 μM for (+)- trans-L (4), respectively. Both (+)-cis-D(2) and (-)-cis-L(4) (Brokamp, R et al. 2014), which were created using cis elements, had almost identical inhibitory effects (IC_{50} = 108±11 μM and 91±2 μM, respectively), but were significantly weaker than the trans products (Mulcahy, S. P et al. 2013).

By altering the tetrahydro-β-carboline scaffold at the C1, N2, and C3 locations, Sharma et al. synthesized two new classes of molecules as shown in Figure 6.6. The first group is a Tetrahydro-β-carboline-quinoline moiety1H-triazole hybridization sequence and an alkyl chain extender (R-H/CH_3 and Linker with Butane to Pentane) of varying chain length (Sharma, B.et al. 2020). Assessments were made of these substances on their efficacy as antiplasmodials against *P. falciparum* strain W2 that is resistant to chloroquine. The length, the functionality as a linker, and the presence of a substituent at C1, the chain of alkyls, all influence the synthesized chemicals' activity. Since then, researchers have implemented a plan, leading to an increase in activity levels. (R-H/CH_3 and Linker with Butane to Pentane)(IC_{50} = 0.49-1.37μM) are more potent than (R-H/CH_3)(IC_{50} = 4.02-9.28μM), illustrating the advantages of a malleable alkyl chain on the aminoquinoline core. The second group involves a Tetrahydro-β-carbolinequinoline moiety hybridizing with an acylhydrazide. Another derivative of R-H/CH3 and Linker with CH2 to Heptane, which are connected to

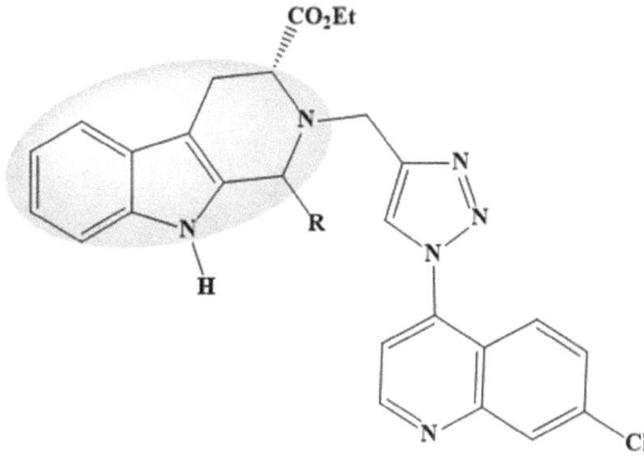

FIGURE 6.6 Modification at N-2 of Tetrahydro-β-carboline Scaffold.

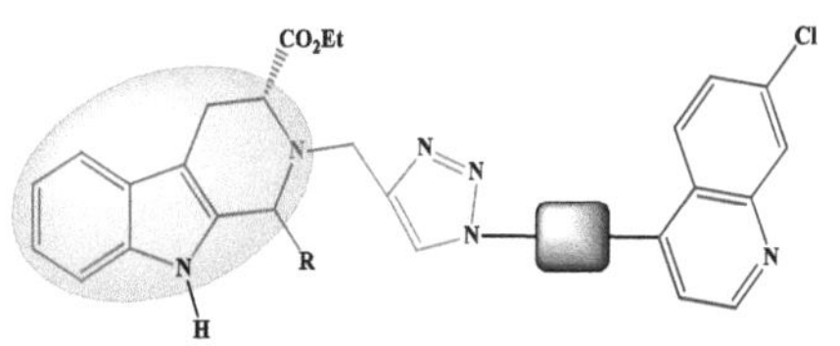
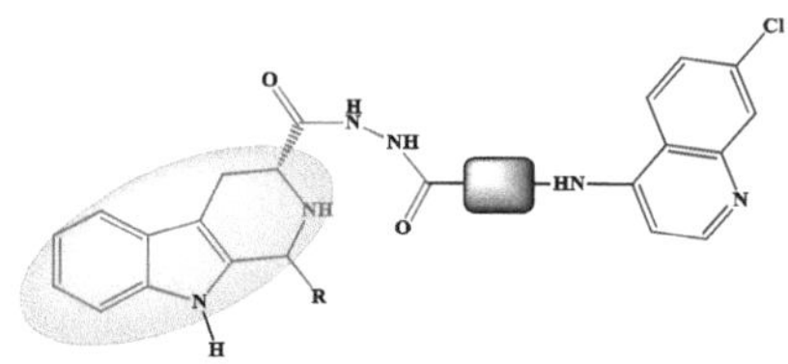

FIGURE 6.7 Modification at N-2 of Tetrahydro-β-carboline Scaffold through linkers.

aliphatic acyl hydrazides, shows the greatest increase as the length of the alkyl chain increases; antiplasmodial action is seen, while the activity profiles do not appear to be affected by tetrahydro-β-carboline substitution at the C1 site. It was found that the activity increased after an aromatic ring replaced an alkyl chain shown by examples R group Hlinkerp-xylene and R groupCH_3linker with p-xylene.

The 2-position of Tetrahydro-β-carboline linked to an isopropyl chain or a propyl chain showed a minimum inhibitory concentration (MIC) of 0.05 μM that is very effective with isomer of D/trans, n =3, R=-$CH(CH_3)_2$, R_1= H and R_2=H. However, unlike compounds, reducing the length of the linker chain at the 2-position did not result in an increase in activity during synthesis by Gupta et.al. CompoundD/trans, n =3, R= -CH_3, R_1= H, and R_2=H in which the isopropyl group was replaced by methyl on the para position of the benzene ring showed good activity with a MIC of 0.06 μM for Tetrahydro-β-carboline, while other compounds increase value of MIC. Also, other compounds had their pKa values reduced when an alkyl chain was replaced with a methoxy group at the para position of the benzene ring. We need to increase the MICs and add three methoxy groups to get very high MIC values. In addition, the para-position chloro group substitutions yielded compounds with modest activity, whereas the big bromo group replacement reduced activity to MIC (Gupta L et al. 2008)

The antimalarial effects of 6-chloro and 6-bromo β-carbolines are widely recognized and widely used. Additional SAR research was conducted by *Liew and colleagues* on indole scaffold-containing β-carbolines (Figure 6.9), 3,4-dihydro

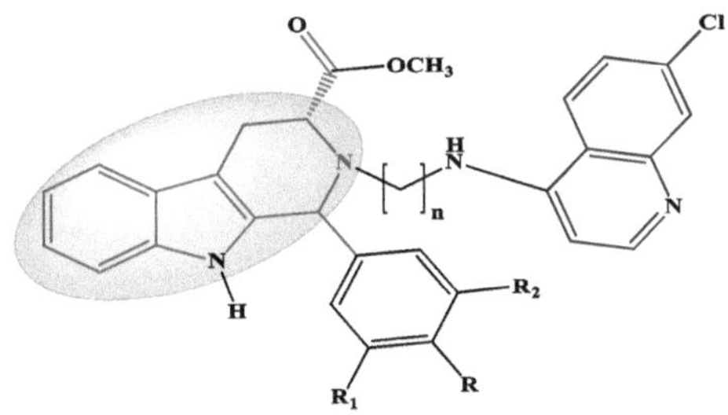

FIGURE 6.8 In vitro antimalarial activity against chloroquine-sensitive strains of *P. falciparum.*

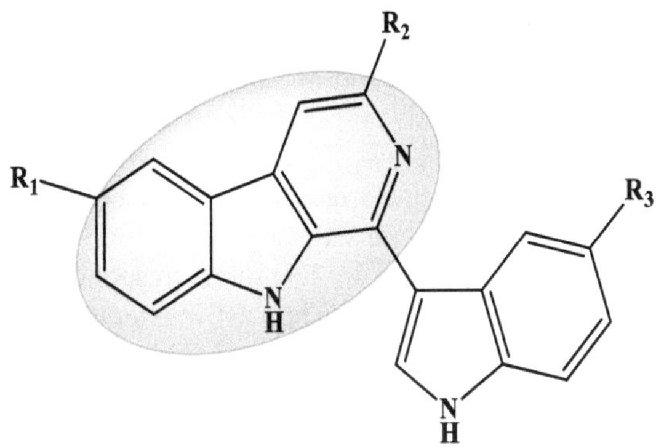

FIGURE 6.9 β-carbolines.

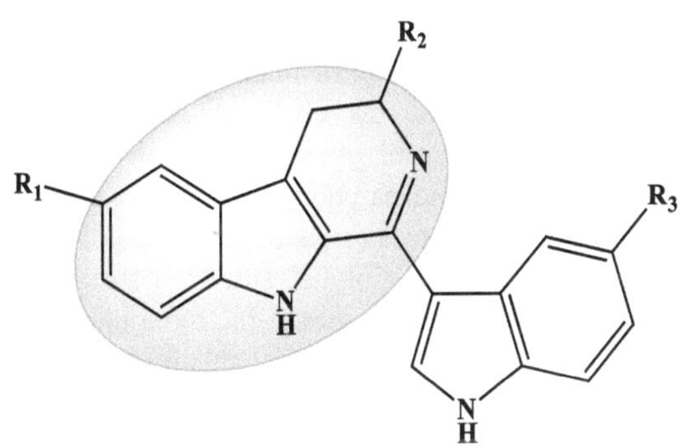

FIGURE 6.10 3,4-dihydro-β-carbolines.

β-carbolines (Figure 6.10), and 1,2,3,4-tetrahydro-β-carbolines (Figure 6.11) (Liew, L.P. et.al. 2014). In the first step of the synthesis process, the authors prepared β-carbolines derivatives with methoxy and hydroxyl groups at the C-6 position of the β-carboline scaffold. It was found that compounds containing an ester or carboxyl group at C-3 were less active than those with a methoxy or hydroxy group. Compound

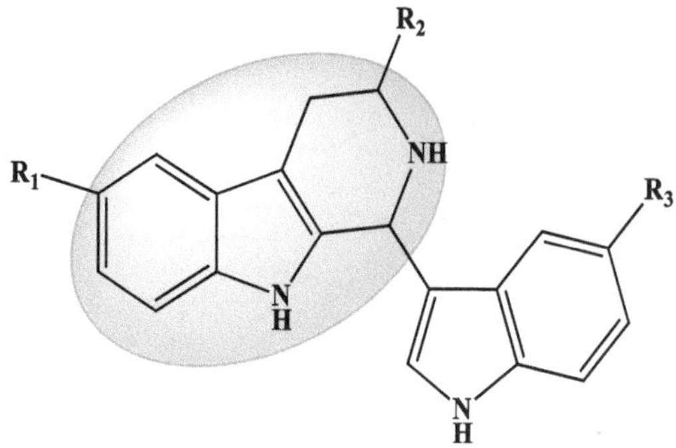

FIGURE 6.11 Tetrahydro-β-carbolines.

R_1= OMe R_2=H R_3=OMe shows the IC50(μM) is 4.7 ± 1.3, which contains a methoxy group at carbon-6, was the most active of all the β-carbolines derivatives tested. The tetrahydro-β-carbolines framework also underwent a similar substitution. These carbolines either have the same antimalarial activity as the corresponding completely aromatic β-carboline compounds with exchanging position it not shoe expected results. The 3,4-dihydro β-carbolines were like the corresponding β-carboline structure in terms of activity, although they were less active overall.

A type of alkaloids known as β-carbolines contains three pyrido[3,4-b] indole rings and has been shown to have antimalarial properties. Different strains of Plasmodium, the bacterium responsible for malaria, have been tested in vitro and in vivo with several β-carboline derivatives that have been produced and assessed for their effectiveness in treating malaria. Another natural β-carboline alkaloid with impressive antimalarial action in vivo is manzamine A, which was discovered in marine sponges. By blocking over 90% of the asexual erythrocytic stages of Plasmodium berghei, a single intraperitoneal injection of manzamine into infected mice greatly extended the median survival time of severely parasitemic animals from 30 days to 60 days. Manzamine A also had a prolonged plasma half-life and caused morphological alterations in the parasite. Researchers have shown that β-carboline possess a diverse array of pharmacological properties effects, including those that are antimalarial, anticonvulsive, hypnotic, anxiolytic, antiviral, antibacterial, inhibitory of topoisomerase-II, and inhibitory of cyclic guanosine monophosphate (cGMP). These findings encourage further research into the mechanism of action and pharmacokinetics of β-carboline as potential therapeutic development candidates.

6.4 CONCLUSION

We have tried to provide a holistic perspective on the various synthetic methods and their utilization in the complete synthesis of β-carboline natural compounds considering the rising importance of these compounds. In addition, we summarized the origins and bioactivities of a selection of representative tetrahydro-β-carbolines alkaloids. While the reaction facilitated by Pictet–Spengler atetrahydro-β-carbolines analogues have been explored thoroughly from a synthetic standpoint, a comprehensive assessment of alternative approaches has not yet been made accessible. More and more, techniques using cyclization, ring-closing metathesis, cycloaddition, condensation techniques, oxidative C-N coupling, rearrangement, and palladiumcatalysedcutting-edge N-heteroannulationare are being adopted. Fischer indolization, a classic technique, is also useful in some situations.

REFERENCES

Ajima, U.; Onah, J. O.; Wannang, N. N., Synthesis, Characterization and biological evaluation of Benzimidazole-Dihydroartemisinin hybrids as potential dual acting antimalarial agents. *Mediterranean Journal of Chemistry* **2019**,*9* (1), 52–64.

Arshad, A. S. M.; Meesala, R.; Hanapi, N. A.; Mordi, M. N., A convenient synthesis of β-carbolines by iron-catalyzed aerobic decarboxylative/dehydrogenative aromatization of tetrahydro-β-carbolines under air. *Tetrahedron* **2021**,*83*, 131960.

Ashley, E. A.; Recht, J.; White, N. J., Primaquine: The risks and the benefits. *Malaria Journal* **2014**,*13* (1), 1–7.

Becher, P. G.; Beuchat, J.; Gademann, K.; Jüttner, F., Nostocarboline: Isolation and synthesis of a new cholinesterase inhibitor from Nostoc 78-12A. *Journal of Natural Products* **2005**,*68* (12), 1793–1795.

Bell, G. J.; Agnandji, S. T.; Asante, K. P.; Ghansah, A.; Kamthunzi, P.; Emch, M.; Bailey, J. A., Impacts of ecology, parasite antigenic variation, and human genetics on RTS, S/AS01e malaria vaccine efficacy. *Current Epidemiology Reports* **2021**,*8*, 79–88.

Brokamp, R.; Bergmann, B.; Müller, I. B.; Bienz, S., Stereoselective preparation of pyridoxal 1, 2, 3, 4-tetrahydro-β-carboline derivatives and the influence of their absolute and relative configuration on the proliferation of the malaria parasite Plasmodium falciparum. *Bioorganic & Medicinal Chemistry* **2014**,*22* (6), 1832–1837.

Chaorattanakawee, S.; Kosaisavee, V.; Bunsermyos, W.; Aonsri, C.; Imaram, W.; Suwannasin, K.; Kunasol, C.; Thamnurak, C.; Boonyalai, N.; Saunders, D., In vitro activity of rhinacanthin analogues against drug resistant Plasmodium falciparum isolates from Northeast Thailand. *Malaria Journal* **2023**,*22* (1), 1–10.

Chu, X.-M.; Wang, C.; Wang, W.-L.; Liang, L.-L.; Liu, W.; Gong, K.-K.; Sun, K.-L., Triazole derivatives and their antiplasmodial and antimalarial activities. *European Journal of Medicinal Chemistry* **2019**,*166*, 206–223.

Chundawat, T. S., Design and strategic synthesis of some β-Carboline-based novel natural products of biological importance. In *Organic Synthesis-A Nascent Relook*, IntechOpen: **2019**.

Condie, G. C.; Bergman, J., Synthesis of some fused β-carbolines including the first example of the pyrrolo [3, 2-c]-β-carboline system. *Journal of Heterocyclic Chemistry* **2004**,*41* (4), 531–540.

Cravo, P.; Napolitano, H.; Culleton, R., How genomics is contributing to the fight against artemisinin-resistant malaria parasites. *Acta Tropica* **2015**,*148*, 1–7.

Dalpozzo, R., The Chiral pool in the Pictet–Spengler reaction for the synthesis of β-Carbolines. *Molecules* **2016**,*21* (6), 699.

Devi, N.; Kumar, S.; Pandey, S. K.; Singh, V., 1 (3)-Formyl-β-carbolines: Potential Aldo-X precursors for the synthesis of β-Carboline-based molecular architectures. *Asian Journal of Organic Chemistry* **2018**,*7* (1), 6–36.

Devi, N.; Rawal, R. K.; Singh, V., Diversity-oriented synthesis of substituted and Fused β-Carbolines from 1-Formyl-9H-β-Carboline scaffolds. *Research Methodology in Chemical Sciences: Experimental and Theoretical Approach* **2017**, 97.

Dimala, C. A.; Kika, B. T.; Kadia, B. M.; Blencowe, H., Current challenges and proposed solutions to the effective implementation of the RTS, S/AS01 Malaria Vaccine Program in sub-Saharan Africa: A systematic review. *PloS One* **2018**,*13* (12), e0209744.

Dhara, S.; Singha, R.; Ahmed, A.; Mandal, H.; Ghosh, M.; Nuree, Y.; Ray, J. K., Synthesis of α, β and γ-Carbolines via Pd-Mediated C sp2-H/N–H Activation. *RSC Advances* **2014**,*4* (85), 45163–45167.

Dhiman, S.; Mishra, U. K.; Ramasastry, S., One-Pot trimetallic relay catalysis: A unified approach for the synthesis of β-Carbolines and other [c]-fused Pyridines. *Angewandte Chemie* **2016**,*128* (27), 7868–7872.

Doerksen, R. S.; Hodik, T.; Hu, G.; Huynh, N. O.; Shuler, W. G.; Krische, M. J., Ruthenium-catalyzed cycloadditions to form five-, six-, and seven-membered rings. *Chemical Reviews* **2021**,*121* (7), 4045–4083.

Gorki, V.; Singh, R.; Walter, N. S.; Bagai, U.; Salunke, D. B., Synthesis and evaluation of antiplasmodial efficacy of β-carboline derivatives against murine malaria. *ACS Omega* **2018**,*3* (10), 13200–13210.

Gupta, L.; Srivastava, K.; Singh, S.; Puri, S.; Chauhan, P. M., Synthesis of 2-[3-(7-Chloroquinolin-4-ylamino)-alkyl]-1-(substituted phenyl)-2, 3, 4, 9-tetrahydro-1H-β-carbolines as a new class of antimalarial agents. *Bioorganic & Medicinal Chemistry Letters* **2008**,*18* (11), 3306–3309.

Huang, Y. Q.; Song, H. J.; Liu, Y. X.; Wang, Q. M., Dehydrogenation of N-Heterocycles by Superoxide Ion Generated through Single-Electron Transfer. *Chemistry–A European Journal* **2018**,*24* (9), 2065–2069.

Kamlah, A.; Lirk, F.; Bracher, F., A new approach to 1-substituted β-carbolines and isoquinolines utilizing tributyl [(Z)-2-ethoxyvinyl] stannane as a C-3, C-4 building block. *Tetrahedron* **2016**,*72* (6), 837–845.

Kim, K.-S., Current challenges in the development of vaccines and drugs against emerging vector-borne diseases. *Current Medicinal Chemistry* **2019**,*26* (16), 2974–2986. www.who.int/news-room/fact-sheets/detail/malaria

Kuo, P.-C.; Shi, L.-S.; Damu, A. G.; Su, C.-R.; Huang, C.-H.; Ke, C.-H.; Wu, J.-B.; Lin, A.-J.; Bastow, K. F.; Lee, K.-H., Cytotoxic and antimalarial β-carboline alkaloids from the roots of Eurycoma longifolia. *Journal of Natural Products* **2003**,*66* (10), 1324–1327.

Lacerda, M. V.; Llanos-Cuentas, A.; Krudsood, S.; Lon, C.; Saunders, D. L.; Mohammed, R.; Yilma, D.; Batista Pereira, D.; Espino, F. E.; Mia, R. Z., Single-dose tafenoquine to prevent relapse of Plasmodium vivax malaria. *New England Journal of Medicine* **2019**,*380* (3), 215–228.

Liew, L. P.; Fleming, J. M.; Longeon, A.; Mouray, E.; Florent, I.; Bourguet-Kondracki, M.-L.; Copp, B. R., Synthesis of 1-indolyl substituted β-carboline natural products and discovery of antimalarial and cytotoxic activities. *Tetrahedron* **2014**,*70* (33), 4910–4920.

Luzzatto, L.; Nannelli, C.; Notaro, R., Glucose-6-phosphate dehydrogenase deficiency. *Hematology/Oncology Clinics* **2016**,*30* (2), 373–393.

Mahmoud, E.; Hayallah, A. M.; Kovacic, S.; Abdelhamid, D.; Abdel-Aziz, M., Recent progress in biologically active indole hybrids: A mini review. *Pharmacological Reports* **2022**,*74* (4), 570–582.

Milen, M.; Ábrányi-Balogh, P., Synthesis of β-carbolines (microreview). *Chemistry of Heterocyclic Compounds* **2016**,*52*, 996–998.

Mulcahy, S. P.; Varelas, J. G., Three-step synthesis of an annulated β-carboline via palladium catalysis. *Tetrahedron Letters* **2013**,*54* (48), 6599–6601.

Nissen, F.; Detert, H., *Total Synthesis of Lavendamycin by a [2+ 2+ 2] Cycloaddition*. Wiley Online Library: **2011**.

Pumphrey, A. L. *Transition Metal-Catalyzed Intramolecular CH Amination Reactions and Applications in Drug Discovery*. University of Illinois at Chicago, **2014**.

Pumphrey, A. L.; Dong, H.; Driver, T. G., RhII2-Catalyzed synthesis of α-, β-, or δ-carbolines from Aryl Azides. *Angewandte Chemie International Edition* **2012**,*51* (24), 5920–5923.

Purohit, P.; Pandey, A. K.; Singh, D.; Chouhan, P. S.; Ramalingam, K.; Shukla, M.; Goyal, N.; Lal, J.; Chauhan, P. M., An insight into tetrahydro-β-carboline–tetrazole hybrids: Synthesis and bioevaluation as potent antileishmanial agents. *Medchemcomm* **2017**,*8* (9), 1824–1834.

Raghavendra, G. M.; Kumar, C. S. P.; Suresha, G. P.; Rangappa, K. S.; Mantelingu, K., T3P catalyzed one pot three-component synthesis of 2, 3-disubstituted 3H-quinazolin-4-ones. *Chinese Chemical Letters* **2015**,*26* (8), 963–968.

Riyazahmed, K. S.; Bora, D.; Shankaraiah, N., Application of transition Metal-Catalyzed C– H activation strategies in the synthesis and functionalization of β-Carbolines. *Asian Journal of Organic Chemistry* **2021**,*10* (5), 1050–1066.

Saha, B.; Sharma, S.; Sawant, D.; Kundu, B., Water as an efficient medium for the synthesis of tetrahydro-β-carbolines via Pictet–Spengler reactions. *Tetrahedron Letters* **2007**,*48* (8), 1379–1383.

Shaikh, A., Application of microwaves in sustainable organic synthesis. In *Green Chemistry*, Elsevier: **2018**; pp. 647–671.

Shang, H.; Tian, Y.; Luo, J.; Li, L.; Tang, Y.; Zou, Z., Rhodium (ii)-catalyzed intramolecular annulation of 1-sulfonyl-1, 2, 3-triazoles with indoles: facile synthesis of functionalized tetrahydro-β-carbolines. *RSC Advances* **2016**,*6* (37), 30835–30839.

Sharma, B.; Kaur, S.; Legac, J.; Rosenthal, P. J.; Kumar, V., Synthesis, anti-plasmodial and cytotoxic evaluation of 1H-1, 2, 3-triazole/acyl hydrazide integrated tetrahydro-β-carboline-4-aminoquinoline conjugates. *Bioorganic & Medicinal Chemistry Letters* **2020**,*30* (2), 126810.

Singh, R.; Jaisingh, A.; Maurya, I. K.; Salunke, D. B., Design, synthesis and bio-evaluation of C-1 alkylated tetrahydro-β-carboline derivatives as novel antifungal lead compounds. *Bioorganic & Medicinal Chemistry Letters* **2020**,*30* (3), 126869.

Singh, V.; Batra, S., 1-Formyl-9H-β-Carboline: A useful scaffold for synthesizing substituted- and fused β-Carbolines. *Current Organic Synthesis* **2012**,*9* (4), 513–528.

Siqueira-Neto, J. L.; Debnath, A.; McCall, L.-I.; Bernatchez, J. A.; Ndao, M.; Reed, S. L.; Rosenthal, P. J., Cysteine proteases in protozoan parasites. *PLoS Neglected Tropical Diseases* **2018**,*12* (8), e0006512.

Suzuki, H.; Unemoto, M.; Hagiwara, M.; Ohyama, T.; Yokoyama, Y.; Murakami, Y., Synthetic studies on indoles and related compounds. Part 46. 1 First total syntheses of 4, 8-dioxygenated β-carboline alkaloids. *Journal of the Chemical Society, Perkin Transactions* **1999**,*1* (12), 1717–1724.

Takasu, K.; Shimogama, T.; Saiin, C.; Kim, H.-S.; Wataya, Y.; Ihara, M., π-Delocalized β-carbolinium cations as potential antimalarials. *Bioorganic & Medicinal Chemistry Letters* **2004**,*14* (7), 1689–1692.

Thanikachalam, P. V.; Maurya, R. K.; Garg, V.; Monga, V., An insight into the medicinal perspective of synthetic analogs of indole: A review. *European Journal of Medicinal Chemistry* **2019**,*180*, 562–612.

Thompson, M. J.; Louth, J. C.; Little, S. M.; Jackson, M. P.; Boursereau, Y.; Chen, B.; Coldham, I., Synthesis and Evaluation of 1-Amino-6-halo-β-carbolines as antimalarial and antiprion agents. *ChemMedChem* **2012**,*7* (4), 578–586.

Varelas, J. G.; Khanal, S.; O'Donnell, M. A.; Mulcahy, S. P., Concise synthesis of annulated pyrido [3, 4-b] indoles via Rh (I)-catalyzed cyclization. *Organic Letters* **2015**,*17* (21), 5512–5514.

Wang, J.; Gong, F.; Liang, T.; Xie, Z.; Yang, Y.; Cao, C.; Gao, J.; Lu, T.; Chen, X., A review of synthetic bioactive tetrahydro-β-carbolines: A medicinal chemistry perspective. *European Journal of Medicinal Chemistry* **2021**,*225*, 113815.

WHO. World malaria report 2017. (2017, November 29). www.who.int/teams/global-malaria-programme/reports/world-malaria-report-2017

WHO. World malaria report 2018. (2018, November 19). www.who.int/teams/global-malaria-programme/reports/world-malaria-report-2018

WHO. World malaria report 2019. (2019, December 4). www.who.int/teams/global-malaria-programme/reports/world-malaria-report-2019

WHO. World Malaria Report 2020. (2020, November 30). www.who.int/teams/global-malaria-programme/reports/world-malaria-report-2020

WHO. World malaria report 2021. (2021, December 6). www.who.int/teams/global-malaria-programme/reports/world-malaria-report-2021

Zhang, M.; Sun, D., Recent advances of natural and synthetic β-carbolines as anticancer agents. *Anti-Cancer Agents in Medicinal Chemistry (Formerly Current Medicinal Chemistry-Anti-Cancer Agents)* **2015**,*15* (5), 537–547.

7 β-Carbolines as Antioxidant

Synthesis and Biological Study

Chetna Ameta, Monika Kumawat, Dharmendra, and Mukesh Kumar

7.1 INTRODUCTION

The synthesis of organic molecules derived from renewable natural resources for diverse applications, particularly in the fields of biology and pharmaceuticals, is a burgeoning trend in contemporary synthetic chemistry. Over the past two decades, an array of alkaloids has been extracted from natural sources, with their significance in biological and pharmaceutical domains under rigorous examination. Research has unveiled that a wide spectrum of natural alkaloids are endowed with bioactivity, and they have been harnessed in the treatment of various pathogenic ailments. One noteworthy class of these natural compounds is the β-carbolines, which have garnered attention due to their potential antioxidant properties. These compounds can be naturally found in a variety of plants, foods, and beverages and have been the subject of investigation regarding their potential health-enhancing attributes. A well-documented member of the β-carboline family is harmine, present in plants such as Peganum harmala (Syrian rue) and Banisteriopsis caapi (a vine used in Ayahuasca preparations).

Antioxidants serve as defenders of our cells against oxidative stress by neutralizing harmful free radicals. Oxidative stress is closely associated with a range of health concerns, including the aging process, inflammation, and chronic diseases. Consequently, antioxidants can play a pivotal role in mitigating oxidative damage and promoting overall well-being. Research into the antioxidant potential of β-carbolines is an ongoing endeavor, and the available evidence suggests that some of these compounds may indeed exhibit antioxidant effects. For instance, harmine has been demonstrated to possess antioxidant and free radical-scavenging properties in various studies. Nonetheless, it is vital to recognize that the effectiveness of these compounds as antioxidants can be influenced by factors such as their concentration, the specific type of β-carboline, and the experimental conditions in which they are assessed.

Carbolines are among the most fascinating alkaloids and are categorized based on their ABC skeleton as α-, β-, γ-, or δ-carbolines (Figure 7.1) (Smirnova et al. 2012).

DOI: 10.1201/9781351058032-7

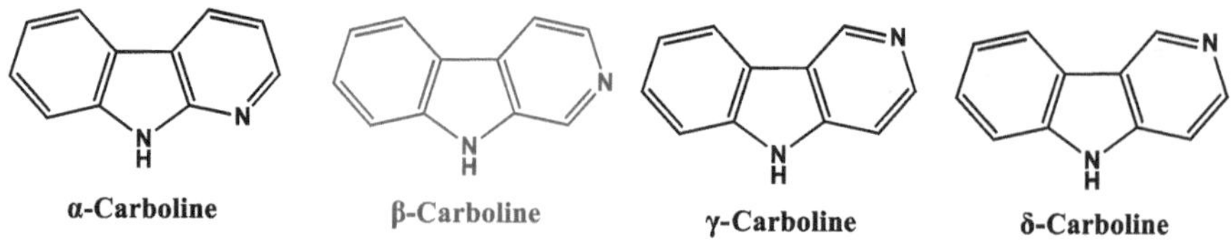

FIGURE 7.1 Structural diversity of carbolines.

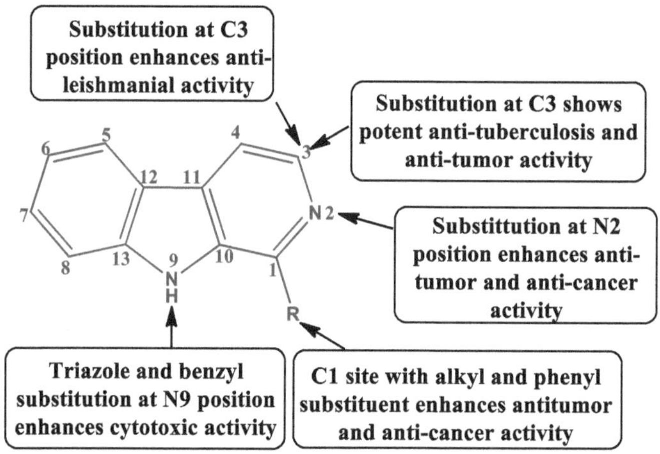

FIGURE 7.2 SARs analysis of β-carbolines.

7.2 PHARMACOLOGICAL ACTIVE β-CARBOLINES

Molecules based on the β-carboline skeleton exhibit remarkable structural diversity. First isolated in 1841, β-carboline (9H-pyrido[3,4-b]indole), also known as nor-harmane, is a nitrogen-containing heterocycle. It falls into the category of indole alkaloids and features a pyridine ring fused with an indole backbone. The structure of β-carboline resembles tryptamine, with the ethylamine chain reconnected to the indole nucleus through an additional carbon atom, resulting in a three-ringed structure (Figure 7.2) (Goebel 1841).

β-carboline alkaloids have been successfully extracted from a diverse array of sources, primarily encompassing plants (particularly in Rutaceae (Domenech-Carbo et al. 2017), Simaroubaceae (Ohishi et al. 2015), Amaranthaceae (Agrawal et al. 2013), Caryophyllaceae (Chen et al. 2010), Rubiaceae (Figueiredo et al. 2014), and Zygophyllaceae (Wang et al. 2016), as well as marine organisms such as hydroids (Skropeta and Wei 2014), bryozoans (Till and Prinsep 2009), soft corals (Gao et al. 2013), and sponges (Tanaka et al. 2014). Furthermore, these alkaloids have been identified in microorganisms (Savi et al. 2015), insects (Marques et al. 2005), various food products (Kim et al. 2016), alcoholic beverages (Tsuchiya et al. 2016), tobacco smoke (Sari and Khalil 2015), and in various human tissues and bodily fluids (Lamounier et al. 2015).

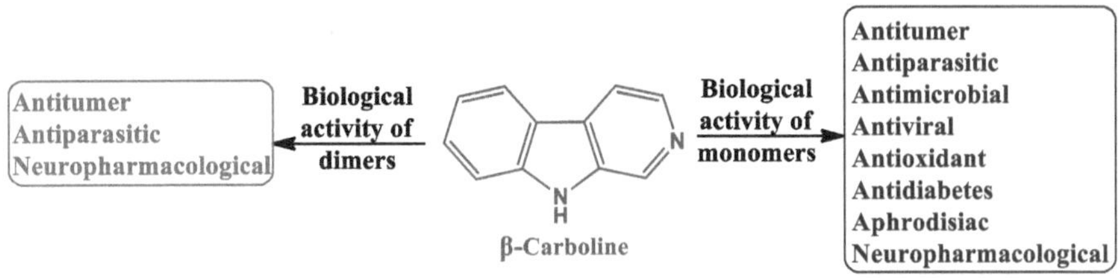

FIGURE 7.3 Biological activities of *β*-carboline alkaloids.

In addition to broad occurrence, (Figure 7.3) they exhibit a wide array of remarkable actions, including sedative (Smith et al. 2013), anxiolytic (Stephens 2013), hypnotic (Baldwin and Tiwari 2015), anticonvulsant (Pogosyan et al. 2007), antineoplastic (Zheng et al. 2014), antiviral (Formagio et al. 2009), antiparasitic (Gooyit et al. 2015), antimicrobial (Zhang et al. 2015), and more. These attributes have garnered significant attention from both the academic and industrial sectors. To date, it is noteworthy that some β-carboline-based remedies (Figure 7.4) have been successfully brought to the commercial market.

Human oxygen consumption is estimated at approximately 250 mL per min, with about 2–4% of this oxygen being altered into various reactive oxygen species (ROS) (Pless et al. 1999). Among these species, the hydroxyl radicals are considered to be the most toxic, and they can be engendered *in vivo* via the Fenton reaction, often originating from the endogenous production of hydrogen peroxide within the brain (Sinet et al. 1980). Hydroxyl radicals can trigger degenerative reactions, including lipid peroxidation in cell membranes. Neurons are particularly vulnerable to damage from free radicals due to their low levels of antioxidant enzymes and limited free radical scavenger concentrations (Kedziora and Bartosz, 1988). This susceptibility has been associated with the development of neurodegenerative diseases, including Alzheimer's disease (Pappolla et al. 1997), Parkinson's disease (Götz et al. 1994), and mental deterioration in Down's syndrome.

An antioxidant treatment is a logical approach, and well-established antioxidant substances like ascorbic acid (vitamin C) and vitamin E are commonly recommended and utilized. However, recent literature suggests that beyond these vitamins, β-caroline derivatives also exhibit antioxidant properties, despite earlier discussions about their potential for producing ROS and phototoxicity (Larson et al. 1988; Pari et al. 2000).

The formation of free radicals and ROS is an inherent aspect of human metabolism. Maintaining a balance between the rate of ROS formation and the activity of antioxidant systems is crucial for homeostasis. This balance hinges on the interplay of low-molecular-weight antioxidants and proteins. Any increase in the rate of ROS formation or inhibition of antioxidant mechanisms can lead to oxidative stress, a pivotal factor in the pathogenesis of various diseases, including neurodegenerative and mental disorders. Research in this area focuses on specific low-molecule components, whose activity or concentration may correlate with the progression of free radical reactions. β-carboline derivatives are potential ligands for 5-HT1A and 5-HT2A

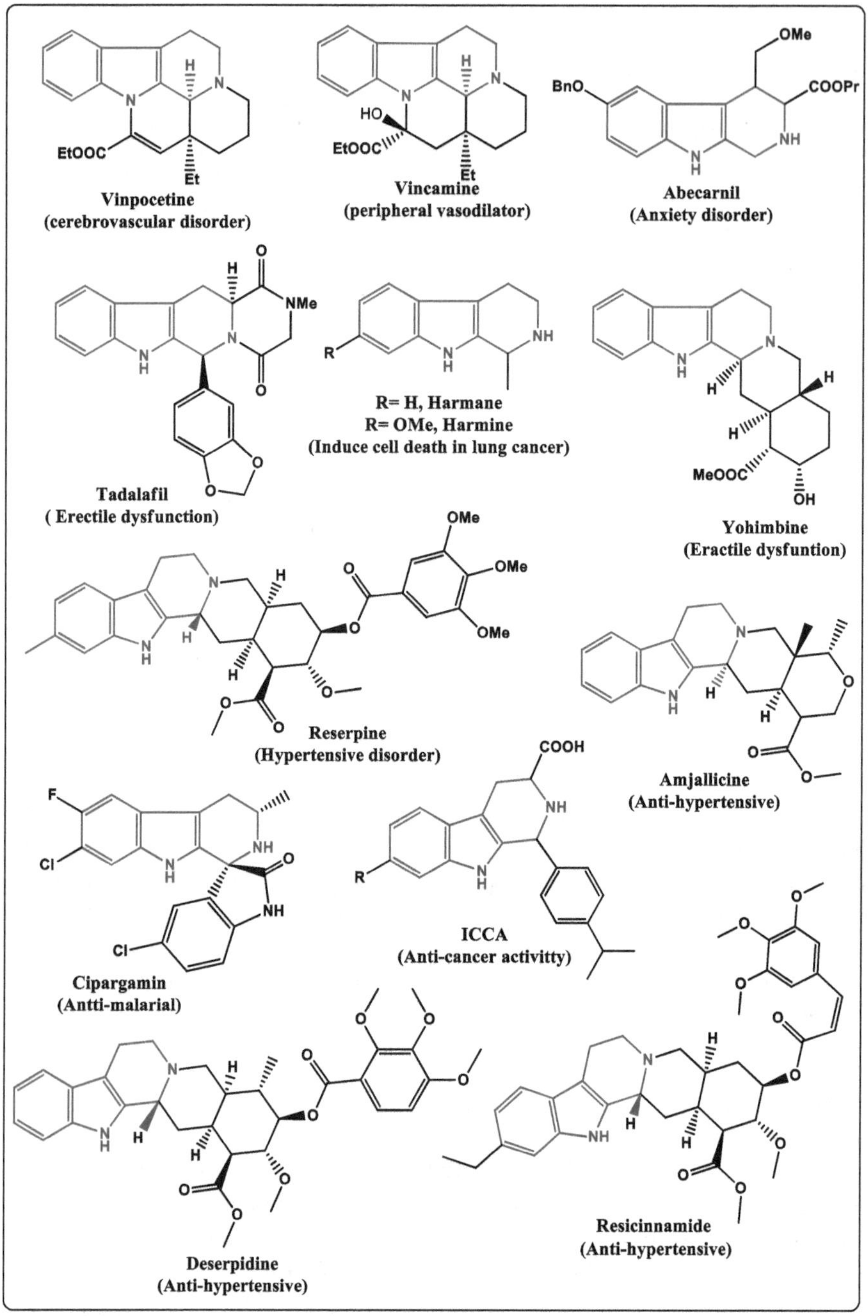

FIGURE 7.4 β-carboline drugs and their pharmacological activity.

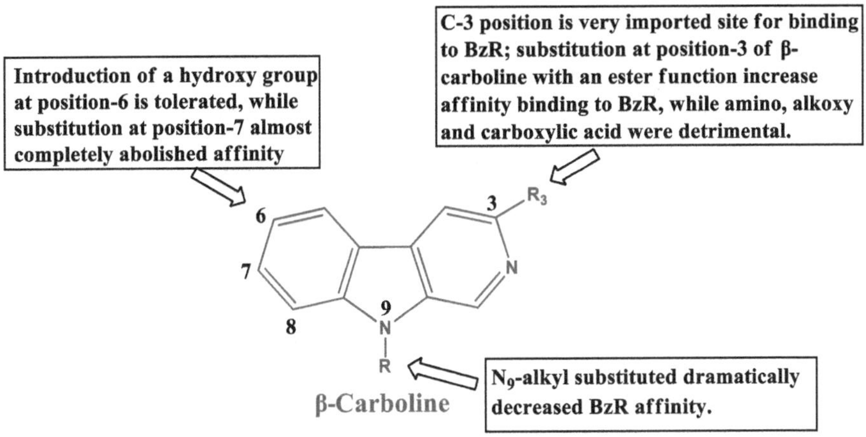

FIGURE 7.5 The Structure-affinity relationship of β-carboline binding to BzR.

receptors. The pro- or antioxidant properties of this compound group may have a direct impact on the overall antioxidant balance. Consequently, it is imperative to investigate whether β-carboline derivatives influence the levels of free radicals or ROS. A inclusive assessment of the antioxidant activity of β-carboline derivatives can shed light on their role in nonenzymatic systems that counteract ROS malfunction or formation (Figure 7.5) (Pless et al. 1999).

7.3 ANTIOXIDANT ACTIVITY OF β-CARBOLINES

7.3.1 Tetrahydro-β-carbolines

Several tetrahydro-β-carboline alkaloids in fruits were identified and the occurrence of these biologically active substances in such products assessed. It has been found that these alkaloids detected in fruit products may act as antioxidants (Herraiz and Galisteo 2003). A role of these compounds as protective agents against free radicals is suggested. Four tetrahydro-β-carboline alkaloids, 1,2,3,4-tetrahydro-β-carboline-3-carboxylic acid, 1-methyl-1,2,3,4-tetrahydro-β-carboline-3-carboxylic acid, 1-methyl-1,2,3,4-tetrahydro-β-carboline, and 6-hydroxy-1-methyl-1,2,3,4-tetrahydro-β-carboline, have bee found as naturally occurring substances in some fruit and fruit juices. These compounds occur in the μg/g level in those products, and a characteristic and distinct profile appears to exist depending on the type of fruit and juice involved. For example, 1-methyl-1,2,3,4-tetrahydro-β-carboline may appear in tomato, tomato juice, and kiwi; 6-hydroxy-1-methyl-1,2,3,4-tetrahydro-β-carboline in bananas, pineapple, tomato, and their corresponding juices; and 1-methyl-1,2,3,4-tetrahydro-β-carboline-3-carboxylic acid in oranges and grapefruits, although it also occurs in most juices. Fruit-occurring tetrahydro-β-carboline alkaloids acted as antioxidants and free radical scavengers in the ABTS assay when compared with ascorbic acid and Trolox. This indicates that tetrahydro-β-carboline alkaloids might act as antioxidants when absorbed and accumulated in the body, contributing to the antioxidant effect of fruit products containing these compounds (Table 7.1).

TABLE 7.1
Tetrahydro-β-carboline derivatives

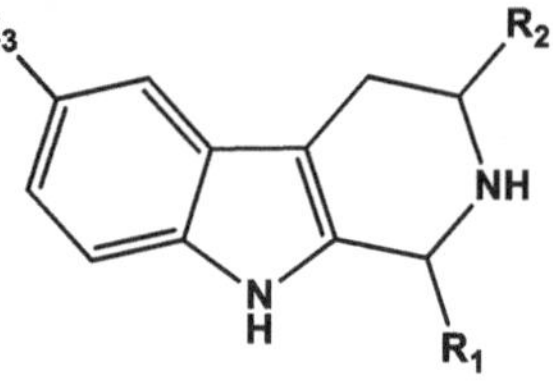

R_1	R_2	R_3	Compound
CH_3	H	OH	6-hydroxy-1-methyl-1,2,3,4-tetrahydro-β-carboline
H	COOH	H	1,2,3,4-tetrahydro-β-carboline-3-carboxylic acid
CH_3	COOH	H	1-metyl-1,2,3,4-tetrahydro-β-carboline-3-carboxylic acid
CH_3	H	H	1-methyl-1,2,3,4-tetrahydro-β-carboline

TABLE 7.2
Tetrahydro derivatives of β-carboline from garlic extracts

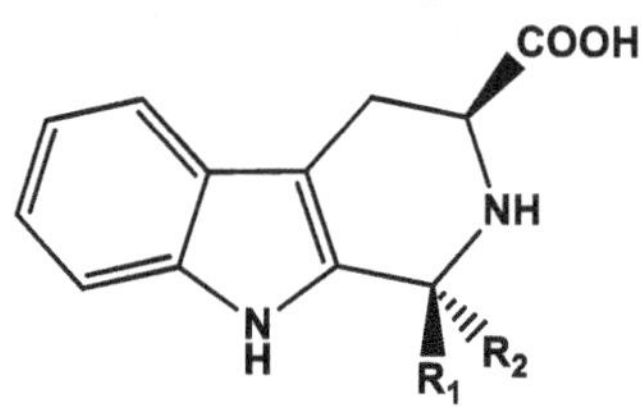

	R_1	R_2
(1R,3S)-MTCC	H	CH_3
(1S,3S)-MTCC	CH_3	H
(1R,3S)-MTCdiC	COOH	CH_3
(1S,3S)-MTCdiC	CH_3	COOH

In their study, Ichikawa et al. (2006) employed a hydrogen peroxide scavenging assay to explore the antioxidant compounds derived from ripened garlic extract. This is made by immersing carved garlic in an ethanolic solution for a period of 10 months. During their investigation, four distinct 1,2,3,4-tetrahydro-β-carboline derivatives (THbCs), namely 1-methyl-1,2,3,4-tetrahydro-β-carboline-3-carboxylic acid and 1-methyl-1,2,3,4-tetrahydro-β-carboline-1,3-dicarboxylic acid (MTCdiC), originating from both diastereoisomers, were successfully isolated and recognized using LCMS (liquid chromatography-mass spectrometry).

Remarkably, these exhibited robust scavenging activities against hydrogen peroxide and effectively inhibited 2,2'-azobis(2-amidinopropane) hydrochloride-induced lipid peroxidation. Notably, among them, (1S,3S)-MTCdiC demonstrated the most potent scavenging activity against hydrogen peroxide, surpassing even ascorbic

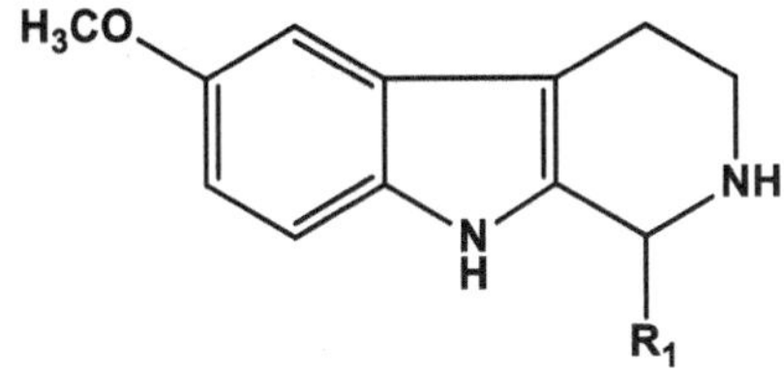

R_1 = -CH_3 (ACE)
R_1 = -C_6H_5 (BEN)
R_1 = -C_7H_8O (ANI)
R_1 = -$C_7H_8O_2$ (VAN)

FIGURE 7.6 Various derivatives of β-carboline.

TABLE 7.3
IC_{50} of DPPH antioxidant assay of pure 6-methoxy scaffold β-carboline v/s BHT and quercetin

	DPPH antioxidantt assay	
Type of standard/pure compound	IC_{50}(ug/mL)	IC_{50} uM
Quercetin	0.20 ± 0.00	0.66
BHT	0.28 ± 0.00	1.27
BEN	2.92 ± 0.04	10.50
ANI	ND	ND
ACE	ND	ND
VAN	1.74 ± 0.02	5.37

acid. Furthermore, (1R,3S)- and (1S,3S)-MTCdiC demonstrated the ability to inhibit LPS-induced nitrite production at concentrations ranging from 50–100 mmol/L and 10–100 mmol/L, respectively.

Notably, THbCs (short for thiosulfinates, a class of compounds) were absent in both raw garlic and various processed garlic products. Surprisingly, they only came into existence and their concentrations rose as part of the inherent aging progression in garlic extraction. These discoveries strongly indicate that THbCs, originating during this natural aging process, manifest themselves as robust antioxidants within aged garlic extract. Consequently, these compounds offer great potential for future applications in the prevention of diseases linked to oxidative harm (Table 7.2 and Figure 7.6).

The scavenging capacities of DPPH (diphenyl picrylhydrazyl) radical at 4 μg/mL concentration were as follows: VAN (58.30 ± 1.65%), BEN (56.43 ± 1.86%), ANI (39.78 ± 1.35%), and ACE (39.23 ± 1.22%) (Table 7.3). The inhibition order ranked is quercetin > BHT > VAN > BEN > ANI > ACE. Significantly, VAN exhibited higher inhibition ($p < 0.05$) compared to BEN, while BEN demonstrated superior DPPH inhibition compared to ANI and ACE. This activity can be ascribed to its hydrogen-donating capability (Herraiz and Galisteo, 2004), with its β-carboline core acting as an efficient free radical inhibitor, promptly reacting with DPPH radicals (Prakash et al. 2012).

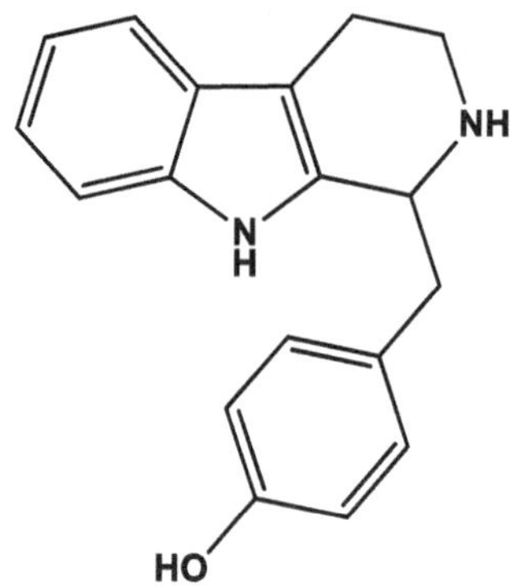

FIGURE 7.7 *dl*-1-(4-hydroxybenzyl)-1,2,3,4-tetrahydro-β-carboline.

Notably, the substances BEN (6-methoxy-1-phenyl-2,3,4,9-tetrahydro-1H-pyrido[3,4-b]indole), ANI (6-methoxy-1-(4-methoxyphenyl)-2,3,4,9-tetrahydro-1H-pyrido[3,4-b]indole), ACE (6-methoxy-1-methyl-2,3,4,9-tetrahydro-1H-pyrido [3,4-b]indole), and VAN (2-methoxy-4-(6-methoxy-2,3,4,9-tetrahydro-1H-pyrido[3,4-b]indol-1-l)phenol) exhibited the most significant antioxidant properties, as evidenced by their IC50 values of 1.74 ± 0.02 and 2.92 ± 0.04 ppm, correspondingly. Conversely, ACE and ANI's IC50 values could not be determined because they displayed less than 50% inhibition even at a concentration of 4 ppm, although antioxidant activity of quercetin (IC50 = 0.20 ± 0.00 ppm) and BHT (IC50 = 0.28 ± 0.00 ppm) was higher than that of VAN (as shown in Table 7.3). Earlier studies have reported IC50 values for various fruits such as orange (IC50 = 5.40 ± 1.30 ppm), guava (IC50 = 2.11 ± 0.63 ppm), and star fruit (IC50 = 3.80 ± 2.10 ppm) (Duh 1998).

In a study by Qais and Jahan, a new tetrahydro-β-carboline (i.e., dl-1-(4-hydroxybenzyl)-1,2,3,4-tetrahydro-β-carboline) was synthesized using the Bischler-Napieralsky reaction (Qais and Jahan 2020). This synthesis involved the usage of $ZnCl_2$ with $POCl_3$ as a dehydrating agent. The product underwent evaluation for its antioxidant and analgesic activities (Figure 7.7). The antioxidant activity of the compound was assessed through the DPPH assess, and the IC50 value obtained was 48.42 ppm, while the standard ascorbic acid showed a significantly lower IC50 value of 2.83 ppm in comparison (Table 7.4).

7.3.2 β-Carboline Alkaloids

Experiments and computer simulations determined that different substances (i.e., harmine, harmaline, harmalol, harmane, and 1,2,3,4-tetrahydroharmane-3-carboxylic acid) can act as antioxidants.

In the laboratory-based research, harmine exhibited the highest hydrogen peroxide scavenging activity at 27.63 ± 1.74%, notably surpassing the performance of ascorbic acid (8.02 ± 0.58%). 1,2,3,4-tetrahydroharmane-3-carboxylic acid also displayed substantial hydrogen peroxide scavenging activity, reaching 11.52 ± 0.82%, and both compounds outperformed the reference standard, ascorbic acid. Additionally,

TABLE 7.4
Antioxidant activity (IC_{50}) of synthesized compound and standard ascorbic acid

Absorbance of blank	Conc. of Standard/test sample (ug/ml)	Absorbance of standard	Aborbance of test sample	Inhibition of standard (%)	Inhibition of test sample (%)	IC_{50} (ug/ml) of standard	IC_{50} (ug/ml) of test sample	P value
0.732	500	0.017	0.046	97.70	93.70	2.83	48.42	0.0169
	250	0.018	0.102	97.50	86.00			
	125	0.020	0.280	97.30	61.70			
	62.5	0.022	0.376	96.90	48.60			
	31.25	0.029	0.521	96.90	28.80			
	15.625	0.054	0.565	92.60	22.80			
	7.813	0.071	0.625	90.30	14.60			
	3.906	0.315	0.634	56.90	13.40			
	1.953	0.505	0.660	31.00	09.80			
	0.977	0.659	0.688	9.90	06.00			

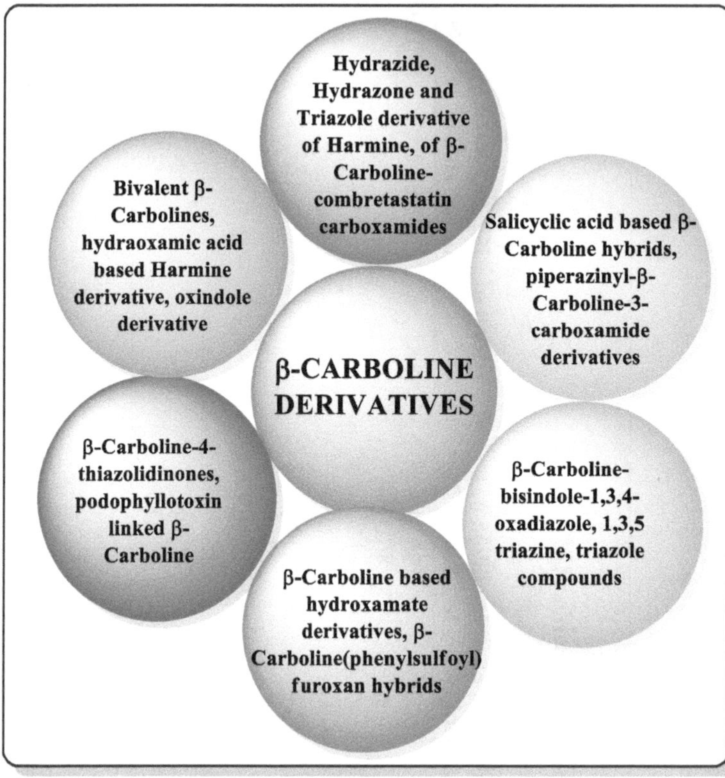

FIGURE 7.8 Examples of β-carboline derivatives.

harmalol displayed outstanding ABTS (2,2-azino-bis-3-ethylbenzothiazoline-6-sulphonic acid) radical scavenging potential, measuring 371.15 ± 1.80 μg TE/mg, while harmaline followed closely at 168 ± 1.30 μg TE/mg. Notably, harmalol showed the highest antioxidant activity in the FRAP test, with a value of 11.30 ± 0.01 μg TE/mg. Harmaline and 1,2,3,4-tetrahydroharmane-3-carboxylic acid also demonstrated significant FRAP values, measuring 5.12 ± 0.38 μg TE/mg and 5.01 ± 0.14 μg TE/mg, respectively.

The medicinally potent aromatic β-carbolines, including β-carboline-1 proponoic acid, Canthi-6-one, 1-Methoxycanthione, 6-Methoxycanthione, Eudistomin-I, Eudistomin-H, Harmane, Harmine, Harmol, and Norharmane, are exemplified in Figure 7.8 (Rajesh and Murugan 2019; Gabriel et al. 2020; Dejos et al. 2014; Xu et al. 2001; Giorgio et al. 2004; Nenaah 2010; Alomar et al. 2013; Ishida et al. 2001; Hayashi et al. 1977; Madle et al. 1981; Hudson et al. 1986: Abe et al. 2011; Mita et al. 1984; Kühn-Velten 1993). Pharmacologically active fully aromatic bet β-carboline marine products, such as Fascalpysin, Hyrtioerectin-A, Plakortamine-A, Plakortamine-B, Plakortamine-C, Plakortamine-D, Manzamine-A, 6-Deoxymanzamine-X, 8-Hydroxymanzamine-A, and 8-Methoxymanzamine-A, are also illustrated in Figure 7.7 (Youssef 2005; Sandler et al. 2002; Ashok et al. 2014; Ashok et al. 2015; Charan et al. 2002).

7.3.3 Dihydro β-Carbolines

3,4-dihydro β-carboline plant products, specifically Harmaline and Harmalol (Sayed et al. 2001; Kaminsky et al. 1991; Chan-Bacab et al. 2001), are pharmacologically superactive. In the marine realm, examples include the pharmacologically active 3,4-dihydro β-carbolines 3,4-dihydro manzamine and Xestomanzamine. β-carboline plant products such as Ajmalicine, Harmalacidine, Pegaharmaline-A, Pegaharmaline-B, Pegaharmine-D, Peganumine-A, Sacleuximine-A, Reserpine, Tangutorine, and Vincaamine (Gendy et al. 2012; Stavrinides et al. 2015; Lamchouri et al. 2013; Wang et al. 2016 and Wang et al. 2014) are also active. Additionally, fully aromatic β-carboline marine products, including Bengacarboline, Callophycin-A, Hytioerectin-B, Maganedin-A, and (+)-Milnamide-C, are also present.

Natural β-carbolines exhibit a wide range of biological effects, encompassing antineoplastic, antimicrobial, antimalarial, antileishmanial, anti-HIV, antitrypanosomal, and antitoxoplasmal properties. Notably, the presence of tetrahydro-β-carbolines in fruit extract suggests consumption of these substances from external sources in one's diet. Once absorbed, these alkaloids have the potential to build up in tissues, leading to their biological activation. Previous researchers have predominantly explored the impact of β-carbolines on the CNS, including their part in inhibiting serotonin uptake, MAO inhibition, and their binding to benzodiazepine-GABA or imidazoline receptors (Samita et al. 2017; Buckholtz 1980; Airaksinen et al. 1981; Glennon et al. 2000), as well as potential toxicological effects (Husbands et al. 2001). The present research, on the other hand, centers on exploring the possible role of tetrahydro-β-carbolines as agents with antioxidant properties and the ability to scavenge free radicals.

β-carbolines present in fruits and juices proved to be highly efficient in combating radicals. In the test, their antioxidant capabilities surpassed those of both ascorbic acid and Trolox. Tetrahydro-β-carbolines possess an indole ring, which could aid in the creation of the indolyl cation or a neutral radical through a single electron transfer mechanism when they function as radical scavengers. A comparable process has been proposed for other indole-based antioxidants, such as melatonin (Turjanski et al. 1998; Gearhart et al. 2002).

The indolyl radical may undergo further oxidation to form aromatic β-carbolines, like harman and norharman, particularly in the case of carboxylic acids derivatives of β-carboline or it may break down into unknown compounds (Poeggeler et al. 1994). These alkaloids, whether obtained from the diet or formed endogenously, can be absorbed and accumulate in tissues and bodily fluids, potentially serving as antioxidants by safeguarding against radicals generated during oxidative stress. Recent suggestions have also pointed to the possible role of endogenous tetrahydro-β-carbolines as antioxidants (Herraiz and Galisteo 2002). However, their contribution to the overall measured antioxidant activity of fruits and fruit juices is expected to be minimal, considering their lower relative concentration compared to vitamins, carotenoids, and phenols. It is important to emphasize that while tetrahydro-β-carbolines have shown their ability to scavenge free radicals, there is still a chance that both dietary and naturally occurring β-carbolines may transform into mutagens or internal toxins as they accumulate in tissues (Pari et al. 2000). Additional research is needed to thoroughly elucidate the biological functions and effects of these alkaloids found in mammals and our diet.

7.3.4 Synthetic β-Carbolines

A series of β-carboline compounds were synthesized, beginning with compound GWC22, and their ability to inhibit lipid peroxidation, a process linked to oxidative damage, was examined. The oxidation of LDL (low-density lipoprotein) was induced using either $CuSO_4$ or 2,20-azobis(2-amidinopropane) dihydrochloride (AAPH). The compounds were also tested for their protective effects against cytotoxicity in bovine aortic endothelial cells (BAECs) by measuring lactate dehydrogenase (LDH) activity and cellular viability, which was assessed through mitochondrial activity using MTT (3-(4,5-dimethylthiazol-2-yl)-2,5-diphenyltetrazolium bromide). Most of the compounds exhibited higher antioxidant activity compared to the GWC22 derivative ($R = 1.6$ for 5 mM $CuSO_4$). The phenolic and benzyloxy derivatives, in particular, demonstrated exceptional antioxidant activity, with R ratios ranging from 1.9 to 2.8 for 1 mM $CuSO_4$. These compounds not only displayed protective effects but also significantly increased cell viability (Albores et al. 1990).

Cervical cancer ranks among the most prevalent cancers specific to the female gender globally, with a high mortality rate (Hadjaz et al. 2011). Its primary causes include human papillomavirus (HPV) infection, smoking, and extended use of oral contraceptives. Initially, cervical cancer may not exhibit noticeable symptoms. However, as the disease progresses, symptoms like abnormal vaginal bleeding (Small et al. 2017), vaginal discharge (Eleje et al. 2015), pelvic pain, or discomfort during sexual intercourse become apparent (Mwaliko et al. 2021). While vaccination has proven to be successful in averting cervical cancer, it continues to pose a substantial challenge in numerous developing and developed regions where effective screening and treatment methods are lacking. Therefore, the creation of affordable and powerful medications for cervical cancer is of paramount significance.

7.4 RECENT ADVANCES

Lately, there has been an increasing awareness of the benefits linked to utilizing active components extracted from herbal remedies, emphasizing their advantages of limited side effects and affordability. Among these compounds, Ethyl β-carboline-3-carboxylate (β-CCE), a β-carboline alkaloid sourced from P. quassioides, has garnered significant attention. Research has highlighted the multifaceted therapeutic properties of β-carboline alkaloids, demonstrating their efficacy in treating a range of conditions, including pulmonary fibrosis in mice (Alappattu 2013), depression (Cui et al. 2019), bacterial infections (Ferraz et al. 2019), inflammatory diseases (Liu et al. 2020), and various forms of cancer (Chen et al. 2022). The anticancer potential of β-carboline alkaloids has been a subject of extensive investigation, encompassing their utility in addressing lung (Ahmad et al. 2020), liver (Abe et al. 2011), gastric (Zhao et al. 2018), colon (Xi et al. 2019), and ovarian cancer (Li et al. 2019). Particularly, β-CCE distinguishes itself as one of the most potent β-carboline alkaloids.

While a substantial portion of research has historically concentrated on the use of β-CCE to stimulate spontaneous hippocampal discharges in rats, lower convulsion induction thresholds, and reduce convulsion incubation periods (Gao et al. 2017;

Petersen et al. 1982; Thiébot et al. 1988; Barrett et al. 1985; Podhorna and Franklin 2000), relatively less attention has been directed toward exploring the anticancer properties of β-CCE. Reactive oxygen species (ROS) play vital roles in controlling cellular growth, differentiation, and development in a range of physiological and pathological situations. When present in small amounts, ROS facilitate functions like cell proliferation, differentiation, and other natural processes. Yet, an excessive increase in ROS levels surpassing the typical threshold causes internal oxidative disruptions, resulting in harm to proteins and lipids, genetic instability, and ultimately contributing to the development of cancer. (Diebold and Chandel 2016; Radak et al. 2011; Tian et al. 2016).

ROS also have a pivotal function in tumor cells, as they stimulate irregular cancer cell proliferation and differentiation, expedite cancer cell spread and the formation of new blood vessels (angiogenesis), and impede the process of apoptosis (cell death) (Tian et al. 2016; Qiu et al. 2019). Elevated intracellular ROS levels disrupt antioxidant defense mechanisms dependent on enzymes like superoxide dismutase (SOD), catalase (CAT), glutathione peroxidase (GPxs), and peroxidase (Prxs). These enzymes typically prevent excessive ROS formation, maintaining redox balance by eliminating ROS (Ismail et al. 2019). Excessive ROS in cancer cells can hinder antioxidant mechanisms and contribute to disease progression. Nevertheless, an excess of ROS production, inadequate removal, and a deficiency of antioxidants can give rise to the buildup of ROS, culminating in oxidative stress and the impairment of diverse cellular organelles. In consequence, ROS are critical for maintaining cellular homeostasis and normal organelle function. Additionally, the presence of various antioxidant enzymes in cells plays a role in regulating redox homeostasis, preventing mitochondria-dependent apoptosis, and mitigating endoplasmic reticulum stress-related signaling pathways that promote cell apoptosis (Cheng et al. 2015; Zhang et al. 2015). Therefore, regulating ROS production levels proves to be an efficacious strategy for treating cancer.

Within the realm of tumor cells, the mitogen-activated protein kinase (MAPK) pathway stands as a key player highly sensitive to oxidative stress, governing processes like gene expression, cell survival, apoptosis, and differentiation. The phosphorylation levels of ERK (extracellular regulated Kinase), JNK (c-Jun N-terminal kinase), and P38 MAPK (mitogen-activated protein kinase) serve as indicators of how intracellular oxidative stress impacts these cells. This research delves into the influence of β-CCE, extracted from P. quassioides, on the apoptosis mechanism within SiHa cervical cancer cells, with the goal of advancing novel treatments for cervical cancer and establishing a theoretical foundation for more potent therapeutics. The integration of therapeutic strategies involving β-CCE may prove to be effective.

β-CCE, an effective biological component, is found widely in P. quassioides, human urine, and the brain (Lee et al. 2021; Braestrup et al. 1980). Since the 1880s, neuro-psychopharmacological studies have observed the beneficial impacts of β-CCE in managing panic disorders, anxiety, and various social behaviors (Skolnick et al. 1984). As a benzodiazepine-derived antagonist, β-CCE is often compared to nordiazepam and has regulatory effects on neurological conditions in humans (Witkin et al. 1986). However, its effects on cancer have remained relatively uncharted. This study uncovers that β-CCE has the potential to impair mitochondria, elevate ROS levels, and trigger apoptosis in SiHa cells. Moreover, it sheds light on the fact that ROS-p38/MAPK signaling plays a pivotal role in governing apoptosis in SiHa cells. Consequently, this research primarily centers on assessing the impact of β-CCE toxicity on cervical cancer SiHa cells.

ROS, acting as second messengers, have the capacity to influence proteins, lipids, and a range of cellular functions, both physiological and pathological (Dandekar et al. 2017; Kattoor et al. 2017). This study demonstrates that β-CCE can induce apoptosis in SiHa cells through the accumulation of ROS and investigates the mechanism behind this apoptosis. It was found that apoptosis is triggered by the accumulation of ROS induced by β-CCE in SiHa cells. Mitochondria, the main producers of energy in cells, are also the primary source of ROS (Sas et al. 2018). Nonetheless, an overabundance of ROS can result in harm to DNA, lipids, and proteins, as well as changes in mitochondrial membrane permeability, ultimately leading to mitochondrial dysfunction. To uphold cellular equilibrium, diverse antioxidant enzymes within the mitochondria manage the redox balance (Sztretye et al. 2019; Bjørklund and Chirumbolo 2017). This research showcases that β-CCE has the capacity to markedly diminish intracellular SOD levels while augmenting cellular MDA production, signifying its influence on cellular antioxidant mechanisms.

Additionally, β-CCE is demonstrated to impact mitochondrial function by instigating the discharge of cytochrome-C, apoptosis-inducing factors, endonuclease G, and other mitochondrial proteins into the outer mitochondrial membrane. Cytochrome-C interacts with apoptotic protease activator 1 to create an apoptotic complex, setting in motion the activation of Cas9, ultimately resulting in cellular harm (Sinha et al. 2013). Furthermore, alterations in ATP production and LDH levels contribute to the promotion of apoptosis. Kinases that are susceptible to oxidative stress include extracellular signal-regulated kinases ½ (ERK1/2), c-Jun amino (N)-terminal kinases ½/3 (JNK1/2/3), p38 isoforms (α, β, γ, and δ), and ERK5 (Cargnello and Roux 2011). Upon stimulation of SiHa cells with β-CCE, a substantial rise in intracellular ROS levels was observed, resulting in the activation of the MAPK signaling pathway. Among these kinases, p38 demonstrated heightened sensitivity to oxidative stress triggers compared to the others (Saeki et al. 2002). The application of p38 phosphorylation inhibitors and NAC (a ROS scavenger) significantly reduced the occurrence of apoptosis compared to ERK and JNK inhibitors. A previous investigation has documented that P. quassioides prompts apoptosis in SiHa cervical cancer cells by triggering the p38/MAPK signaling pathway. This provides a solid theoretical basis for β-CCE, a monomeric compound derived from P. quassioides, inducing apoptosis in SiHa cells (Gong et al. 2020). It is crucial to acknowledge that this study comes with certain limitations. We did not provide direct molecular evidence illustrating the interaction between p38 MAPK and ROS, nor did we thoroughly elucidate the mitochondrial damage induced by β-CCE in SiHa cell apoptosis. These aspects necessitate further investigation at the molecular level, with an emphasis on uncovering the underlying mechanisms in greater detail.

The induction of apoptosis in SiHa cervical cancer cells by β-CCE is driven by the accumulation of both cytoplasmic and mitochondrial ROS levels. Simultaneously, this process triggers the p38/MAPK and mitochondrial-dependent pathways for apoptosis. Our research findings hold the potential to expand the utility of β-CCE in scientific studies and introduce new avenues for cervical cancer treatment.

Reverse transcriptase plays a central role in the process of viral reverse transcription. Studies have shown its potential to impede the growth of cancer cells and reduce the activity of HIV-1 RT. An antifungal protein derived from P. harmala seed extracts underwent testing to evaluate its impact on the proliferation of cancer cells, based on a cytotoxicity curve (Ma et al. 2013). Using an HIV-1 RT kit, the inhibitory effect on HIV-1 RT was measured, revealing that this novel antifungal protein could hinder the growth of esophageal cancer, cervical cancer, gastric cancer, and melanoma cells, as well as diminish the activity of HIV-1 RT.

Nucleic acid, encompassing both deoxyribonucleic acid (DNA) and ribonucleic acid (RNA), serves as the central information carrier within cells. In a specific experiment, PhAMP underwent incubation with Pseudomonas aeruginosa and Staphylococcus aureus, being continuously shaken for 16 hours. Subsequently, the real-time PCR method was utilized to assess the gene expression levels associated with biofilm formation. The findings demonstrated a significant reduction in the expression of genes responsible for the flagellum (flgK), fimbriae protein (pilA), and fimbriae (cupA1) in P. aeruginosa. Conversely, the expressions of the capsular polysaccharide synthesis gene (CPS5) and the intercellular adhesion gene (icaA) in S. aureus were upregulated (Mirza et al. 2019). The PI3K/Akt/GSK-3β/ROS/eIF2B pathway promoted breast cancer growth and metastasis via suppression of NK cell cytotoxicity and tumor cell susceptibility (Jin et al. 2019)

In another instance, the extract derived from Peganum harmala seeds was subjected to cultivation alongside Ralstonia solanacearum inocula in a broth medium. This experiment led to several notable observations, including cell wall thickening, cytoplasmic disarray, extensive cellular damage, and genome condensation, which were documented using transmission electron microscopy (Shaheen et al. 2020).

Furthermore, in an independent research endeavor, a blend of eight plant extracts was combined with yeast extract sucrose (YES), and the expression levels of genes associated with aflatoxin B1 synthesis were investigated using RT-PCR. The results demonstrated a reduction in aflatoxin B1 production, which corresponded with the decreased expression or inhibition of genes such as aflR, aflM, and aflP by the plant extracts, including P. harmala (Safari et al. 2020).

Furthermore, it was observed that harmine could impact virus replication. When Madin-Darby bovine kidney (MDBK) cells infected with Bovine herpesvirus-1 (BoHV-1) were exposed to harmine at diverse steps, a significant reduction in viral production was detected. This effect was particularly pronounced during both the early and later stages of viral replication (Jiang et al. 2017).

7.5 CONCLUSION

In conclusion, the pharmacologically active compounds derived from Peganum harmala and other natural sources, including β-carbolines and related alkaloids, have demonstrated diverse biological activities and potential therapeutic applications. These compounds have shown promise in fields such as cancer treatment, antimicrobial activities, inhibition of biofilm-related genes, and modulation of aflatoxin biosynthesis. Additionally, these compounds have exhibited antioxidant properties, influencing the redox balance within cells and impacting the development of various diseases.

Furthermore, the studies discussed emphasize the role of ROS in both normal cellular processes and pathological conditions, particularly in cancer. The induction of apoptosis in cervical cancer SiHa cells by β-carboline-3-carboxylate (β-CCE) from P. quassioides highlights the potential of these compounds as effective anticancer agents. The involvement of ROS in promoting cell differentiation, proliferation, and various physiological functions, as well as their role in intracellular oxidation imbalance, underscores the significance of regulating ROS levels in cancer therapy.

The studies also shed light on the involvement of oxidative stress-sensitive kinases, such as the p38 mitogen-activated protein kinase (MAPK), in mediating apoptosis in response to increased ROS levels. This research contributes to a better understanding of the mechanisms behind β-CCE-induced apoptosis and its potential as a novel therapeutic approach for cervical cancer.

In summary, the findings discussed in these studies expand our knowledge of the multifaceted properties of β-carbolines and related alkaloids, making them promising candidates for future research and drug development in various fields, including oncology, microbiology, and antioxidant therapy. However, further investigations are necessary to explore their precise molecular interactions and potential clinical applications fully.

7.6 FUTURE ASPECTS

The future prospects of β-carbolines as antioxidants encompass the identification of novel compounds, a deeper understanding of their underlying mechanisms, rigorous clinical validation for therapeutic use, determination of optimal dosages and efficient delivery systems, exploration of synergistic combinations with other antioxidants, comprehensive safety assessments, regulatory approval pathways, personalized antioxidant therapies, formulation of nutraceuticals, interdisciplinary research collaboration, investigation into their role in aging and age-related diseases, and potential environmental applications. These avenues of research are expected to not only expand the repertoire of effective antioxidants but also provide tailored approaches to combat oxidative stress-related diseases and promote healthy aging, contributing to advancements in overall health and environmental protection.

REFERENCES

Abe, A., Yamada, H., Moriya, S., & Miyazawa, K. (2011). The β-carboline alkaloid harmol induces cell death via autophagy but not apoptosis in human non-small cell lung cancer A549 cells. *Biological & Pharmaceutical Bulletin*, 34(8), 1264–1272. https://doi.org/10.1248/bpb.34.1264

Agrawal, R., Sethiya, N.K., & Mishra, S.H. (2013). Antidiabetic activity of alkaloids of Aerva lanata roots on streptozotocin-nicotinamide induced type-II diabetes in rats. *Pharmceutical Biology*, 51(5):635–642. doi: 10.3109/13880209.2012.761244

Ahmad, I., Fakhri, S., Khan, H., Jeandet, P., Aschner, M., & Yu, Z.L. (2020). Targeting cell cycle by β-carboline alkaloids in vitro: Novel therapeutic prospects for the treatment of cancer. *Chemical Biology Interactions,* 330, 109229. DOI: 10.1016/j.cbi.2020.109229

Airaksinen, M.M., & Kari, I. (1981). Beta-carbolines, psychoactive compounds in the mammalian body. Part I: Occurrence, origin, and metabolism. *Medical Biology*, 59(1), 21–34.

Alappattu, M.J. (2013). Pain and psychological outcomes after rehabilitative treatment for a woman with chronic pelvic pain with stage III cervical cancer: A case report. *Journal of Women's Health Physical Therapy*, 37(3), 97–102. DOI: 10.1097/JWH.0000000000000004

Albores, R., Neafsey, E.J., Drucker, G., Fields, J.Z., & Collins, M.A. (1990). Mitochondrial respiratory inhibition by N-methylated beta-carboline derivatives structurally resembling N-methyl-4-phenylpyridine. *Proceedings of the National Academy of Sciences of the United States of America*, 87(23), 9368–9372. https://doi.org/10.1073/pnas.87.23.9368

Alomar, M.L., Rasse-Suriani, F.A., Ganuza, A., Cóceres, V.M., Cabrerizo, F.M., & Angel, S.O. (2013). In vitro evaluation of β-carboline alkaloids as potential anti-Toxoplasma agents. *BMC Research Notes,* 6, 193. https://doi.org/10.1186/1756-0500-6-193.

Ashok, P., Ganguly, S., & Murugesan, S. (2014). Manzamine alkaloids: Isolation, cytotoxicity, antimalarial activity, and SAR studies. *Drug Discovery Today*, 19(11), 1781–1791. https://doi.org/10.1016/j.drudis.2014.06.010

Ashok, P., Lathiya, H., & Murugesan, S. (2015). Manzamine alkaloids as antileishmanial agents: A review. *European Journal of Medicinal Chemistry,* 97, 928–936. https://doi.org/10.1016/j.ejmech.2014.07.006

Baldwin, D.S., & Tiwari, N. (2015). Anxiolytic and hypnotic drugs. *Fundamentals of Clinical Psychopharmacology*, 1501, 125.

Barrett, J.E., Brady, L.S., Witkin, J.M., Cook, J.M., & Larscheid, P. (1985). Interactions between the benzodiazepine receptor antagonist Ro 15-1788 (flumazepil) and the inverse agonist beta-CCE: Behavioral studies with squirrel monkeys. *Life Sciences*, 36(14), 1407–1414. DOI: 10.1016/0024-3205(85)90047-5

Bjørklund, G., & Chirumbolo, S. (2017). Role of oxidative stress and antioxidants in daily nutrition and human health. *Nutrition*, 33, 311–321. DOI: 10.1016/j.nut.2016.07.018

Braestrup, C., Nielsen, M., & Olsen, C.E. (1980). Urinary and brain beta-carboline-3-carboxylates as potent inhibitors of brain benzodiazepine receptors. *Proceedings of the National Academy of Sciences of the United States of America*, 77(4), 2288–2292. DOI: 10.1073/pnas.77.4.2288

Buckholtz, N.S. (1980). Neurobiology of tetrahydro-beta-carbolines. *Life Sciences*, 27(11), 893–903. https://doi.org/10.1016/0024-3205(80)90098-3

Cargnello, M., & Roux, P.P. (2011). Activation and function of the MAPKs and their substrates, the MAPK-activated protein kinases. *Microbiology and Molecular Biology Reviews*, 75(1), 50–83. DOI: 10.1128/MMBR.00031-10

Chan-Bacab, M.J., & Peña-Rodríguez, L.M. (2001). Plant natural products with leishmanicidal activity. *Natural Product Reports*, 18(6), 674–688. https://doi.org/10.1039/b100455g

Charan, R.D., McKee, T.C., Gustafson, K.R., Pannell, L.K., & Boyd, M.R. (2002). Thorectandramine, a novel β-carboline alkaloid from the marine sponge Thorectandra sp. *Tetrahedron Letters,* 43, 5201–5204.

Chen, Y.F., Kuo, P.C., Chan, H.H., Kuo, I.J., Lin, F.W., Su, C.R., Yang, M.L., Li, D.T., & Wu, T.S. (2022). β-carboline alkaloids from Stellaria dichotoma var. lanceolata and their anti-inflammatory activity. *Journal of Natural Products*, 73(12), 1993–1998. DOI: 10.1021/np1003627

Cheng, D., Zhao, L., Xu Y, Ou, R., Li, G., Yang, H., & Li, W. (2015). K-Ras promotes the survival of non-small lung cancer cells by cooperating with sirtuin 1 and p27 under ROS stimulation. *Tumour Biology*, 36(9), 7221–7232. DOI: 10.1007/s13277-015-3429-8

Cui, Y., Jiang, L., Yu, R., Shao, Y., Mei, L., & Tao, Y. (2019). β-carboline alkaloids attenuate bleomycin induced pulmonary fibrosis in mice through inhibiting NF-kb/p65 phosphorylation and epithelial-mesenchymal transition. *Journal of Ethnopharmacology,* 243, 112096. DOI: 10.1016/j.jep.2019.112096

Dandekar, A., Mendez, R., & Zhang, K. (2017). Cross-talk between ER stress, oxidative stress, and inflammation in health and disease. *Methods in Molecular Biology*, 1292, 205–214. DOI: 10.1007/978-1-4939-2522-3_15

Dejos, C., Voisin, P., Bernard, M., Regnacq, M., & Berges, T. (2014). Canthin-6-one displays antiproliferative activity and causes accumulation of cancer cells in the G2/M phase. *Journal of Natural Products,* 77, 2481–2487.

De Oliveira Figueiredo, P., Perdomo, R.T., Garcez, F.R., de Fatima Cepa Matos, M., de Carvalho, J.E., & Garcez, W.S. (2014). Further constituents of Galianthe thalictroides (Rubiaceae) and inhibition of DNA topoisomerases I and IIα by its cytotoxic β-carboline alkaloids. *Bioorganic & Medicinal Chemistry Letters*, Mar 1;24(5):1358–1361. DOI: 10.1016/j.bmcl.2014.01.039

Di Giorgio, C., Delmas, F., Ollivier, E., Elias, R., Balansard, G., & Timon-David, P. (2004). In vitro activity of the beta-carboline alkaloids harmane, harmine, and harmaline toward parasites of the species Leishmania infantum. *Experimental Parasitology,* 106(3-4), 67–74. https://doi.org/10.1016/j.exppara.2004.04.002

Diebold, L., & Chandel, N.S. (2016). Mitochondrial ROS regulation of proliferating cells. *Free Radical Biology and Medicine,* 100, 86–93. DOI: 10.1016/j.freeradbiomed.2016.04.198

Domenech-Carbo, A., Rodrigo, R., Maréchal, J.D., Poupon, E., Fournet, A., Figadère, B., & Cebrián-Torrejón, G. (2017). Bioelectrochemical monitoring of soluble guanylate cyclase inhibition by the natural β-carboline canthin-6-one. *Journal of Molecular Structure,* 1134, 661–667. DOI: https://doi.org/10.1016/j

Duh, P.D. (1998). Antioxidant activity of burdock (Arctium lappa Linne): Its scavenging effect on free radicals and active oxygen. *Journal of the American Oil Chemists' Society*, 75(4), 455–461.

El Gendy, M.A., Soshilov, A.A., Denison, M.S., & El-Kadi, A.O. (2012). Harmaline and harmalol inhibit the carcinogen-activating enzyme CYP1A1 via transcriptional and posttranslational mechanisms. *Food and Chemical Toxicology: An International Journal Published for the British Industrial Biological Research Association*, 50(2), 353–362. https://doi.org/10.1016/j.fct.2011.10.052

El Sayed, K.A., Kelly, M., Kara, U.A., Ang, K.K., Katsuyama, I., Dunbar, D.C., Khan, A.A., & Hamann, M.T. (2001). New manzamine alkaloids with potent activity against infectious diseases. *Journal of the American Chemical Society*, 123(9), 1804–1808. https://doi.org/10.1021/ja002073o

Eleje, G.U., Eke, A.C., Igberase, G.O., Igwegbe, A.O., & Eleje, L.I. (2015). Palliative interventions for controlling vaginal bleeding in advanced cervical cancer. *Cochrane Database of Systematic Reviews*, (5). DOI: 10.1002/14651858.CD011000.pub2

Ferraz, C.A.A., de Oliveira Júnior, R.G., Picot, L., da Silva Almeida, J.R.G., Nunes, X.P. (2019). Pre-clinical investigations of β-carboline alkaloids as antidepressant agents: A systematic review. *Fitoterapia,* 137, 104196. DOI: 10.1016/j.fitote.2019.104196

Gabriel, R.S., Amaral, A.C.F., Lima, I.C., Cruz, J.D., Garcia, A.R., Souza, H.A.S., Rodrigues, I.A. (2020). β-Carboline-1-propionic acid alkaloid: Antileishmanial and cytotoxic effects. *Revista Brasileira de Farmacognosia.* 29, 755–762. https://doi.org/10.1016/j.bjp.2019.08.002

Gao, C., Yi, X., Huang, R., Yan, F., He, B., & Chen, B. (2013). Alkaloids from corals. *Chemistry & Biodiversity*, 10(8), 1435–1447. DOI: 10.1002/cbdv.201100276. PMID: 23939792

Gao, J., Zhu, H., Wan, H., Zou, X., Ma, X., & Gao, G. (2017). Harmine suppresses the proliferation and migration of human ovarian cancer cells through inhibiting ERK/CREB pathway. *Oncology Reports*, 38(5), 2927–2934. PMID: 28901502. DOI: 10.3892/or.2017.5952

Gearhart, D.A., Neafsey, E.J., & Collins, M.A. (2002). Phenylethanolamine N-methyltransferase has beta-carboline 2N-methyltransferase activity: Hypothetical relevance to Parkinson's disease. *Neurochemistry International*, 40(7), 611–620. https://doi.org/10.1016/s0197-0186(01)00115-2

Glennon, R.A., Dukat, M., Grella, B., Hong, S., Costantino, L., Teitler, M., Smith, C., Egan, C., Davis, K., & Mattson, M.V. (2000). Binding of beta-carbolines and related agents at serotonin (5-HT(2) and 5-HT(1A)), dopamine (D(2)) and benzodiazepine receptors. *Drug and Alcohol Dependence*, 60(2), 121–132. https://doi.org/10.1016/s0376-8716(99)00148-9

Goebel, F. (1841). Über das harmalin. *European Journal of Organic Chemistry*, 38 363–366.

Gong, Y.X., Liu, Y., Jin, Y.H., Jin, M.H., Han, Y.H., Li, J., Shen, G.N., Xie, D.P., Ren, C.X., Yu, L.Y., Lee, D.S., Kim, J.S., Jo, Y.J., Kwon, J., Lee, J., Park, Y.H., Kwon, T., Cui, Y.D., & Sun, H.N. (2020). Picrasma quassioides extract elevates the cervical cancer cell apoptosis through ROS-mitochondrial axis activated p38 MAPK signaling pathway. *In Vivo*, 34(4), 1823–1833.

Gooyit, M., Tricoche, N., Javor, S., Lustigman, S., & Janda, K.D. (2015). Exploiting the polypharmacology of β-carbolines to disrupt O. volvulus molting. *ACS Medicinal Chemistry Letters*, 6(3), 339–343. DOI: 10.1021/ml500516r

Götz, M.E., Künig, G., Riederer, P., & Youdim, M.B. (1994). Oxidative stress: Free radical production in neural degeneration. *Pharmacology & Therapeutics*, 63(1), 37–122. https://doi.org/10.1016/0163-7258(94)90055-8

Hadjaz, F., Besret, S., Martin-Nizard, F., Yous, S., Dilly, S., Lebegue, N., Chavatte, P., Duriez, P., Berthelot, P., & Carato, P. (2011). Antioxidant activity of β-carboline derivatives in the LDL oxidation model. *European Journal of Medicinal Chemistry*, 46(6), 2575–2585. https://doi.org/10.1016/j.ejmech.2011.03.048

Hayashi, K., Nagao, M., & Sugimura, T. (1977). Interactions of norharman and harman with DNA. *Nucleic Acids Research*, 4(11), 3679–3685. https://doi.org/10.1093/nar/4.11.3679

Herraiz, T., & Galisteo, J. (2002). Tetrahydro-beta-carboline alkaloids that occur in foods and biological systems act as radical scavengers and antioxidants in the ABTS assay. *Free Radical Research*, 36(8), 923–928. https://doi.org/10.1080/1071576021000005762

Herraiz, T., & Galisteo, J. (2003). Tetrahydro-beta-carboline alkaloids occur in fruits and fruit juices. Activity as antioxidants and radical scavengers. *Journal of Agricultural and Food Chemistry*, 51(24), 7156–7161. https://doi.org/10.1021/jf030324h

Herraiz, T., & Galisteo, J. (2004). Endogenous and dietary indoles: A class of antioxidants and radical scavengers in the ABTS assay. *Free Radical Research*, 38(3), 323–331. https://doi.org/10.1080/10611860310001648167.

Hudson, J.B., Graham, E.A., Fong, R., Hudson, L.L., & Towers, G.H. (1986). Further studies on the antiviral activity of harmine, a photoactive beta-carboline alkaloid. *Photochemistry and Photobiology*, 44(4), 483–487. https://doi.org/10.1111/j.1751-1097.1986.tb04696.x

Husbands, S.M., Glennon, R.A., Gorgerat, S., Gough, R., Tyacke, R., Crosby, J., Nutt, D.J., Lewis, J.W., & Hudson, A.L. (2001). Beta-carboline binding to imidazoline receptors. *Drug and Alcohol Dependence*, 64(2), 203–208. https://doi.org/10.1016/s0376-8716(01)00123-5

Ichikawa, M., Yoshida, J., Ide, N., Sasaoka, T., Yamaguchi, H., & Ono, K. (2006). Tetrahydro-beta-carboline derivatives in aged garlic extract show antioxidant properties. *The Journal of Nutrition*, 136(3 Suppl), 726S–731S. https://doi.org/10.1093/jn/136.3.726S

Ishida, J., Wang, H.K., Oyama, M., Cosentino, M.L., Hu, C.Q., & Lee, K.H. (2001). Anti-AIDS agents. 46. Anti-HIV activity of harman, an anti-HIV principle from Symplocos setchuensis, and its derivatives. *Journal of Natural Products*, 64(7), 958–960. https://doi.org/10.1021/np0101189

Ismail, T., Kim, Y., Lee, H., Lee, D.S., & Lee, H.S. (2019). Interplay between mitochondrial peroxiredoxins and ROS in cancer development and progression. *International Journal of Molecular Sciences,* 20(18), 4407. DOI: 10.3390/ijms20184407

Jiang, X., Zou, J., Zhuang, Y., Yuan, W., Zhu, L., & Zhu, G. (2017). The antiviral effects of harmine against BoHV-1 infection in vitro. *Letters in Drug Design & Discovery*, 14(11), 1303–1307.

Jin, F., Wu, Z., Hu, X., Zhang, J., Gao, Z., Han, X., Qin, J., Li, C., & Wang, Y. (2019). The PI3K/Akt/GSK-3β/ROS/eIF2B pathway promotes breast cancer growth and metastasis via suppression of NK cell cytotoxicity and tumor cell susceptibility. *Cancer Biology & Medicine*, 16(1), 38–54. DOI: 10.20892/j.issn.2095-3941.2018.0253

Kaminsky, S.M., Levy, O., Garry, M.T., & Carrasco, N. (1991). Inhibition of the Na+/I- symporter by harmaline and 3-amino-1-methyl-5H-pyrido(4,3-b)indole acetate in thyroid cells and membrane vesicles. *European Journal of Biochemistry*, 200(1), 203–207. https://doi.org/10.1111/j.1432-1033.1991.tb21068.x

Kattoor, A.J., Pothineni, N.V.K., Palagiri, D., & Mehta, J.L. (2017). Oxidative stress in atherosclerosis. *Current Atherosclerosis Reports*, 19, 1–11. DOI: 10.1007/s11883-017-0678-6

Kedziora, J., & Bartosz, G. (1988). Down's syndrome: A pathology involving the lack of balance of reactive oxygen species. *Free Radical Biology & Medicine*, 4(5), 317–330. https://doi.org/10.1016/0891-5849(88)90052-4

Kim, D.C., Quang, T.H., Yoon, C.S., Ngan, N.T.T., Lim, S.I., Lee, S.Y., Kim, Y.C., & Oh, H. (2016) Anti-neuroinflammatory activities of indole alkaloids from kanjang (Korean fermented soy source) in lipopolysaccharide-induced BV2 microglial cells. *Food Chemistry,* 213, 69–75. https://doi.org/10.1016/j.foodchem.2016.06.068

Kühn-Velten W.N. (1993). Norharman (beta-carboline) as a potent inhibitory ligand for steroidogenic cytochromes P450 (CYP11 and CYP17). *European Journal of Pharmacology*, 250(1), R1–R3. https://doi.org/10.1016/0014-2999(93)90648-2

Lamchouri, F., Zemzami, M., Jossang, A., Abdellatif, A., Israili, Z.H., & Lyoussi, B. (2013). Cytotoxicity of alkaloids isolated from Peganum harmala seeds. *Pakistan Journal of Pharmaceutical Sciences*, 26(4), 699–706.

Lamounier, A., Mateus, N., da Cunha, A., Luna, A., & Aucélio, R. (2015) Determination of six β-carboline alkaloids in urine and phytotherapic extracts using micellar liquid chromatography with fluorimetric detection. *Journal of Liquid Chromatography & Related Technologies,* 38, 997–1006.

Larson, R.A., Marley, K.A., Tuveson, R.W., & Berenbaum, M.R. (1988). Beta-carboline alkaloids: Mechanisms of phototoxicity to bacteria and insects. *Photochemistry and Photobiology*, 48(5), 665–674. https://doi.org/10.1111/j.1751-1097.1988.tb02879.x

Lee, J., Gong, Y.X., Jeong, H., Seo, H., Xie, D.P., Sun, H.N., & Kwon, T. (2021). Pharmacological effects of Picrasma quassioides (D. Don) Benn for inflammation, cancer, and neuroprotection. *Experimental and Therapeutic Medicine*, 22(6), 1357. DOI: 10.3892/etm.2021.10792

Li, J.M., Huang, Y.C., Kuo, Y.H., Cheng, C.C., Kuan, F.C., Chang, S.F., Lee, Y.R., Chin, C.C., & Shi, C.S. (2019). Flavopereirine suppresses the growth of colorectal cancer cells through P53 signaling dependence. *Cancers (Basel)*, 11(7), 1034. DOI: 10.3390/cancers11071034

Liu, H.W., Ji, Q.T., Ren, G.G., Wang, F., Su, F., Wang, P.Y., Zhou, X., Wu, Z.B., Li, Z, & Yang, S. (2020). Antibacterial functions and proposed modes of action of novel 1,2,3,4-tetrahydro-β-carboline derivatives that possess an attractive 1,3-diaminopropan-2-ol pattern against rice bacterial blight, kiwifruit bacterial canker, and citrus bacterial canker. *Journal of Agricultural and Food Chemistry*, 68(45), 12558–12568. DOI: 10.1021/acs.jafc.0c02528

Ma, X., Liu, D., Tang, H., Wang, Y., Wu, T., Li, Y., Yang, J., Yang, J., Sun, S., & Zhang, F. (2013). Purification and characterization of a novel antifungal protein with antiproliferation and anti-HIV-1 reverse transcriptase activities from Peganum harmala seeds. *Acta Biochimica et Biophysica Sinica*, 45, 87–94. DOI: 10.2174/1389203720666191119095624

Madle, E., Obe, G., Hansen, J., & Ristow, H. (1981). Harman and norharman: Induction of sister-chromatid exchanges in human peripheral lymphocytes in vitro and interaction with isolated DNA. *Mutation Research*, 90(4), 433–442. https://doi.org/10.1016/0165-1218(81)90065-3

Marques, M.R., Mendes, M.A., Tormena, C.F., Souza, B.M., Cesar, L.M., Rittner, R., & Palma, M.S. (2005). Structure determination of a tetrahydro-beta-carboline of arthropod origin: A novel alkaloid-toxin subclass from the web of spider Nephila clavipes. *Chemistry & Biodiversity*, 2(4), 525–534. https://doi.org/10.1002/cbdv.200590034

Mirza, R., Azeem, M., & Qaisar, U. (2019). Influence of Peganum harmala peptides on the transcriptional activity of biofilm-related genes in sensitive and resistant strains of Pseudomonas aeruginosa and Staphylococcus aureus. *Pakistan Journal of Pharmaceutical Sciences*, 32(5, Supplementary), 2341–2345.

Mita, S., Kamataki, T., & Kato, R. (1984). Effect of norharman on DNA strand breaks and mutation of Chinese hamster V79 cells by chemical carcinogens. *Carcinogenesis*, 5(6), 715–718. https://doi.org/10.1093/carcin/5.6.715

Mwaliko, E., Van Hal, G., Bastiaens, H., Van Dongen, S., Gichangi, P., Otsyula, B., Naanyu, V., & Temmerman, M. (2021). Early detection of cervical cancer in western Kenya: Determinants of healthcare providers performing a gynaecological examination for abnormal vaginal discharge or bleeding. *BMC Family Practice,* 22(1), 52. DOI: 10.1186/s12875-021-01395-y

Nazari Formagio, A.S., Santos, P.R., Zanoli, K., Ueda-Nakamura, T., Düsman Tonin, L.T., Nakamura, C.V., & Sarragiotto, M.H. (2009). Synthesis and antiviral activity of beta-carboline derivatives bearing a substituted carbohydrazide at C-3 against poliovirus and herpes simplex virus (HSV-1). *European Journal of Medicinal Chemistry*, 44(11), 4695–4701. https://doi.org/10.1016/j.ejmech.2009.07.005

Nenaah, G. (2010). Antibacterial and antifungal activities of (beta)-carboline alkaloids of Peganum harmala (L) seeds and their combination effects. *Fitoterapia*, 81(7), 779–782. https://doi.org/10.1016/j.fitote.2010.04.004

Ohishi, K., Toume, K., Arai, M.A., Koyano, T., Kowithayakorn, T., Mizoguchi, T., Itoh, M., & Ishibashi, M. (2015). 9-Hydroxycanthin-6-one, a β-carboline alkaloid from eurycoma longifolia, Is the first Wnt signal inhibitor through activation of glycogen synthase kinase 3β without depending on casein kinase 1α. *Journal of Natural Products*, 22, 78(5), 1139–1146. DOI: 10.1021/acs.jnatprod.5b00153

Pappolla, M.A., Sos, M., Omar, R.A., Bick, R.J., Hickson-Bick, D.L., Reiter, R.J., Efthimiopoulos, S., & Robakis, N.K. (1997). Melatonin prevents death of neuroblastoma cells exposed to the Alzheimer amyloid peptide. *The Journal of Neuroscience: The*

Official Journal of the Society for Neuroscience, 17(5), 1683–1690. https://doi.org/10.1523/JNEUROSCI.17-05-01683.1997

Pari, K., Sundari, C.S., Chandani, S., & Balasubramanian, D. (2000). Beta-carbolines that accumulate in human tissues may serve a protective role against oxidative stress. *The Journal of Biological Chemistry*, 275(4), 2455–2462. https://doi.org/10.1074/jbc.275.4.2455

Petersen, E.N., Paschelke, G., Kehr, W., Nielsen, M., & Braestrup, C. (1982). Does the reversal of the anticonflict effect of phenobarbital by beta-CCE and FG 7142 indicate benzodiazepine receptor-mediated anxiogenic properties? *European Journal of Pharmacology*, 82(3-4), 217–221. DOI: 10.1016/0014-2999(82)90517-9

Pless, G., Frederiksen, T.J., Garcia, J.J., & Reiter, R.J. (1999). Pharmacological aspects of N-acetyl-5-methoxytryptamine (melatonin) and 6-methoxy-1,2,3,4-tetrahydro-beta-carboline (pinoline) as antioxidants: Reduction of oxidative damage in brain region homogenates. *Journal of Pineal Research*, 26(4), 236–246. https://doi.org/10.1111/j.1600-079x.1999.tb00589.x

Podhorna, J., & Franklin, K.B. (2000). Pontine tegmentum lesions increase anxiety-like behavior in rats: A comparison with anxiety produced by beta-CCE. *Pharmacology, Biochemistry, and Behavior*, 65(2), 267–273. DOI: 10.1016/s0091-3057(99)00200-2

Poeggeler, B., Saarela, S., Reiter, R.J., Tan, D.X., Chen, L.D., Manchester, L.C., & Barlow-Walden, L.R. (1994). Melatonin--a highly potent endogenous radical scavenger and electron donor: New aspects of the oxidation chemistry of this indole accessed in vitro. *Annals of the New York Academy of Sciences,* 738, 419–420. https://doi.org/10.1111/j.1749-6632.1994.tb21831.x

Pogosyan, S.A., Grigoryan, N.P., & Paronikyan, R.G. (2007). Synthesis and anticonvulsant activity of dihydrochlorides of indoline-3′spiro-1-(1,2,3,4-tetrahydro-β-carboline derivatives. *Pharmaceutical Chemistry Journal*, 41(10), 527–528. https://doi.org/10.1007/s11094-008-0002-4

Prakash, D., Upadhyay, G., Gupta, C., Pushpangadan, P., & Singh, K. (2012). Antioxidant and free radical scavenging activities of some promising wild edible fruits. *International Food Research Journal*, 19(3).

Qais, N., & Jahan, N. (2020). Synthesis of dl-1-(4-Hydroxybenzyl)-1,2,3,4-Tetrahydro-β-carboline and evaluation of its biological activity, 9(6), 203–211.

Qiu, J., Zhang, T., Zhu, X., Yang, C., Wang, Y., Zhou, N., Ju, B., Zhou, T., Deng, G., & Qiu, C. (2019). Hyperoside induces breast cancer cells apoptosis via ROS-mediated NF-κB signaling pathway. *International Journal of Molecular Sciences*, 21(1), 131. DOI: 10.3390/ijms21010131

Radak, Z., Zhao, Z., Goto, S., & Koltai, E. (2011). Age-associated neurodegeneration and oxidative damage to lipids, proteins, and DNA. *Molecular Aspects of Medicine*, 32(4-6), 305–315. DOI: 10.1016/j.mam.2011.10.010

Rajesh, R.P., & Murugan, S. (2019). Spectroscopic identification of brominated, non-brominated alkaloids and evaluation of antimicrobial activity of Eudistomin-I, Eudistomin H, from green ascidian Eudistoma viride. *Journal of Applied Pharmaceutical Science,* 9, 116–123. https://doi.org/10.7324/JAPS.2019.90216

Saeki, K., Kobayashi, N., Inazawa, Y., Zhang, H., Nishitoh, H., Ichijo, H., Saeki, K., Isemura, M., & Yuo, A. (2002). Oxidation-triggered c-Jun N-terminal kinase (JNK) and p38 mitogen-activated protein (MAP) kinase pathways for apoptosis in human leukemic cells stimulated by epigallocatechin-3-gallate (EGCG): A distinct pathway from those of chemically induced and receptor-mediated apoptosis. *Biochemical Journal*, 368(Pt 3), 705–720. DOI: 10.1042/BJ20020101

Safari, N., Mirabzadeh Ardakani, M., Hemmati, R., Parroni, A., Beccaccioli, M., & Reverberi, M. (2020). The potential of plant-based bioactive compounds on inhibition of aflatoxin B1 biosynthesis and down-regulation of aflR, aflM, and aflP genes. *Antibiotics,* 9(11). DOI: 10.3390/antibiotics9110728

Samita, F., Ochieng, C.O., Owuor, P.O., Manguro, L.O., & Midiwo, J.O. (2017). Isolation of a new β-carboline alkaloid from aerial parts of Triclisia sacleuxii and its antibacterial and cytotoxicity effects. *Natural Product Research*, 31(5), 529–536. https://doi.org/10.1080/14786419.2016.1201666

Sandler, J.S., Colin, P.L., Hooper, J.N., & Faulkner, D.J. (2002). Cytotoxic beta-carbolines and cyclic peroxides from the Palauan sponge Plakortis nigra. *Journal of Natural Products*, 65(9), 1258–1261. https://doi.org/10.1021/np020228v

Sari, Y., & Khalil, A. (2015). Monoamine oxidase inhibitors extracted from tobacco smoke as neuroprotective factors for potential treatment of Parkinson's disease. *CNS & Neurological Disorders Drug Targets,* 14(6), 777–785. https://doi.org/10.2174/1871527314666150325235608

Sas, K., Szabó, E., & Vécsei, L. (2018). Mitochondria, oxidative stress and the kynurenine system, with a focus on aging and neuroprotection. *Molecules*, 23(1), 191. DOI: 10.3390/molecules23010191

Savi, D.C., Shaaban, K.A., Vargas, N., Ponomareva, L.V., Possiede, Y.M., Thorson, J.S., Glienke, C., & Rohr, J. (2015) Microbispora sp. LGMB259 endophytic actinomycete isolated from Vochysia divergens (Pantanal, Brazil) producing β-carbolines and indoles with biological activity. *Current Microbiology*, 70(3), 345–354. DOI: 10.1007/s00284-014-0724-3

Shaheen, H.A., & Issa, M.Y. (2020). In vitro and in vivo activity of Peganum harmala L. alkaloids against phytopathogenic bacteria. *Scientia Horticulturae,* 264, 108940. DOI: 10.1016/j.scienta.2019.108940

Sinet, P.M., Heikkila, R.E., & Cohen, G. (1980). Hydrogen peroxide production by rat brain in vivo. *Journal of Neurochemistry*, 34(6), 1421–1428. https://doi.org/10.1111/j.1471-4159.1980.tb11222.x

Sinha, K., Das, J., Pal, P.B., & Sil, P.C. (2013). Oxidative stress: The mitochondria-dependent and mitochondria-independent pathways of apoptosis. *Archives of Toxicology*, 87(7), 1157–1180. DOI: 10.1007/s00204-013-1034-4

Skolnick, P., Ninan, P., Insel, T., Crawley, J., & Paul, S. (1984). A novel chemically induced animal model of human anxiety. *Psychopathology*, 17(Suppl 1), 25–36. PMID: 6143342. DOI: 10.1159/000284074

Skropeta, D., & Wei, L. (2014). Recent advances in deep-sea natural products. *Natural Product Reports*, Aug;31(8), 999–1025. DOI: 10.1039/c3np70118b

Small, W. Jr., Bacon, M.A., Bajaj, A., Chuang, L.T., Fisher, B.J., Harkenrider, M.M., Jhingran, A., Kitchener, H.C., Mileshkin, L.R., Viswanathan, A.N., & Gaffney, D.K. (2017). Cervical cancer: A global health crisis. *Cancer*, 123(13), 2404–2412. DOI: 10.1002/cncr.30667

Smirnova, O.B., Golovko, T.V., & Granik, V.G. (2012). Carbolines. Part 2. comparison of some of the properties of α-, γ-, and δ-carbolines. *ChemInform*, 43. DOI: 10.1002/CHIN.201209258

Smith, K.L., Ford, G.K., Jessop, D.S., & Finn, D.P. (2013). Behavioral, neurochemical, and neuroendocrine effects of the endogenous β-carboline harmane in fear-conditioned rats. *Journal of Psychopharmacology (Oxford, England)*, 27(2), 162–170. https://doi.org/10.1177/0269881112460108

Stavrinides, A., Tatsis, E.C., Foureau, E., Caputi, L., Kellner, F., Courdavault, V., & O'Connor, S.E. (2015). Unlocking the diversity of alkaloids in Catharanthus roseus: Nuclear localization suggests metabolic channeling in secondary metabolism. *Chemistry & Biology*, 22(3), 336–341. https://doi.org/10.1016/j.chembiol.2015.02.006

Stephens, D.N. (2013). Anxiolytic β-carbolines: From molecular biology to the clinic. *Springer Science & Business Media* (Vol. 11, pp. 230). ISBN-13 978-3540571445.

Sztretye, M., Dienes, B., Gönczi, M., Czirják, T., Csernoch, L., Dux, L., Szentesi, P., & Keller-Pintér, A. (2019). Astaxanthin, a potential mitochondrial-targeted antioxidant treatment in diseases and with aging. *Oxidative Medicine and Cellular Longevity*, 2019, 3849692. DOI: 10.1155/2019/3849692

Tanaka, N., Momose, R., Takahashi-Nakaguchi, A., Gonoi, T., Fromont, J., & Kobayashi, J. (2014) Hyrtimomines, indole alkaloids from Okinawan marine sponges Hyrtios spp. *Tetrahedron,* 70, 832–837. DOI: 10.1002/chin.201428215

Thiébot, M.H., Soubrié, P., & Sanger, D. (1988). Anxiogenic properties of beta-CCE and FG 7142: A review of promises and pitfalls. *Psychopharmacology (Berl)*, 94(4), 452–463. DOI: 10.1007/BF00212837

Tian, H., Gao, Z., Wang, G., Li, H., & Zheng, J. (2016). Estrogen potentiates reactive oxygen species (ROS) tolerance to initiate carcinogenesis and promote cancer malignant transformation. *Tumor Biology*, 37(1), 141–150. DOI: 10.1007/s13277-015-4370-6

Till, M., & Prinsep, M. (2009). 5-Bromo-8-methoxy-1-methyl-β-carboline, an Alkaloid from the New Zealand Marine Bryozoan Pterocella Wesiculosa. *Journal of Natural Products,* 72, 796–798. https://doi.org/10.1021/np8007655

Tsuchiya, H. (2016) Anesthetic effects changeable in habitual drinkers: Mechanistic drug interactions with neuro-active indoleamine–aldehyde condensation products associated with alcoholic beverage consumption. *Medical Hypotheses,* 92, 62–66. https://doi.org/10.1016/j.mehy.2016.04.038

Turjanski, A.G., Rosenstein, R.E., & Estrin, D.A. (1998). Reactions of melatonin and related indoles with free radicals: A computational study. *Journal of Medicinal Chemistry*, 41(19), 3684–3689. https://doi.org/10.1021/jm980117m

Wang, K.B., Di, Y.T., Bao, Y., Yuan, C.M., Chen, G., Li, D.H., Bai, J., He, H.P., Hao, X.J., Pei, Y.H., Jing, Y.K., Li, Z.L., & Hua, H.M. (2014). Peganumine A, a β-carboline dimer with a new octacyclic scaffold from Peganum harmala. *Organic Letters*, 16(15), 4028–4031. https://doi.org/10.1021/ol501856v

Wang, K.B., Li, D.H., Hu, P., Wang, W.J., Lin, C., Wang, J., Lin, B., Bai, J., Pei, Y.H., Jing, Y.K., Li, Z.L., Yang, D, & Hua, H.M. (2016). A series of β-carboline alkaloids from the seeds of peganum harmala show G-Quadruplex interactions. *Organic Letters*, Jul 15, 18(14), 3398–3401. DOI: 10.1021/acs.orglett.6b01560

Wang, K.-B., Yuan, C.-M., Xue, C.-M., Li, D.-H., Jing, Y.-K., He, H.-P., Hao, X.-J., Di, Y.-T., Li, Z.-L., & Hua, H.-M. (2014). Pegaharmalines A and B, two novel β-carboline alkaloids with unprecedented carbon skeletons from Peganum harmala. *RSC Advances*, 4(96), 53725–53729. https://doi.org/10.1039/C4RA07985J

Witkin, J.M., Barrett, J.E., Cook, J.M., & Larscheid, P. (1986). Differential antagonism of diazepam-induced loss of the righting response. *Pharmacology, Biochemistry, and Behavior*, 24(4), 963–965. DOI: 10.1016/0091-3057(86)90443-0

Xi, T., Xia, H., Fan, Y.X., Cao, Y.C., & Zhang, H.L. (2019). Mechanism of β-carboline alkaloids inhibiting migration and invasion of SGC7901 cells. *Chinese Journal of Clinical Medicine*, 44(1), 119–124. DOI: 10.19540/j.cnki.cjcmm.20180914.001

Xu, Z., Chang, F.-R., Wang, H.-K., Kashiwada, Y., McPhail, A., Bastow, K., Tachibana, Y., Cosentino, M., & Lee, K.-H. (2001). Anti-HIV agents 45 1 and antitumor agents 205. 2 two new sesquiterpenes, leitneridanins A and B, and the cytotoxic and anti-HIV principles from leitneria floridana. *Journal of Natural Products,* 63, 1712–1715. https://doi.org/10.1021/np000260u

Youssef, D.T. (2005). Hyrtioerectines A-C, cytotoxic alkaloids from the red sea sponge hyrtioserectus. *Journal of Natural Products*, 68(9), 1416–1419. https://doi.org/10.1021/np050142c

Zhang, J., Li, L., Dan, W., Li, J., Zhang, Q., Bai, H., & Wang, J. (2015). Synthesis and antimicrobial activities of 3-Methyl-β-Carboline derivatives. *Natural Product Communications*, 10(6), 899–902.

Zhao, W.Y., Shang, X.Y., Zhao, L., Yao, G.D., Sun, Z., Huang, X.X., & Song, S.J. (2018). Bioactivity-guided isolation of β-Carboline alkaloids with potential anti-hepatoma effect from Picrasma quassioides (D. Don) Benn. *Fitoterapia,* 130, 66–72. DOI: 10.1016/j.fitote.2018.08.004

Zheng, C., Fang, Y., Tong, W., Li, G., Wu, H., Zhou, W., Lin, Q., Yang, F., Yang, Z., Wang, P., Peng, Y., Pang, X., Yi, Z., Luo, J., Liu, M., & Chen, Y. (2014). Synthesis and biological evaluation of novel tetrahydro-β-carboline derivatives as antitumor growth and metastasis agents through inhibiting the transforming growth factor-β signaling pathway. *Journal of Medicinal Chemistry*, 57(3), 600–612. https://doi.org/10.1021/jm401117t

8 β-Carbolines as Anti-inflammatory and Antidepressant Agents
Synthesis and Biological Study

Nidhi Jangir, Surendra Kumar Bagaria, Surbhi Dhadda, Shikha Agarwal, Dinesh Kumar Jangid

8.1 INTRODUCTION OF *β*-CARBOLINES

Carbolines are also known as nor-harmane, which is an extraordinary class of heterocyclic compounds found naturally having exceptional pharmacological properties. These are tricyclic, pyridine-fused indole structures (where the rings are recognized as A, B, and C) and are categorized as per the position of N-atom in the C-ring skeleton as α-, β-, γ- or δ-carbolines and degree of saturation such as 1,2,3,4-tetrahydro carboline (fully saturated), 3,4-dihydro carboline (partially saturated), and β-carbolines (unsaturated), respectively (Figure 8.1) (Smirnova et al. 2011; Singh et al. 2012).

Carbolines are among the indole alkaloids that have a pyridine ring fused to an indole moiety. Various forms of carbolines are produced by fusing the pyridine ring with the indole moiety. Depending on where the nitrogen is located in the ring, these carbolines are categorized as α, β, γ, or δ-carbolines. Compounds with β-carbolines scaffolds are pharmacologically effective (Figure 8.2).

β-carbolines are indole alkaloids that have a tricyclic 9H-pyrido[3,4-b] indole ring. The indole alkaloids' nor-harmane, which has an indole molecule attached to a pyridine ring, is another name for it (Smirnova et al. 2011). Plants, food, microorganisms, and insects all contain β-carbolines (Singh et al. 2012). Many of these alkaloids have been isolated from plants in the family, including the Rutaceae, Zygophyllaceae, Simaroubaceae, Amaranthaceae, Caryophylaceae, and Rubiaceae, in addition to marine sources such soft hydroids, corals, and bryozoans. (Piechowska et al. 2019; Dai et al. 2018). For the production of β-carbolines, tryptophan and tryptamine are regarded as precursors (Casal et al. 2019).

 DOI: 10.1201/9781351058032-8

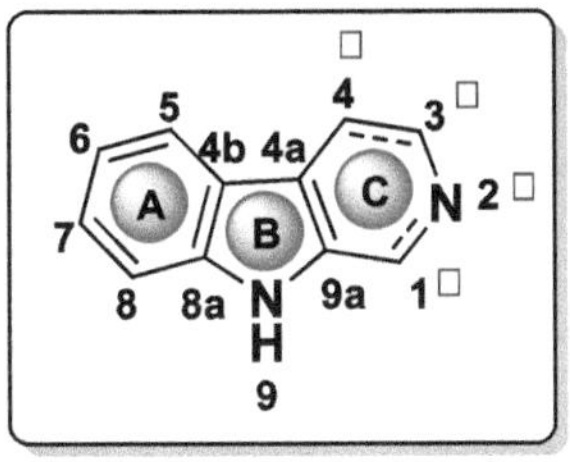

FIGURE 8.1 Skeleton of carboline derivatives.

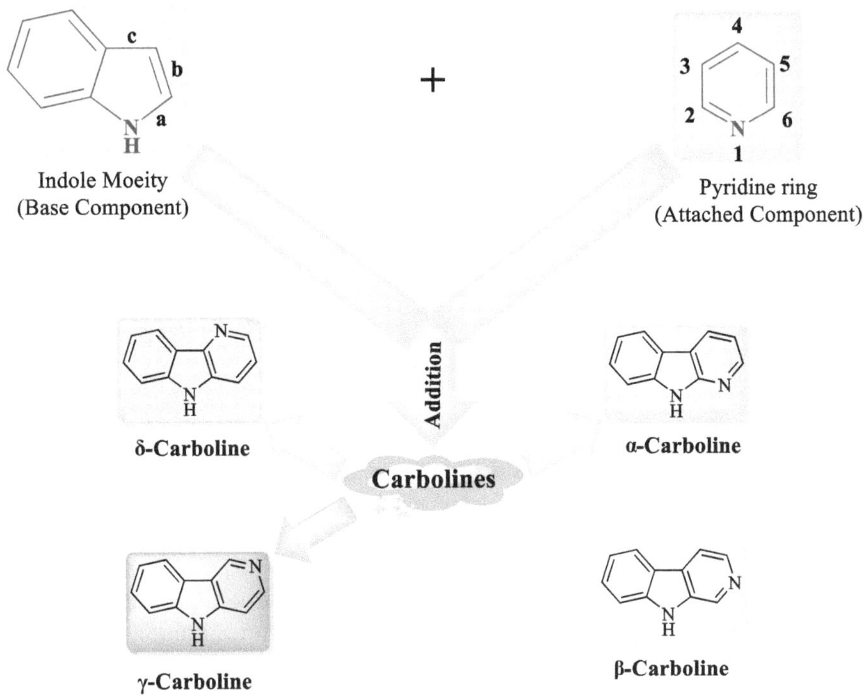

FIGURE 8.2 Chemical composition of several carboline kinds.

β-carbolines are grouped based on the degree of pyridine ring saturation. According to Cao et al. (2007), β-carbolines with a fully saturated aromatic pyridine ring are referred to as tetrahydro-β-carbolines (THβCs), β-carbolines with a partially saturated aromatic pyridine ring as 3,4 dihydro-β-carbolines (DHβCs), and β-carbolines with a total unsaturated aromatic pyridine ring as fully aromatic beta carbolines (FAβCs) (Figure 8.3).

β-carbolines are excessively disseminated in flora and fauna and are found in numerous foodstuffs, plants, insects, marine creatures, and mammals in addition to human tissues and body fluids. A number of derivatives of this family demonstrate countless medicinal activities (Piechowska et al. 2019; Dai et al. 2018; Casal et al.

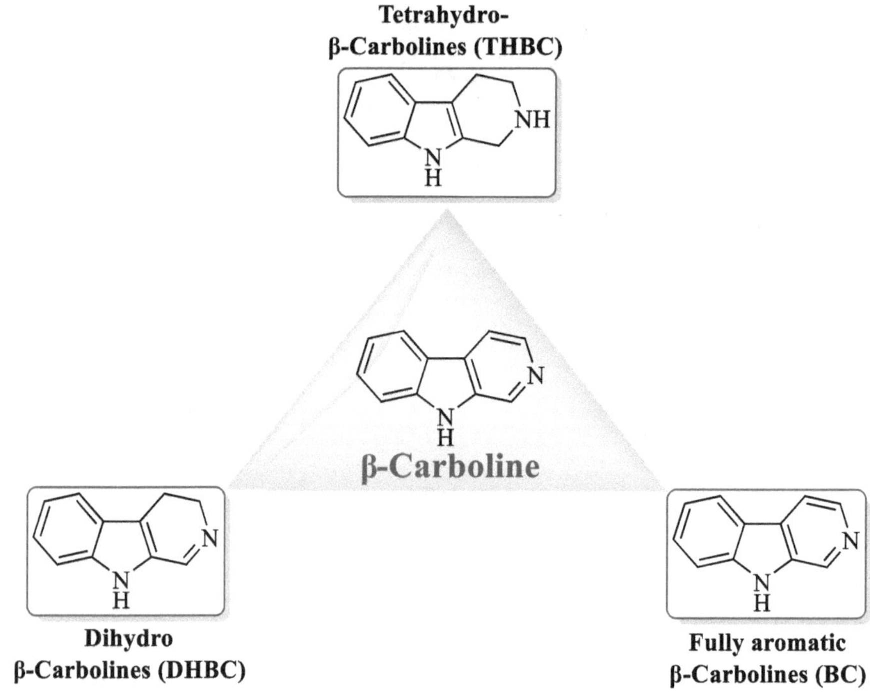

FIGURE 8.3 Various β-carboline kinds' chemical structures.

2019). The structural diversity and therapeutic potential (Cao et al. 2007) integrated in them inspire various scientists to synthesize β-carboline derivatives or separate them from natural products (Singh et al. 2012; Dai et al. 2018; Maity et al. 2019; Domínguez et al. 2011; Devi et al. 2018; Singh et al. 2017). Some commercially available drug motifs include Reserpine, Vinpocetine, Abecarnil, Brovincamine, Lurbinectedin, Tadalafil, and Cipargamin. A few of the reported drugs are mentioned in Figure 8.4.

In the year 1911, Ame Pictet and Theodor Spengler initially reported the synthesis of β-carboline derivatives (Royer et al. 2004). The β-carboline and tetra hydro β-carboline show significance in pharmacological activity upon substitution at different positions (i.e., 1, 2, 3, 6, 7, and 9) with a range of medicinal potential. The β-carbolines have been reported previously as antimalerial agent (Ang et al. 2000), antimicrobial agent (Sheng et al. 2023), anticancer agent (Abinaya et al. 2022), anti-HIV agent (Thatikayala et al. 2022), etc. (Figure 8.5).

Natural products have been utilized as traditionally available herbal medicines to preclude or medicate innumerable sicknesses for thousands of years (Coussens et al. 2002) and also assist as prime fragments for the anti-inflammatory drug advancement (Zhao et al. 2012). Anti-inflammatory medicines are applicable to relieve pain or ease swelling (i.e., inflammation). A drug or treatment that decreases inflammation or swelling is said to be anti-inflammatory or antiphlogistic. Furthermore, temperament and emotional dysregulation play a role in depression and anxiety. These conditions

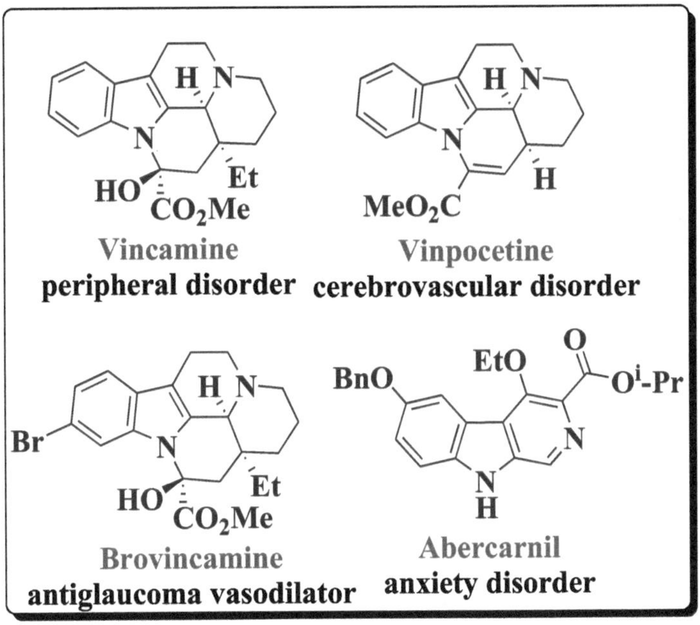

FIGURE 8.4 Structure of drugs containing β-carboline motif.

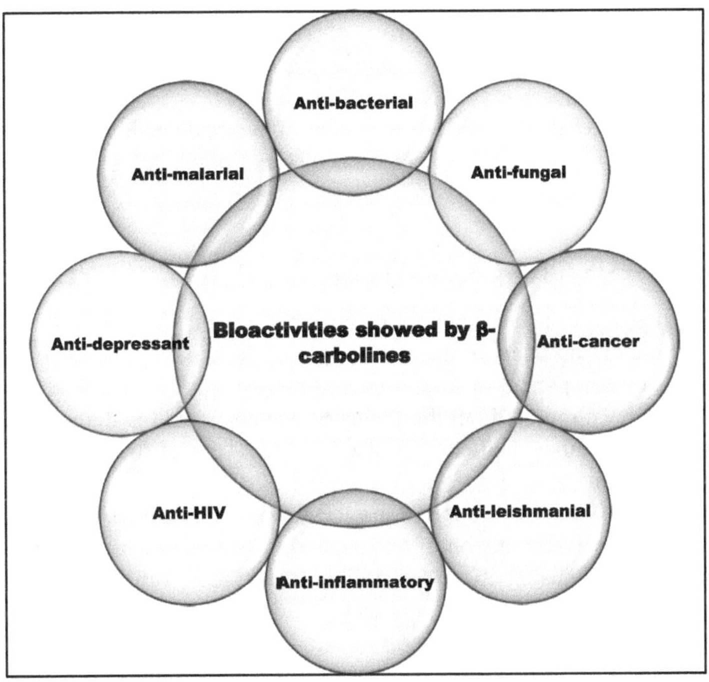

FIGURE 8.5 Representation of pharmaceutical activities of β-carbolines.

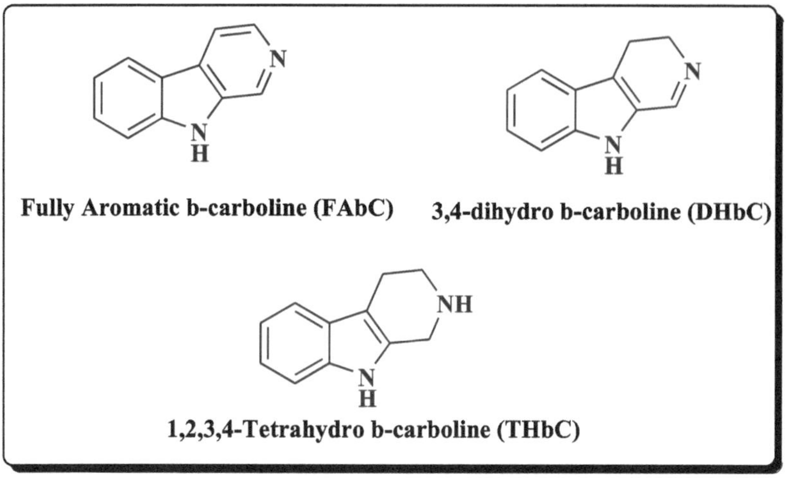

FIGURE 8.6 Structure of different types of β-carbolines.

can range from dysthymia and a chronic low mood syndrome to serious depression, which is typically accompanied by additional mental health issues like anxiety, sleeplessness, and obsessive behaviors and social phobia (Gold et al. 2015). The antidepressant effect of β-carbolines received attention from the scientific community due to the actions on the monoamine neurotransmission, particularly the serotoninergic system (Heshmati et al. 2014). β-carboline derivatives (Figure 8.6) for intense canthin-6-ones and bis-β-carbolines were separated from *Picrasma. quassioides* displayed promising anti-inflammatory properties (Yang et al. 2011).

In this chapter, we discuss the advancement of synthetic pathways for production of β-carboline derivatives and their pharmacological potential as anti-inflammatory and antidepressant agents.

8.2 SYNTHESIS OF *β*-CARBOLINE AND ITS DERIVATIVES

Many literature studies have been done on β-carbolines and their derivatives based on their role in natural products in the past couple of decades. β-carboline has been to be a found a key component in several natural products and drugs. Several synthetic protocols have been reported but the Pictet–Spengler method is the most promising protocol; however, other methods are also producing novel β-carbolines that have great importance due to theeir heterocyclic moiety.

Viswanathan et al. developed a protocol for the synthesis of β-carboline derivatives that involved reaction of an appropriate amount of tryptamine and aldehyde that were heated using microwave radiation, and the resulting mixture was then precipitated as salt of either a trifluoroacetic acid or a hydrochloride (Viswanathan et al. 2023). This tetrahydro-β-carboline salt is extracted by filtration, then it is dissolved in dimethyl formamide and heated in the existence of lithium and silver carbonates to produce the product *β*-carboline, which is then refined using flash chromatography (Scheme 8.1).

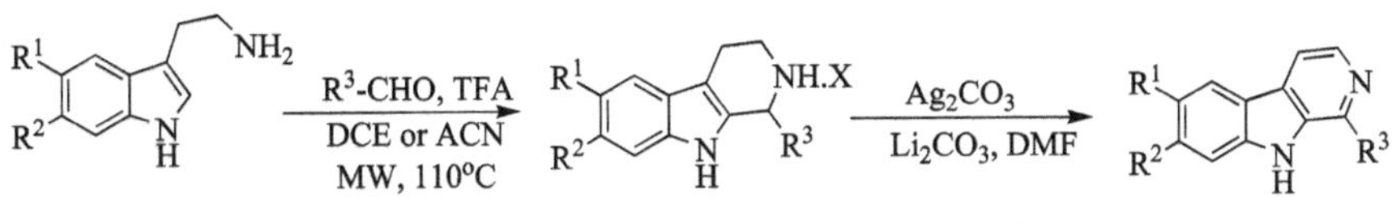

SCHEME 8.1 Microwave-assisted preparation of derivatives of *β*-carbolines.

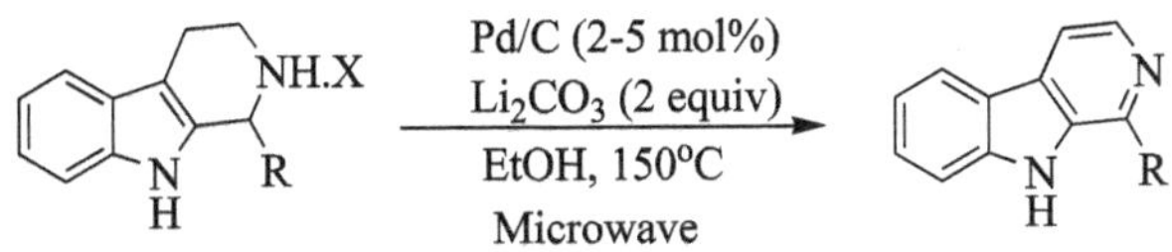

SCHEME 8.2 Microwave-assisted preparation of *β*-carboline derivatives.

Eagon et al. demonstrated the microwave-assisted preparation of β-carboline within 20 minutes (Eagon et al. 2014). In this synthesis tryptamine served as the starting material for the microwave-assisted Pictet–Spengler reaction, which produced a yield of up to 99% utilizing 1,2-dichloroethane (DCE) and trifluoroacetic acid (TFA) as solvents (Scheme 8.2). Following the initial THBC synthesis, Pd/C in EtOH was used to produce tetrahydro-β-carboline salts, which were then aromatized to produce β-carboline salts.

The synthesis of phenyl piperazine derivatives of β-carboline was reported by Ashok et al. in which DL-tryptophan was used as a starting ingredient (Ashok et al. 2019). Following this, aldehyde conjugation and esterification with thionyl chloride produced 1-substituted THBC ester. Following that, the β-carboline ester was produced by oxidizing it with potassium permanganate in THF, and alkali hydrolysis produced the β-carboline carboxylic acid. Finally, the preparation of diverse 3-substituted-β-carboline phenyl piperazines (Scheme 8.3) was achieved by acid amide coupling with a variety of substituted phenyl piperazines in the existence of EDC, HCl, and HOBt.

Maestri et al. studied the use of a palladium catalyst for the chemo- and regioselective synthesis of tetrahydro β-carbolines. Co-catalysis of palladium and carboxylic acid at 100°C using phosphine as a ligand in toluene, with complete configuration control in every scenario, results in the creation of C-C bonds via indolic $C(SP^2)$-H bond activation of N-propylated tryptamines, recovering products as a single E-isomer (Cera et al. 2018) (Scheme 8.4).

Another new, extremely effective protocol for the preparation of aromatic β-carbolines from 2-acyl-3-bromoindoles employing tributyl [(Z)-2-ethoxyvinyl] stannane as a C2 building block via Stille cross-coupling was developed by Kamlah et al. (2015) The synthesis of β-carbolines occurred in modest yields in the first Pd-catalyzed step, which was succeeded by the ring closure in glacial acetic acid using NH_4OAc (Scheme 8.5).

One-pot bimetallic relay catalysis and one-pot triple-orthogonal metal relay catalysis were devised by Dhiman et al. for the preparation of 1,3-disubstituted 4-hydroxy β-carbolines and 1,3-disubstituted β-carbolines (Dhiman et al. 2016). Both cases began with the easily available 3-(2-aminophenyl)-5-hexenyn-3-ols as raw ingredients (Scheme 8.6).

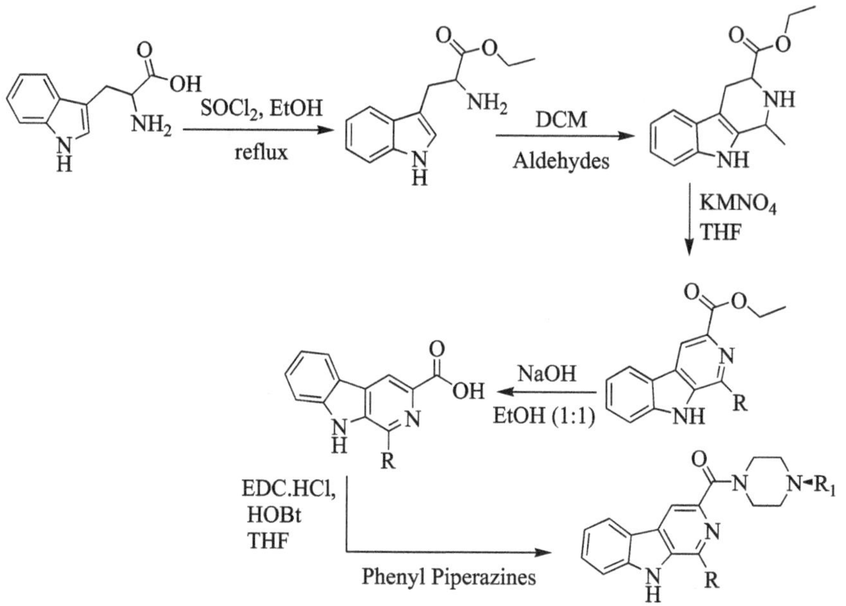

SCHEME 8.3 Preparation of *β*-carbolines derivatives.

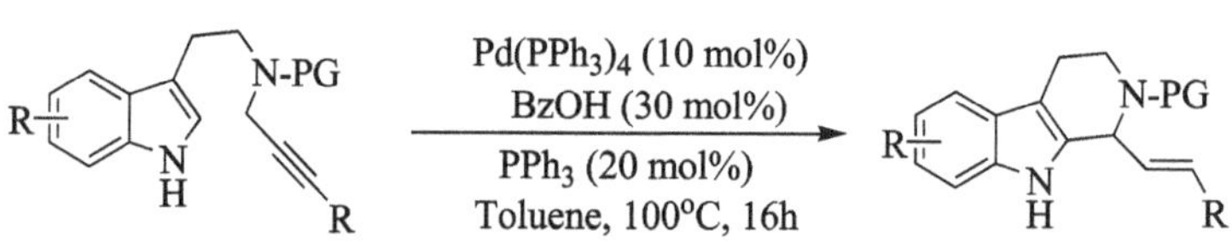

SCHEME 8.4 Preparation of *β*-carbolines with palladium assistance.

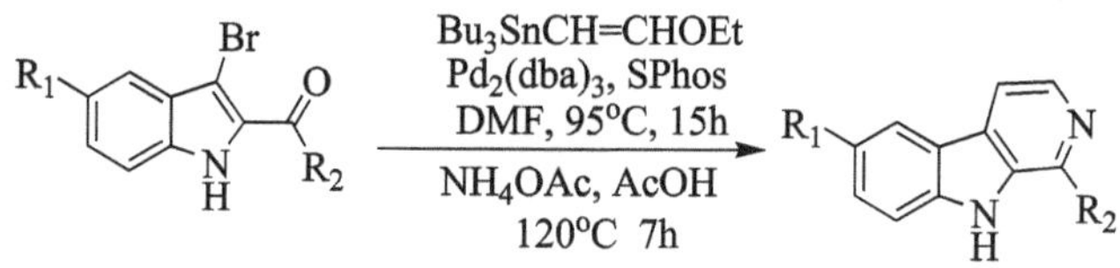

SCHEME 8.5 Pd-catalyzed preparation of *β*-carboline derivatives.

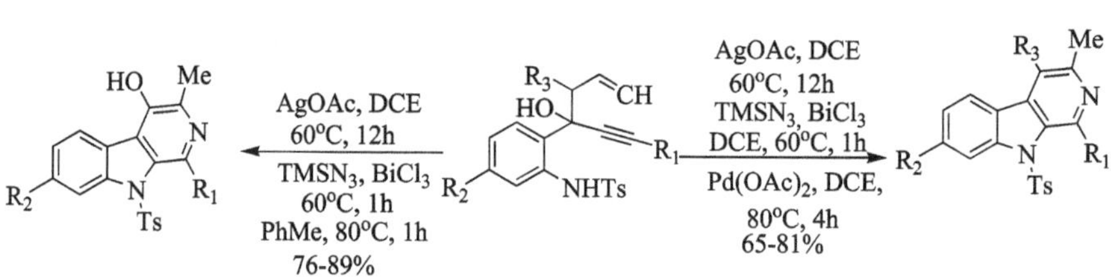

SCHEME 8.6 A method of making *β*-carbolines in a single pot.

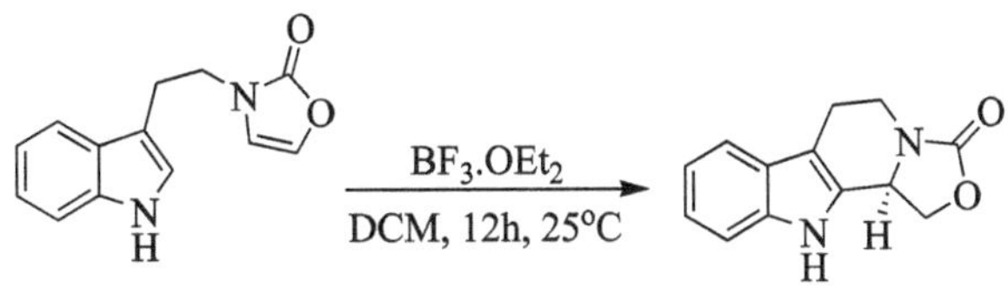

SCHEME 8.7 $BF_3.OEt_2$-catalyzed synthetic approach for the synthesis of *β*-Carboline derivative.

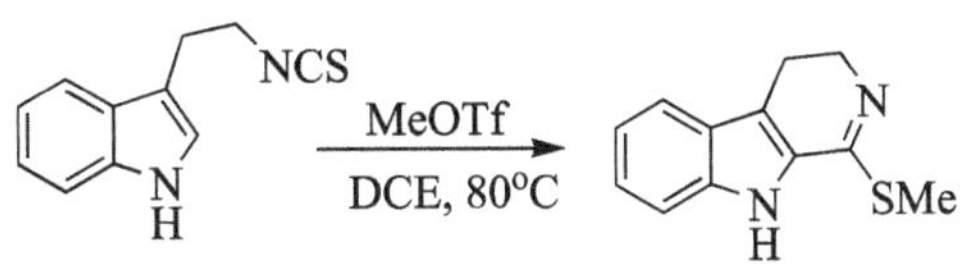

SCHEME 8.8 MeOTf-catalysed synthetic pathway for the synthesis of *β*-Carboline derivative.

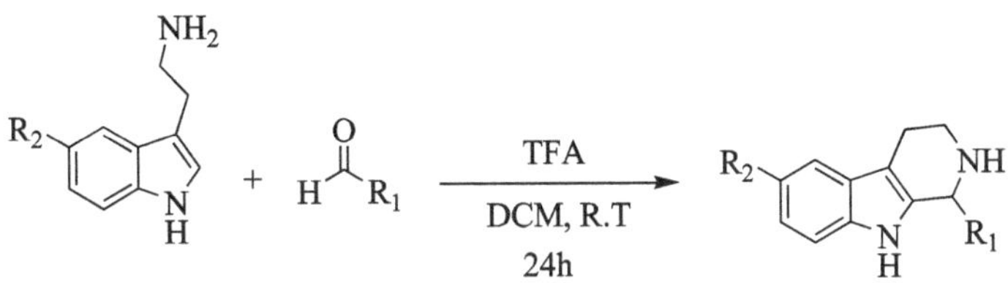

SCHEME 8.9 TFA catalyzed synthetic route for the synthesis of *β*-Carboline derivatives.

In another protocol, He et al. described a facile synthesis of β-carboline derivatives at 25°C for 12 hours with a good yield (80%) (He et al. 2016). In this protocol, the intended product was produced using dichloro methane (DCM) as a polar aprotic solvent and borontrifluoride etherate ($BF_3.OEt_2$) as a Lewis acid catalyst during the indole oxazolone cyclization process (Scheme 8.7). However, this simple one-pot method produces 80% of the therapeutically important tetrahydroisoquinoline alkaloids found in nature.

In 2017, Wen et al. demonstrated a simple and efficient protocol for the formation of β-carboline derivatives in a non-polar solvent like DCE at 80°C with 90% yield of product (Wen et al. 2017) under catalytic media of isothiocyanate and methyltrifluoro methanesulfonate (MeOTf) (Scheme 8.8). This synthetic process is used to make the traditional Chinese medicine drug rutacarpine (an alkaloid quinazolinone), which is useful for treating diseases associated with inflammation.

Spindler et al. presented a straightforward, all-purpose, capable, and practical route at room temperature for the synthesis of 1-substituted tetrahydro-β-carbolines with a superior yield range (18–97%) (Spindler et al. 2016). The Pictet–Spengler reaction using tryptamine derivatives and aldehyde, together with a suitable catalyst such trifluoroacetic acid (TFA), can be used to carry out this method over the course of 24 hours in a polar DCM solvent (Scheme 8.9).

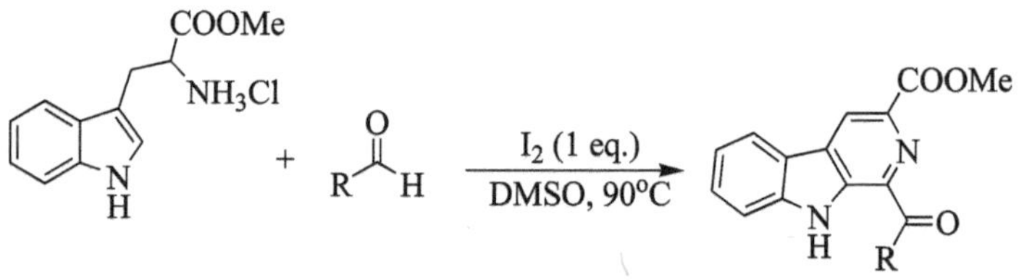

SCHEME 8.10 *β*-Carboline derivative synthetically prepared using I_2 catalysis.

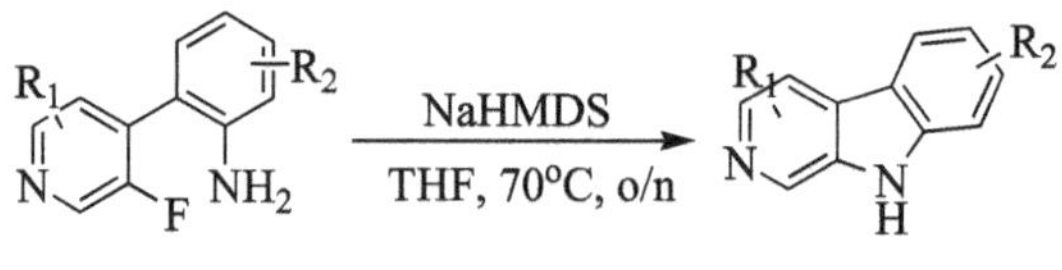

SCHEME 8.11 Synthesis of *β*-carbolines via bases.

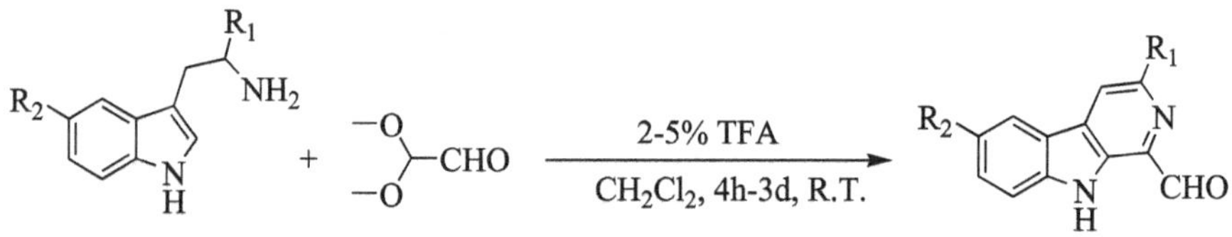

SCHEME 8.12 The use of TFA in the synthesis of a derivative of *β*-carboline.

An efficient protocol for the synthesis of β-carboline derivative based on Pictet-Spengler condition with moderate yield range (60–85%) was developed by Battini et al. (2014). In this protocol, tryptophan methyl ester was reacted with an aldehyde derivative at a temperature of 90°C in the presence of molecular iodine (I_2) as a catalyst and polar aprotic solvent dimethylsulfoxide (DMSO) (Scheme 8.10). The demonstrated catalytic pathway is competent, cost-effective, and incredibly selective; it yields a high spectrum of β-carboline derivatives, ranging from 60–85%, and concludes without the need for any protective groups for the separation and processing of these compounds.

Purified 2-aminobiaryls produced the best results, yielding β-carbolines in an excess of 18–66% yield reported by Shainthavaan et al. when they were combined with an excess of NaHMDS. As far as we are aware, this reaction process, which begins with commercially available 3-fluoropyridines, offers the fastest way to synthesize the naturally occurring carbolines norharmane (Sathiyalingama et al. 2022) (Scheme 8.11).

A straightforward and effective approach for the synthesis of β-carboline derivatives with an appreciable yield range of 81–96% was developed by Singh et al. (2012). This method can be carried out at room temperature by reacting tryptamine with an aldehyde derivative for 3 to 4 days in the existence of TFA as an acid catalyst and DCM as a solvent to produce a derivative of β-carboline (Scheme 8.12). This protocol's benefits include simplicity, cost-effectiveness, and a typically important

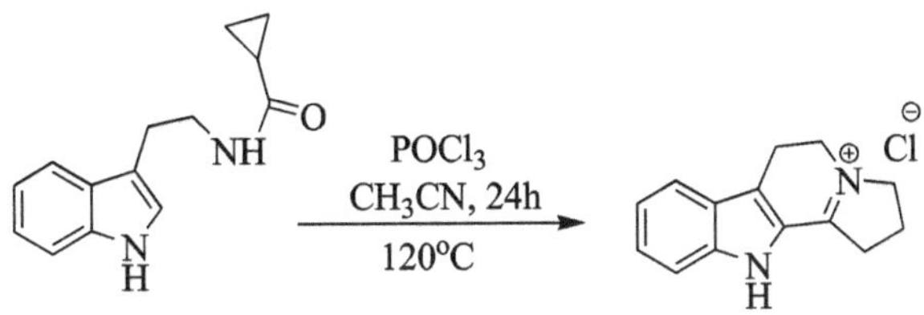

SCHEME 8.13 $POCl_3$ catalyzed synthetic path for *β*-Carboline derivative in acetonitrile solvent.

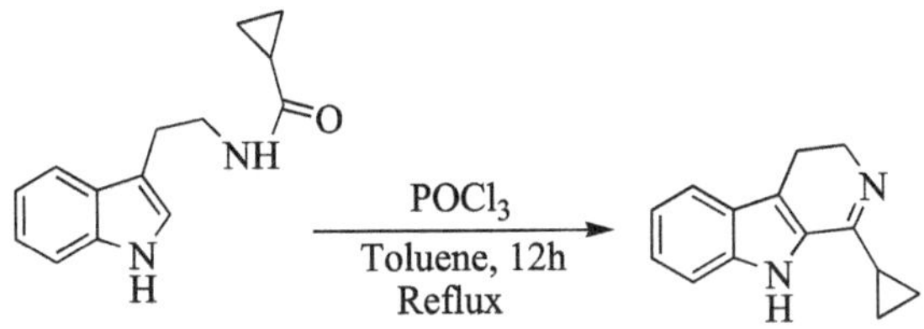

SCHEME 8.14 Synthesis for *β*-Carboline derivative in toluene, accelerated by $POCl_3$.

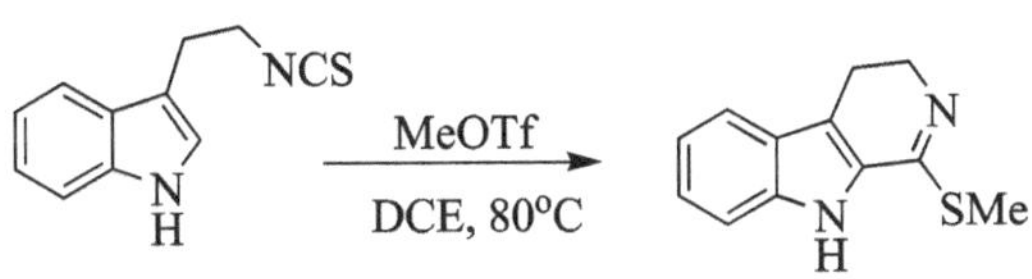

SCHEME 8.15 A synthetic process for producing derivatives of *β*-carboline using the MeOTf catalyst.

reaction with as superb yield (81–96%). Additionally, the derivative encourages a variety of fused and substituted-β-carbolines that operate as bioactive agents.

Saha et al. developed an efficient synthetic route of Bischler–Napieralski cyclized β-carboline derivative (Saha et al. 2011). Using phosphoryl chloride ($POCl_3$) as a Lewis acid catalyst and a polar aprotic solvent such acetonitrile, tryptamine amide was cyclized using this approach for a period of 24 hours at a temperature of 120°C with an excellent yield (70%) (Scheme 8.13).

A subsequent synthesis of β-carboline derivative was reported by Saha et al. (2011). Using $POCl_3$ and tryptamine amide for 12 hours under reflux conditions in a nonpolar solvent such as toluene yielded a product with a respectable percentage of 45% (Scheme 8.14). The benefit of this $POCl_3$ promoted synthetic technique is the straightforward, affordable, and one-pot procedure.

Wen et al. established a protocol at 80°C for the synthesis of an isothiocyanate-derived β-carboline derivative using methyltrifluoro methanesulfonate (MeOTf) as a catalyst in a nonpolar solvent such as DCE (Wen et al. 2017). This method produced an outstanding yield of 90%. In traditional Chinese medicine, the quinazolinone alkaloid rutaecarpine is used to treat inflammation-related illnesses and is synthesized using this method (Scheme 8.15).

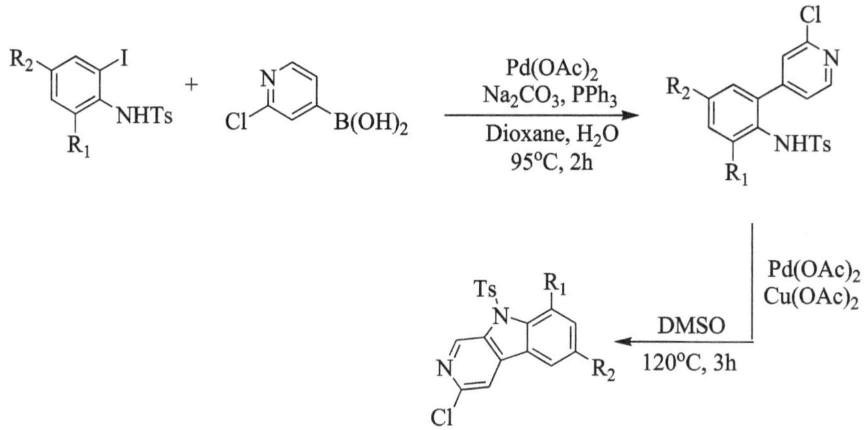

SCHEME 8.16 Economically beneficial process for making *β*-carbolines.

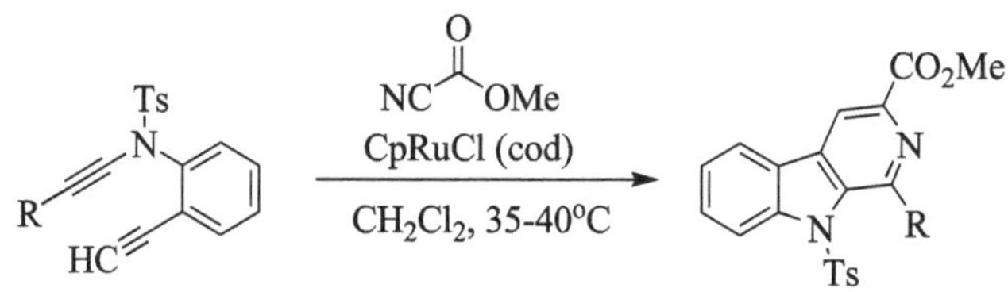

SCHEME 8.17 Ruthenium-catalyzed synthesis of *β*-carboline derivatives.

A two-step, economically advantageous method for producing β-carbolines was described by Dhara et al. (2014). N-tosylated 2-iodoanilines were coupled with 2-chloropyridin-4-yl-boronic acid precursors in the Suzuki reaction. Following ring closure brought about by Pd-mediated C-H/N-H activation, a modest yield of the intended product was obtained (Scheme 8.16).

Witulski et al. created a quick ruthenium-catalyzed process for creating the skeleton of the β-carboline (Witulski et al. 2011). Starting with readily accessible 2-iodoaniline, the necessary yne-ynamides were made in five steps (Scheme 8.17). The corresponding β-carbolines were produced by the [2+2+2] cyclo addition of yneynamides with methyl cyanoformate under the influence of CpRuCl(cod). The marine alkaloid eudistomin U was successfully synthesized using this method.

A simple procedure to synthesize β-carboline compounds with palladium support was described by Ding et al. (2010). This can be achieved by directly dehydrogenating internal alkynes and tert-butylimines of N-substituted indole carboxaldehydes in an annulation reaction (Scheme 8.18). In this procedure, elemental oxygen was employed to activate the C-H bond.

In another protocol, Wu et al. demonstrated one-pot preparation of β-carboline derivative at 110°C temperature for 5–10 h of duration time with excellent yield (86%) (Zhu et al. 2013). This procedure can be carried out by reacting tryptamine and aceophenone with molecular iodine (I_2), hydrogen peroxide (H_2O_2), and a polar

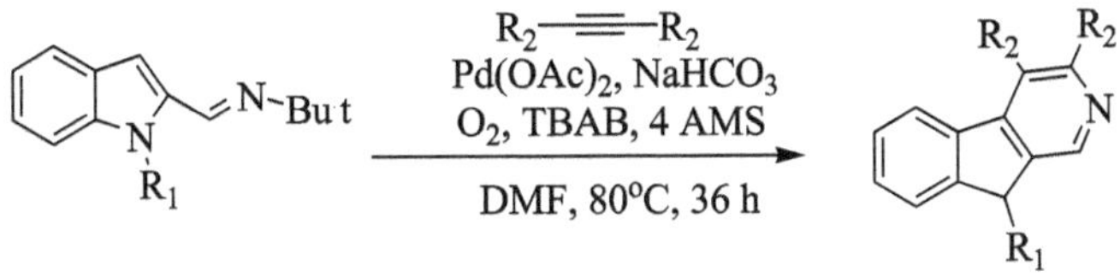

SCHEME 8.18 Pd catalysed synthesis of *β*-carboline compounds.

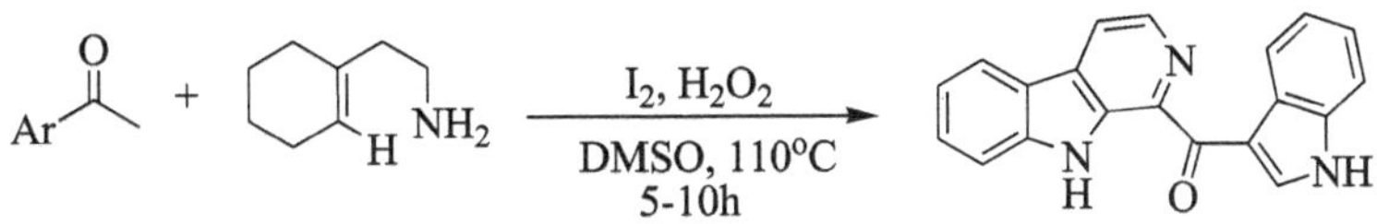

SCHEME 8.19 Preparation of a derivative of *β*-Carboline using of H_2O_2 and I_2.

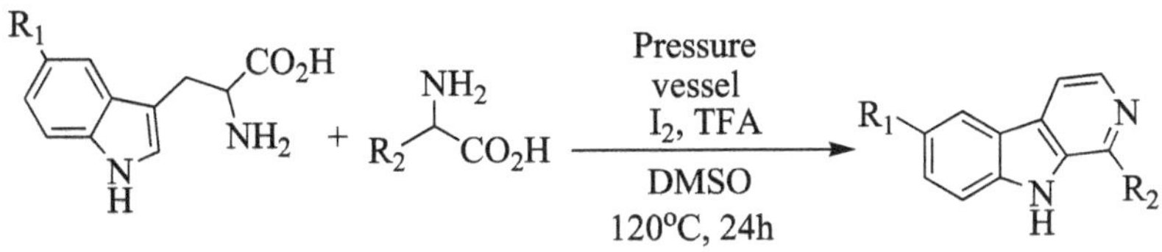

SCHEME 8.20 A single-pot synthesis of derivatives *β*-carbolines.

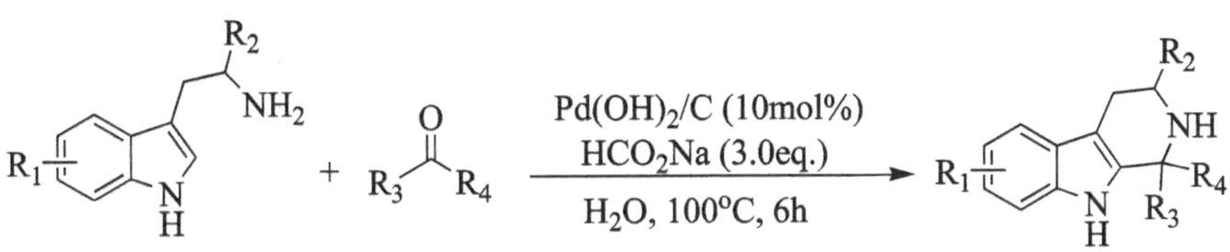

SCHEME 8.21 Synthetic route for the synthesis of functionalized tetrahydro *β*-carbolines derivatives.

aprotic solvent like DMSO (Scheme 8.19). The result is pityriacitrin (a derivative of the amino acid carboline).

Wang et al. reported a unique one-pot preparation of β-carbolines (Wang et al. 2018). The planned synthesis, which begins with racemic tryptophan and different amino acids, progresses in a series of stages that include decarboxylation, deamination, the Pictet–Spengler reaction, and oxidation (Scheme 8.20). These steps are driven by a reaction with I_2 and TFA.

Zeng et al. effectively developed a "two-in-one" approach to synthesize functionalized tetrahydro β-carbolines by palladium-catalyzed C-H bond activation of tryptamines with carbonyl compounds in water, which involves in situ imine production as a guiding group (Zeng et al. 2019) (Scheme 8.21) to develop a transformative protocol that is more atom-economic and environmentally friendly.

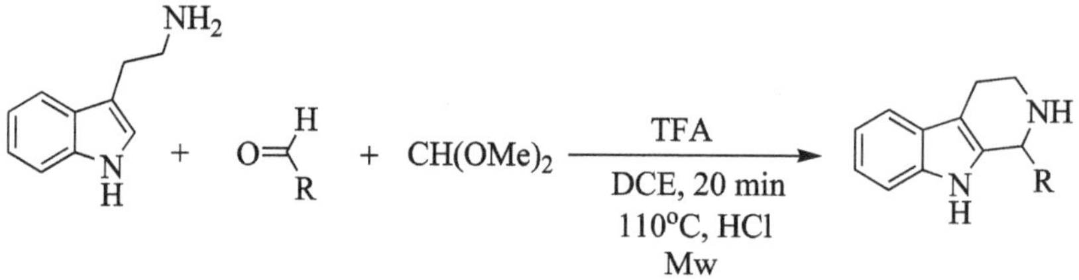

SCHEME 8.22 Synthesis of *β*-carbolines with microwave assistance, supported by TFA.

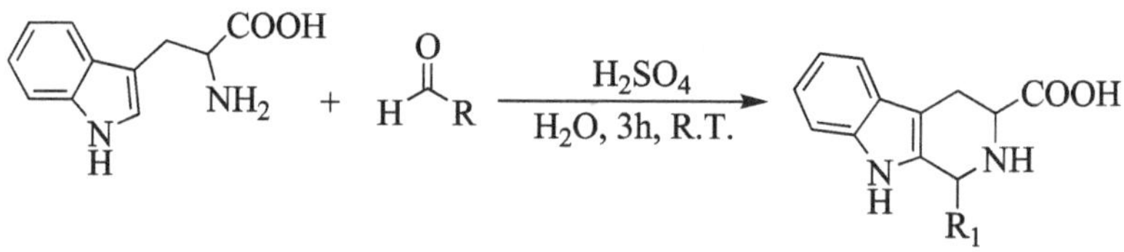

SCHEME 8.23 By virtue of water synthetic route for a derivative of *β*-carboline.

A microwave-assisted protocol designed by Anderson et al. involves Pictet–Spengler synthesis to obtain β-carboline (Anderson et al. 2014). In order to obtain the (70–99%) yield range of the product, tryptamine is allowed to react with substituted aldehyde and 2,2-dimethoxypropane (CH(OME)2) in the presence of TFA as an acid catalyst in a mobile solvent like 1,2-dichloroethane (DCE) under acidic conditions containing hydrochloric acid (HCl) at 110°C temperature for 20 min (Scheme 8.22). The primary advantage of this method is its ability to quickly and easily manufacture pure crystalline products on a milligram to gram scale through precipitation after simple filtration, with a high yield range of 70–99%, all without the need for liquid liquid extraction or column chromatography.

An efficient and greener methodology for the synthesis of β-carboline derivatives was developed by Buxi et al. which yielded a good yield range (50–83%) using the Pictet–Spengler condensation reaction of L-tryptophan with aldehyde derivative in water as a green solvent under reflux conditions for 3 hours (Buxi et al. 2013) (Scheme 8.23). This method is cost-effective, effective, and uses water as a green solvent.

Ame Pictet and Theodor Spengler developed a process for producing substitutions or fused beta-carbolines by artificial synthetic methods (Royer et al. 2004). They produced a protocol to synthesize 1,2,3,4-tetrahydroisoquinoline by heating phenylethylamine with aldehyde in the existence of acidic media. An imine is produced by the Pictet-Spengler reaction, which then cycles electron-rich aryl or heteroaryl groups onto imine or iminium ion electrophiles to yield an iminium ion in an acidic environment (Scheme 8.24). The same method has been used for the preparation of a variety of aza-heterocyclic compounds (Tatsui et al. 1928).

The biosynthesis of β-carboline alkaloids was achieved by the formation of Schiff base of tryptamine and then subsequent intramolecular Mannich reaction, in which the C2-carbon of indole ring behave as a nucleophile (França et al. 2014). Further the aromaticity of the nucleus was regenerated via abstraction of a proton from the C2.

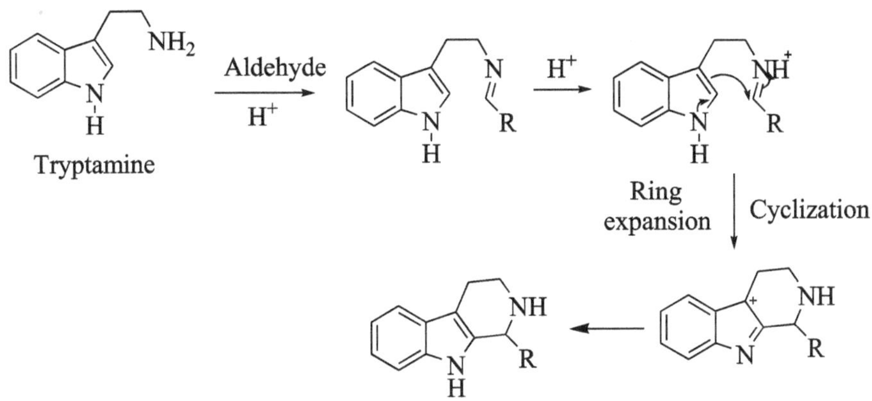

SCHEME 8.24 Pictet-Spengler reaction for the synthesis of β-carboline.

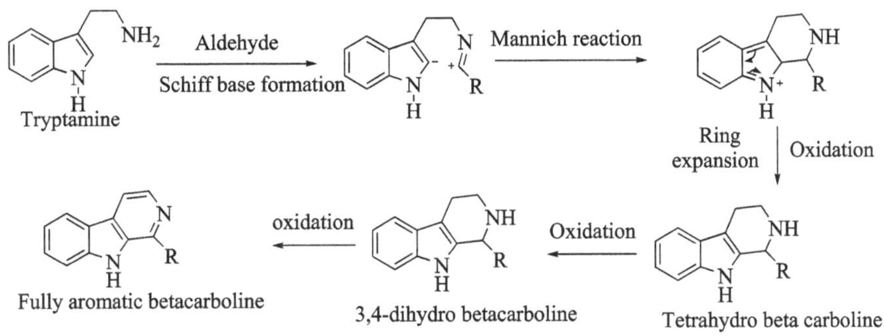

SCHEME 8.25 Biosynthesis of fully aromatic beta-carboline from tryptamine.

Finally, the desired compound (Rajesh et al. 2019) i.e. fully aromatic *β*-carboline is obtained by the oxidation of 3,4-dihydro- *β*-carboline (Scheme 8.25).

Yang et al. described the synthesis of β-carboline by the condensation of pyruvic aldehyde with L-tryptophan or L-tryptophanylamide (Yang et al. 2006). Similar products could be obtained by the coupling reaction between 4-methoxyphenylglyoxal and tryptophan methyl ester. Changing the functionalities from carboxylic acid group to amide (Mei-L et al. 2011) and ester significantly increased the yields of the desired products (Scheme 8.26).

A facile and environmentally friendly method has been developed in a single step from tryptamine and aldehydes for the preparation of tetrahydro-β-carbolines (tryptolines) (Hong-J et al. 2020). The preparation of diverse tryptolines derivatives was catalyzed by L-tartaric acid, a natural existing compound used to achieve the desired products in form of colourless crystals. This new protocol of synthesis using water and L-tartaric acid is easy, safe, and inexpensive, and the products can be easily separated using basic filtration method as they are obtained in crystals (Scheme 8.27).

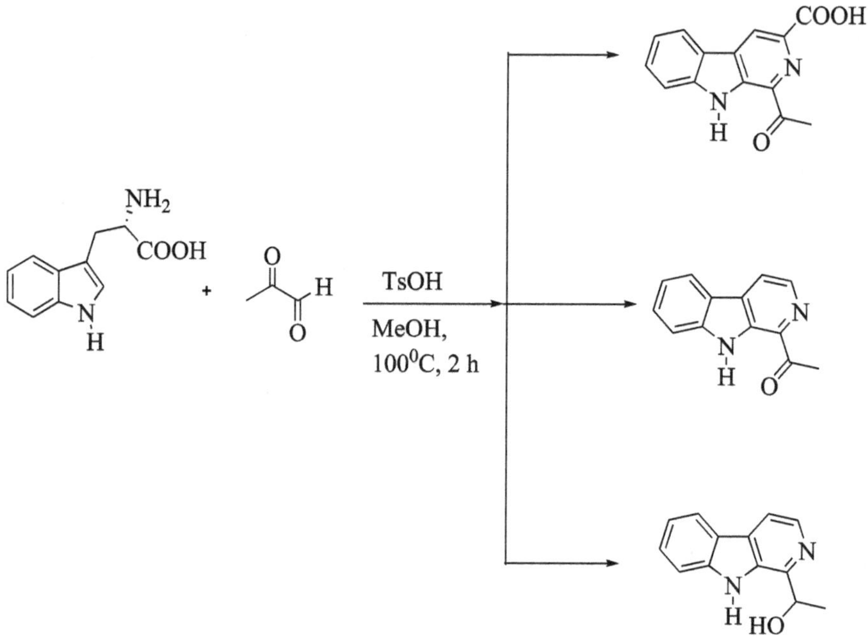

SCHEME 8.26 Synthesis of β-carboline analogs.

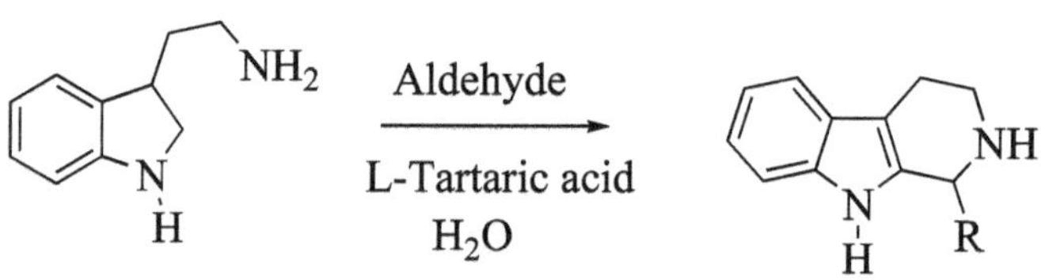

SCHEME 8.27 Tetrahydro-β-carbolines (tryptolines).

Bischler–Napieralski designed a novel method for the preparation of 1,3-diarylated 3,4-dihydro-β-carbolines. Using this process, benzophenone imimine is selectively hydrolyzed, then classic Bischler-Napieralski (Ji et al., 2021) cyclization is performed to produce 1,3-diarylated DHBCs. Corresponding β-carbolines are then obtained through oxidation (Scheme 8.28).

Tetrahydro-β-carbolines were synthesized using an environmentally safe and effective approach that involved the dehydrogenative coupling of alcohols with tryptamines in the presence of a weak base and a catalytic quantity of Mn catalyst (Zhihui et al. 2023). This procedure produces good to outstanding yields of the desired products for a range of benzylic and aliphatic alcohols with distinct functional groups. Numerous medicinal compounds, including harman, harmaline, and harmine, were effectively synthesized using this method (Scheme 8.29).

Wu and his co-workers devised a microwave-assisted technique to synthesize 1,2,3,4-tetrahydro-β-carbolines using a soluble polymer support system (Wu et al.

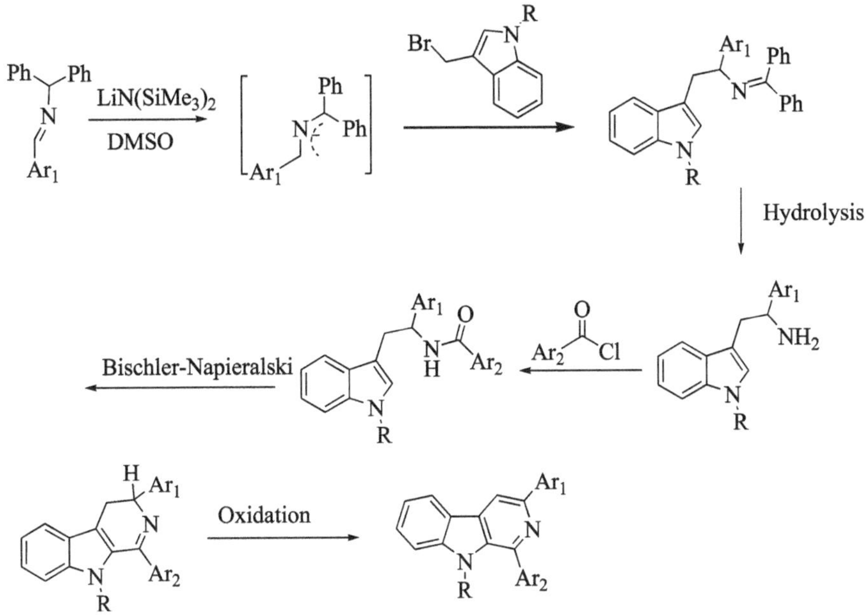

SCHEME 8.28 Proposed Strategy for the Construction of 1,3-Diaryl DHBCs and β-Carbolines.

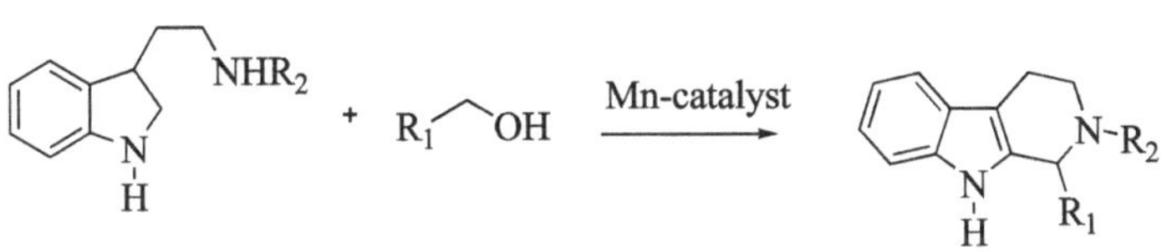

SCHEME 8.29 Synthesis of tetrahydro-β-carbolines by coupling of alcohols.

2002). The loading of F-moc protected L-tryptophan on the soluble polymer support in the presence of the coupling agent DCC/DMAP marked the beginning of this synthetic approach. First, the F-moc group was deprotected in the presence of piperidine, and then the Pictet–Spengler cyclization with carbonyl compounds produced 1,2,3,4-tetrahydro-β-carbolines. Treatment of the cyclized product with 1% potassium cyanide in methanol was used to separate the polyethylene glycol support, resulting in the production of 1,2,3,4-tetrahydro-β-carbolines (Scheme 8.30).

An innovative technique for creating stereoselective tetrahydro-carbolines with fused hydantoin under microwave irradiation on the molecular support of soluble biarm hydroxy polymer (polyethylene glycol-400) was reported by Chang et al. (2006). The easily available NHBoc protected L-tryptophan was reacted with polyethylene glycol-400 in the presence of DCC/ DMAP to produce the Boc (tert-Butyloxycarbonyl) protected intermediate. The following stage involves deprotecting

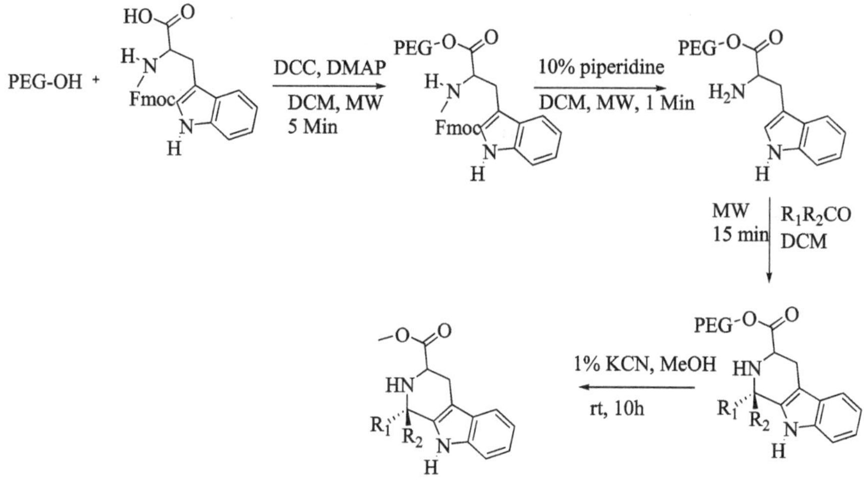

SCHEME 8.30 Microwave assisted synthesis of 1,2,3,4-tetrahydro-β-carbolines.

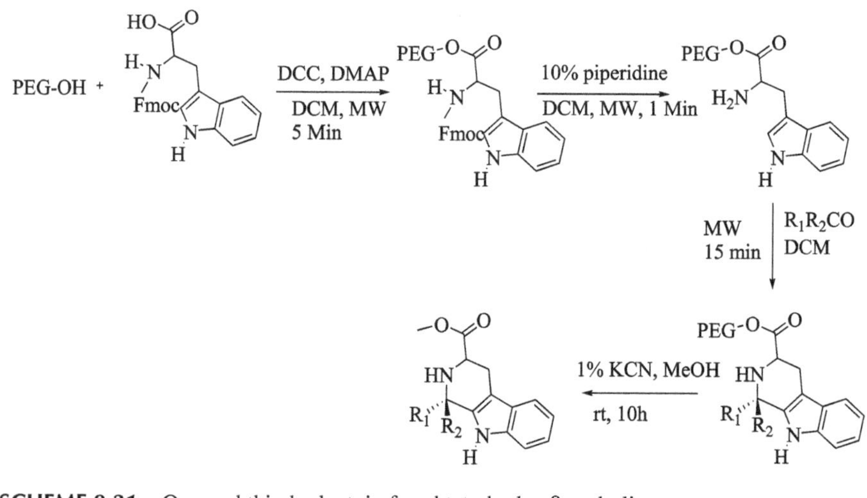

SCHEME 8.31 Oxo and thio hydantoin fused tetrahydro-β-carboline.

the Boc group under microwave irradiation, followed by cyclization with aldehydes to produce tetrahydro-β-carbolines in a cis/trans isomer ratio of 1:1 Yeh et al. (2007) reported the second diversity, which was developed by applying for the terminal thiohydantoin moiety through the N-2/C-3 bond of the tetrahydro-β-carboline frame. The process of cyclization takes place in between the polymer-attached tetrahydro-β-carboline and substituted isothiocyanates to produce tetrahydro-β-carbolines (Scheme 8.31).

Balogh and co-workers developed a novel one-pot protocol for the synthesis of substituted 3,4-dihydro-β-carbolines (DHβCs) in the presence of T_3P (Abranyi-Balogh et al. 2016). The synthesis of isoeudistomin alkaloid, which involves the

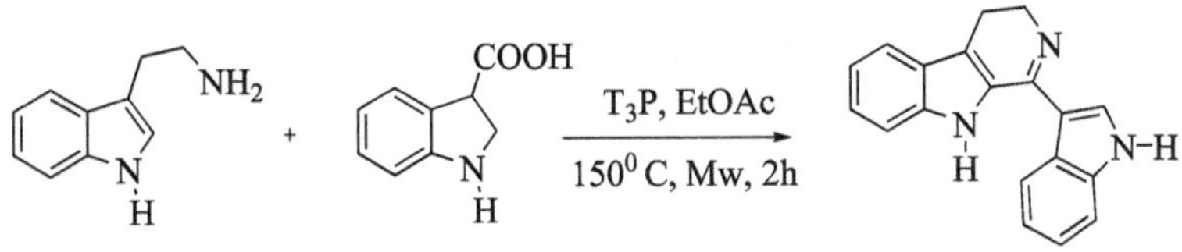

SCHEME 8.32 Synthesis of substituted 3,4-dihydro-beta carboline.

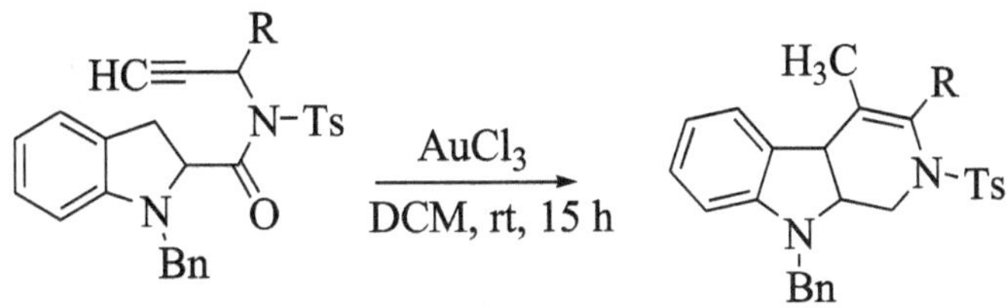

SCHEME 8.33 Synthesis of N-Tosyl-β-carbolinones.

reaction of tryptamine and indole-3-carboxylic acid, was also carried out using this approach (Scheme 8.32).

In another protocol, Verniest and colleagues devised a procedure for N-Tosyl-β-carboline synthesis that involves the reaction of indole substituted N-propargylamides in the $AuCl_3$ and followed by the 6-exo-dig cyclization under milder reaction reactions (Verniest et al. 2010). Easily available as starting materials for the aforesaid synthesis is the acylation of N-tosylpropargylamines, which is generated in situ using N-benzyl-1H-2-indolecarbonyl chloride (Scheme 8.33).

8.3 *β*-CARBOLINE AS ANTIDEPRESSANT AGENTS

Both the chemistry and functionality of the brain are affected by depression. Depression is a major mental disorder, which often manifests as exhaustion, apathy, low spirits/thoughts, lack of concentration, anorexia, sleeplessness. etc. (Bakim et al. 2012). Patients with long-term illnesses like Alzheimer's, Parkinson's, and HIV frequently exhibit depressive symptoms (Skalisz et al. 2002). Mortality and morbidity are brought on by major depressive illnesses, which include post-traumatic stress disorder (PTSD), generalized anxiety, obsessive-compulsive disorder (OCD), and irritability (Muhammad et al. 2013). Many factors are involved in depression such as genetic, psychosocial, physical, biological, etc. Imbalance in the concentration levels of various monoamine neurotransmitters such as serotonin/5-hydroxytryptamine (5-HT), norepinephrine/noradrenaline (NA), dopamine, and their metabolites is the major cause of depression (Ferraz et al. 2019). Medically, tricyclic antidepressants (TCAs), monoamine oxidase inhibitors (MAOIs), selective serotonin reuptake inhibitors (SSRIs), serotonin-norepinephrine reuptake inhibitors (SNRIs), and other serotonergic synthetic or herbal remedies are frequently used to maintain the pathophysiological effects of depression (Hamid et al. 2017). The primary cause of depression is monoaminergic activity in the brain. The concentration levels of amine neurotransmitters in the central nervous system are modulated by MAO enzyme, which serves as a catalyst for neurotransmitter oxidative deamination (Herraiz et al. 2018). Thus, inhibition of MAO is essential to maintain balanced levels of 5-HT, dopamine, and norepinephrine/noradrenaline (NA) neurotransmitters in the brain by preventing their degradation and reuptake. An imbalance of dopamine in the brain causes symptoms of depression such as feelings of sadness and hopelessness, while an imbalance of serotonin affects the processing of emotions. Dopamine, norepinephrine, and serotonin are the three neurotransmitters linked to depression. The main purpose of an antidepressant is to treat depression. By changing the networks and chemicals that transmit signals through nerve pathways to the brain, antidepressants are helpful in treating depression The five main categories of antidepressants are as follows:

[a] Selective serotonin reuptake inhibitors (SSRIs)
[b] Serotonin and nor-epinephrine reuptake inhibitors (SNRIs)
[c] Tricyclic antidepressants (TCAs)
[d] Monoamine oxidase inhibitors (MAOIs)
[e] Atypical antidepressants

The neurochemical effects of antidepressants vary depending on the moieties such as aromaticity, functional groups, chain conformations, etc., with varying degrees of SAR (Ayoob et al. 2017). β-Carboline analogues have distinctive pharmacophore characteristics that express their interactions with the depressive syndromes-related processes. The beta-carboline alkaloids like harmine, harman, and nor-harman (e) are MAO inhibitors present in a wide variety of plant products, including foods, medicines, and narcotics. For example, 6-methoxy-1,2,3,4-tetrahydro-β-Carboline (pinoline) lowers depression symptoms by competitively blocking the function of the MAO-A and by directly binding to the serotonin transporter, avoiding serotonin absorption into the brain's synaptosomes and platelet (Jiang et al. 2009). Through the restriction of MAO-A and 5-HT reuptake and related selective serotonin reuptake inhibitor-binding process, pinoline functions as an antidepressant (Pähkla et al. 2000). The possible antidepressant effects of β-carboline derivatives are outlined in Table 8.1, along with the mechanisms of action shown in both in vivo and in vitro investigations.

The β-Carboline scaffolds have a significant role in biological function. When different positions 1, 3, 5, 6, 7, 8, and 9 are substituted, several physiologically active molecules with a broad spectrum of pharmacological actions are produced (Thatikayala et al. 2022). (Figure 8.7).

The main biochemical theory of depression disorders is the monoamine hypothesis. It was first put forth in the 1950s (Freis et al. 1954) and explains depression as the result of a change in the levels of one or more monoamines, such as serotonin (5-HT), dopamine (DA), and noradrenaline (NA). Noradrenergic, serotoninergic, and dopaminergic neurons are widely distributed throughout the central nervous system.

TABLE 8.1
Pharmacological movements of β-Carboline scaffolds against depression

S.N.	β-Carbolines Compound	Mechanism of action	Ref.
1.	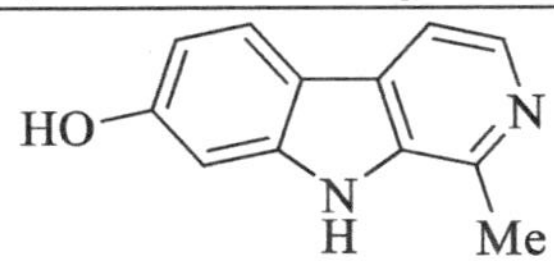	Encourage of hippocampal neurotransmitter generation.	(Morales-García et al. 2017)
2.	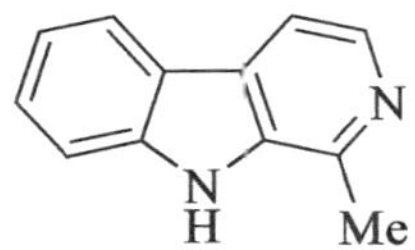	Obstruct of MAO-A and Inverse agonism on benzodiazepine receptors.	(Aricioglu et al. 2003; Farzin et al. 2006)
3.	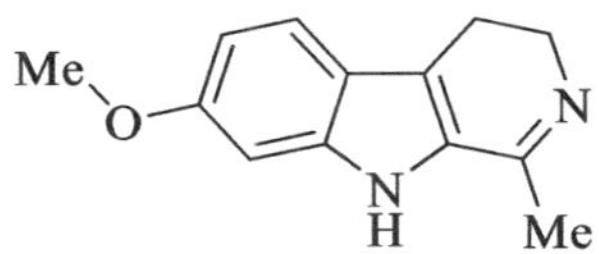	Prevention of MAO-B and inverse agonism on benzodiazepine receptors to modulate glutamate/GABA levels.	(Ebrahimi-Ghiri et al. 2019)
4.	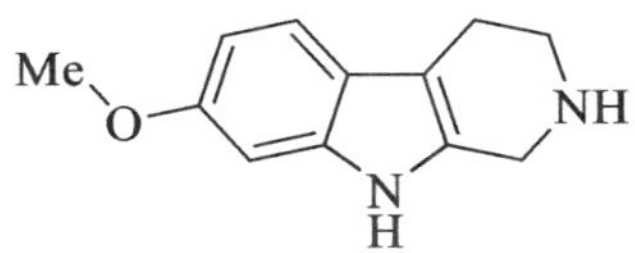	Inhibition of MAO-A, 5-HT reuptake, and affinity to SSRI binding sites.	(Pahkla et al. 2000)

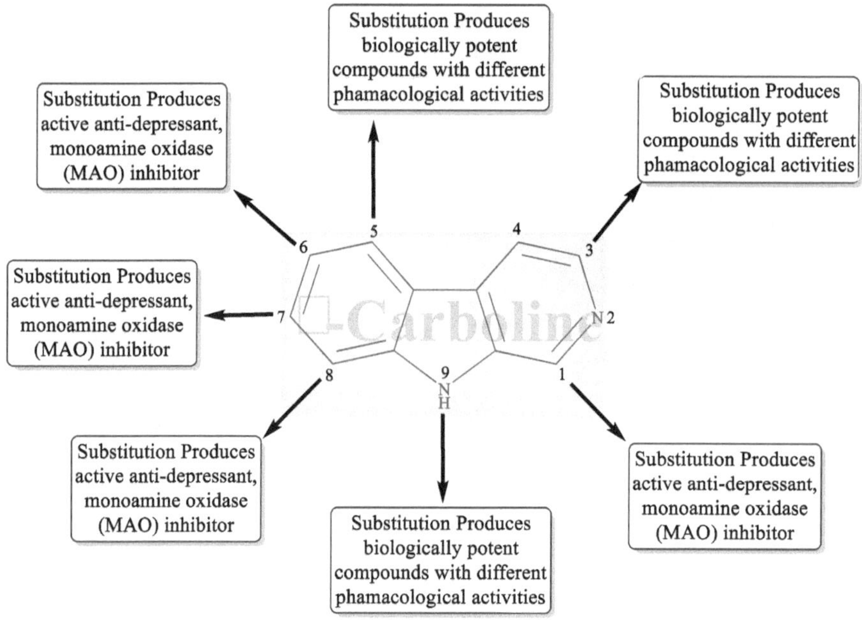

FIGURE 8.7 Structure activity relationship (SAR) of β-Carboline scaffolds.

In general, NA affects how the prefrontal cortex works, how working memory is processed, and how behavior and attention are controlled. The most cohesive network of neurotransmitters in the brain is 5-HT, which is essential for controlling mood, emotion, and motor function. Depressive behaviors may be influenced by hypo- or hyperserotoninergic states, indicating that this monoamine plays a critical role in regulating other neurotransmitters that are also connected to depressive disorders. Lastly, DA regulates working memory, attention, reward and motivation processes, and motor coordination (Jesulola et al. 2018; Hamon et al. 2013; Belzung et al. 2015; Dean et al. 2017; Goldberg et al. 2014). This means that many behavioral signs of depression, including low mood, alertness, decreased motivation, exhaustion, aggressiveness, impaired focus, and cognitive performance, may be caused by monoaminergic system failure (Jesulola et al. 2018). Because of this, the majority of antidepressant drugs function by raising monoamine levels in synaptic clefts and subsequently regulating the corresponding neurotransmission systems. Numerous investigations have shown that by inhibiting MAO (Kim et al. et al. 1997) and 5-HT reuptake (McIsaac et al. 1972), β-carbolines raise the extracellular concentration levels of these monoamines in various brain areas. The oxidative deamination of monoamines is catalyzzed by MAOs, a family of flavin-containing amine oxidoreductases that are typically found within the outer mitochondrial membrane.

MAO-A and MAO-B are two isoforms of MAO that have different substrate preferences. MAO-A primarily breaks down 5-HT molecules, while MAO-B has a stronger affinity for phenylethylamine and phenylamine. DA and NA are normally metabolized by both isoforms (Corbineau et al. 2017; Fisar et al. 2016). Previous studies have shown that β-carboline increases 5-HT levels in the synapse by preferentially inhibiting MAO-A.

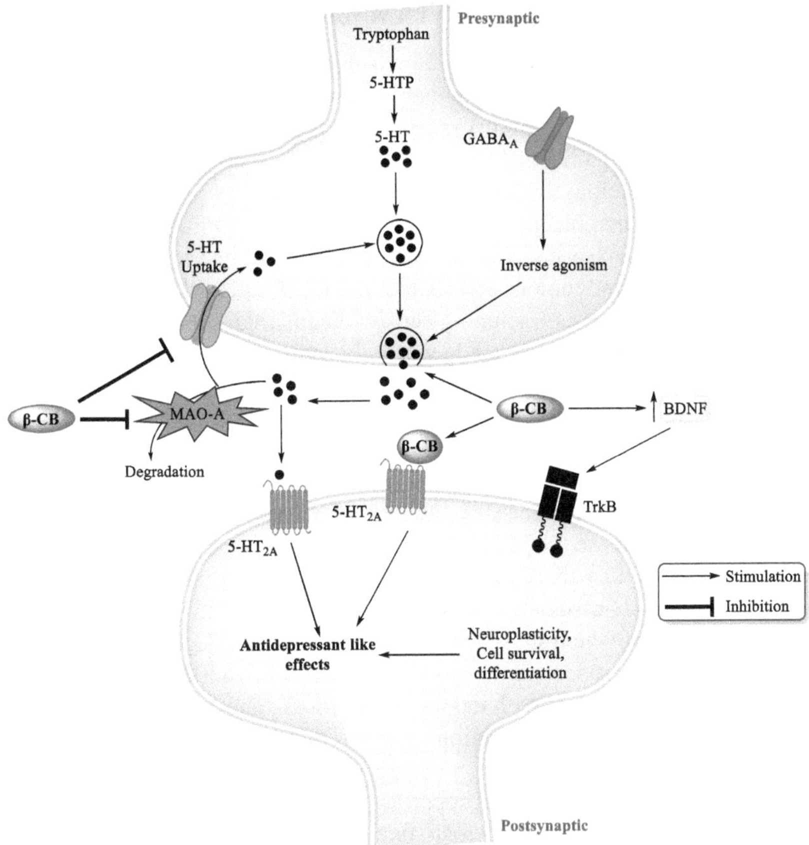

FIGURE 8.8 Molecular mechanisms involved in the antidepressant effects of β-carbolines (βCB).

It is interesting to note that these alkaloids can interact with a variety of cell-surface receptors, including 5-HT$_{2A}$ receptors (Glennon et al. 2000), which are connected to the pharmacology of antidepressants. 5-HT$_{2A}$, which are G-protein coupled receptors, are found in the brain's thalamus, amygdala, hippocampal, and cortical areas.

Once activated, these receptors can regulate neuronal excitation implicated in depression through modulation of glutamatergic and GABAergic postsynaptic neurons, thereby producing antidepressant effects (Guiard et al. 2015). β-carbolines interact with GABAA receptors in addition to serotoninergic receptors. In this situation, an inverse-agonistic mechanism is at work, which lowers inhibitory postsynaptic signals and subsequently modifies the antidepressant-like effects of monoamines, such as 5-HT. Since certain carbolines serve as reverse agonists of GABAA receptors they directly affect the 5-HT system by inhibiting MAO-A and 5-HT reuptake process (Herraiz et al. 2010).

Figure 8.8 demonstrate the primary chemical processes that underlie the antidepressant-like effects of β-carbolines.

8.4 *β*-CARBOLINE AS ANTI-INFLAMMATORY AGENTS

The β-carboline derivatives with diversified substitution at positions like 1, 2, 3, 6, 7, and 9 showcase a variety of biological activities. According to Medzhitov et al. (2008), inflammation is an adaptive reaction brought on by unpleasant stimuli and circumstances including infection and tissue damage. The majority of the cellular and molecular pathophysiology of inflammation is mediated by proinflammatory cells, primarily activated macrophages, which produce a variety of inflammatory mediators, such as reactive oxygen species (ROS), nitric oxide (NO), interleukin-6 (IL-6), prostaglandin E_2 (PGE_2), and tumor necrosis factor-a (TNF-a) (Laskin et al. 2011; Kang et al. 2012). Many diseases, including sepsis, rheumatoid arthritis, atherosclerosis, and even cancer, are mediated or made worse by an excess or uncontrolled synthesis of these mediators (Tabas et al. 2013). Nuclear factor erythroid 2-related factor 2 (Nrf_2) is a transcription factor involved in the regulation of inflammation. Through consensus sequences known as antioxidant-response elements, Nrf_2, a member of the basic-leucine zipper transcription factor, controls the expression of many detoxifying and antioxidant genes, including heme oxygenase-1 (HO-1), NAD(P)H:quinone oxidoreductase (NQO1), glutathione S-transferases, and g-glutamyl cysteine synthetase catalytic subunit (GCLC) (Jaramillo et al. 2013). When Nrf_2 is bound to the Kelch-like ECH-associated protein (Keap1), it enters the Keap1-Cul_3-E_3 ubiquitin ligase complex, which subsequently ubiquitinates and degrades Nrf_2 (Jaramillo et al.). In response to oxidative stress, Nrf_2 is released from Keap1, moves to the nucleus, forms a heterodimer with the tiny Maf protein, and binds to sites associated with antioxidants in the promoter regions of genes that are cytoprotective and antioxidant (Jaramillo et al. 2013).

β-carboline's anti-inflammatory properties were attained by inhibiting the generation of nitric oxide and tumor necrosis factor (TNF)-alpha in lipopolysaccharide-stimulated murine RAW264 and human THP-1 cells. In RAW264 cells, β-carboline exhibited more potent TNF-alpha suppression effects compared to butein and reference polyphenol. It was also discovered that β-carboline inhibited the production of interleukin-6 in RAW264 cells (Yamazaki et al. 2011). Furthermore, another way that Nrf_2 might be activated is by phosphorylation. It has been demonstrated that several kinases are involved in the phosphorylation of Nrf2, including mitogen-activated protein kinases (MAPKs) including extracellular signal-regulated kinases (ERK), c-jun N-terminal kinases (JNK), and p38 kinases (Alfieri et al. 2008). Neurodegenerative disorders, aging, diabetes, photooxidative stress, cardiovascular disease, inflammation, pulmonary fibrosis, acute pulmonary damage, and cancer have all been demonstrated to be protected against by activating the Nrf_2 defense response (Boutten et al. 2011).

According to Paine et al. (2010), HO-1 is the main antioxidative and anti-inflammatory enzyme that is controlled by activating Nrf_2. The rate-limiting enzymatic phase of heme breakdown is catalyzed by heme oxygenase (HO), which also generates carbon monoxide, ferrous iron, and biliverdin. Biliverdin reductase then converts biliverdin into bilirubin (Paine et al. 2010). HO-1 and HO-2 are two known genetically different HO isozymes (Paine et al. 2010). The constitutive isoform is represented by HO-2 (Paine et al. 2010). On the other hand, a variety of triggers,

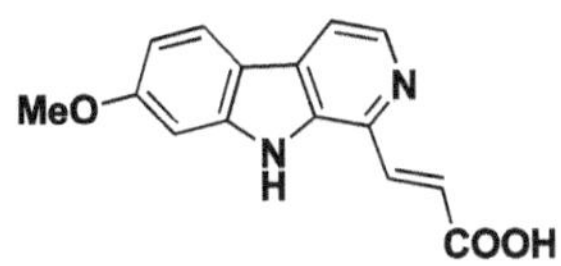

(*E*)-3-(7-methoxy-9*H*-pyrido[3,4-*b*]indol-1-yl)acrylic acid

FIGURE 8.9 Anti-inflammatory drug.

such as inflammatory mediators and factors associated with oxidative stress, significantly upregulate the inducible isoform HO-1 (Maines et al. 2001). According to Abraham et al. (2008), HO-1 preserves cellular homeostasis and is crucial for tissue protection by lowering oxidative damage and lowering the inflammatory response. In activated macrophages, the expression of HO-1 or carbon monoxide therapy inhibits the production of proinflammatory cytokines and chemokines (Choo et al. 2015). Accordingly, using natural phytochemicals to target the Nrf_2/HO-1 pathway may be a useful tactic for the management or prevention of inflammatory illnesses (Motterlini et al. 2014).

In Southeast Asian nations, Eurycoma longifolia Jack (Simaroubaceae) is a significant medicinal plant (Bhat et al. 2010). Quassinoids, canthin-6-one, β-carboline alkaloids, and other bioactive components found in the plant are responsible for E. longifolia's efficacy (Bhat et al. 2010). According to a recent study (Tran et al. 2014), some quassinoids and β-carboline alkaloids from E. longifolia exhibit strong NF-kB inhibitory effects when TNF-a is present. Nevertheless, not much research has been done on E. longifolia's anti-inflammatory properties. Searching further for Nrf2 activators in herbal medicinal plants, they discovered a strong Nrf2 activator in 7-MCPA, a β-carboline alkaloid isolated from *E. longifolia* hairy-root cultures. In this work, we used RAW264.7 cells to examine the anti-inflammatory properties of 7-MCPA as well as its underlying processes. Here, we demonstrated how 7-MCPA prevented mice from developing septic shock from lipopolysaccharide (LPS)-induced inflammation in vivo by activating the Nrf2/HO-1 pathway through a ROS-dependent p38 MAPK pathway. Strongly inhibiting the generation of NO, the anti-inflammatory chemical (E)-3-(7-methoxy-9H-pyrido[3,4-b]indol-1-yl)acrylic acid was isolated from the hairy-root cultures of Eurycoma longifolia (Ngoc et al. 2016) (Figure 8.9).

8.5 CONCLUSION

This current review chapter summarizes the modern synthetic routes and in vitro and in vivo anti-inflammatory, antidepressant properties of β-carbolines. β-carbolines have also been investigated for other biological purposes, but most of the clinical studies are superficial and did not employ detailed experimental studies or molecular mechanisms. More and more techniques are being discovered that are based on ring-closing metathesis, cyclization, condensation, oxidative C-N coupling, and palladium catalyzed reductive N-hetero annulation for the synthesis of β-carbolines. The synthetic diversity of the proposed mechanisms and the action of these compounds suggests that β-carbolines are multitargeted useful for the clinical treatment of many disorders.

REFERENCES

Abinaya, R., Srinath, S., Soundarya, S., Sridhar, R., Balasubramanian, K.K. and Baskar, B. 2022. Recent developments on synthesis strategies, SAR studies and biological activities of β-carboline derivatives–an update. *J. Mol. Struct.* 1261:132750.

Abraham, N.G. and Kappas, A. 2008. Pharmacological and clinical aspects heme oxygenase. *Pharmcol. Rev.* 60:79–127.

Ábrányi-Balogh, P., Földesi, T., Grün, A., Volk, B., Keglevich, G. and Milen, M. 2016. Synthetic study on the T_3P®-promoted one-pot preparation of 1-substituted-3,4-dihydro-β-carbolines by the reaction of tryptamine with carboxylic acids. *Tetrahedron Lett.* 57:1953–1957.

Alfieri, A., Srivastava, S., Siow, R.C., Modo, M.P., Fraser, A. and Mann, G.E. 2008. Targeting the Nrf2-Keap1 antioxidant defense pathway for neurovascular protection in stroke. *J. Physiol.* 589:4125–4136.

Ang, K.K., Holmes, M.J., Higa, T., Hamann, M.T. and Kara, U.A. 2000. In vivo antimalarial activity of the beta-carboline alkaloid manzamine A. *Antimicrob. Agents Chemother.* 44(6):1645–1649.

Aricioglu, F., Yillar, O., Korcegez, E. and Berkman, K. 2003. Effect of harmane on the convulsive threshold in epilepsy models in mice. *Ann. N. Y. Acad. Sci* 1009(1):190–195.

Ashok, P., Chander, S., Smith, T.K., Singh, R.P., Jha, P.N. and Sankaranarayanan, M. 2019. Biological evaluation and structure activity relationship of 9-methyl-1-phenyl-9H-pyrido[3,4-b]indole derivatives as anti-leishmanial agents. *Bioorg. Chem.* 84:98–105. doi.org/10.1016/j.bioorg.2018.11.037

Ayoob, I., Hazari, Y.M., Lone, S.H., Shakeel-u-Rehman, Khuroo, M.A., Fazili, K.M. and Bhat, K.A. 2017. Phytochemical and cytotoxic evaluation of peganum harmala: Structure activity relationship studies of harmine. *ChemistrySelect*, 2(10):2965–2968.

Bakim, B., Sertcelik, S. and Tankaya, O. 2012. A case of serotonin syndrome with antidepressant treatment and concomitant use of the Herbal Remedy (Peganum Harmala). *BCP* 22(4):359–361.

Battini, N., Padala, A.K., Mupparapu, N., Vishwakarma, R.A. and Ahmed, Q.N. 2014. Unexplored reactivity of 2-Oxoaldehydes towards Pictet–Spengler conditions: Concise approach to *β*-Carboline based marine natural products," *RSC Adv.* 4(50):26258–26263.

Belzung, C., Willner, P. and Philippot, P. 2015. Depression: From psychopathology to pathophysiology. *Curr. Opin. Neurobiol.* 30:24–30.

Bhat, R. and Karim, A.A. 2010. Tongkat Ali (Eurycoma longifolia Jack): A review on its ethnobotany and pharmacological importance. *Fitoterapia* 81:669–679.

Boutten, A., Goven, D., Artaud-Macari, E., Boczkowski, J. and Bonay, M. 2011. NRF2 targeting: A promising therapeutic strategy in chronic obstructive pulmonary disease. *Trends. Mol. Med.* 17:363–371.

Buxi, S., Rihui, C., Wenxi, F., Liang, G., Qin, M., Xuemei, C., Guoxian, Z., Liqin, Q. and Huacan, S. 2013. Design, synthesis and in vitro and in vivo antitumor activities of novel bivalent Bcarbolines. *Eur. J. Med. Chem.* 60:10–22.

Cao, R., Peng, W., Wang, Z. and Xu, A. 2007. β-Carboline alkaloids: Biochemical and pharmacological functions. *Curr. Med. Chem.* 14:479–500.

Casal, S. Chapter 20. 2019. Potential effects of β-carbolines on human health. In *Coffee: Consumption and Health Implications*. The Royal Society of Chemistry: Croydon, UK, 461–468.

Cera, G., Lanzi, M., Balestri, D., Della, C.N., Maggi, R., Bigi, F., Malacria, M. and Maestri, G. 2018. Synthesis of Carbolines via Palladium/Carboxylic acid joint catalysis. *Org. Lett.* 20:3220–3224.

Chang, W.J., Kulkarni, M.V. and Sun, C.M. 2006. Traceless and stereoselective synthesis of tetrahydro-β-carbolinethiohydantoins by microwave irradiation. *J. Comb. Chem.* 8:141–144.

Choo, Y.Y., Lee, S., Nguyen, P.H., Lee, W., Woo, M.H., Min, B.S. and Lee, J.H. 2015. Caffeoylglycolic acid methyl ester, a major constituent of sorghum, exhibits anti-inflammatory activity via the Nrf2/heme oxygenase-1 pathway. *RSC Adv.* 5:17786–17796.

Corbineau, S., Breton, M., Mialet-Perez, J. and Costemale-Lacoste, J.-F. 2017. Major depression and heart failure: Interest of monoamine oxidase inhibitors. *Int. J. Cardiol.* 247:1–6.

Coussens, L.M. and Werb, Z. 2002. Inflammation and cancer. *Nature* 420(6917):860–867.

Dai, J., Dan, W., Schneider, U. and Wang, J. 2018. β-Carboline alkaloid monomers and dimers: Occurrence, structural diversity, and biological activities. *Eur. J. Med. Chem.* 157:622–656.

Dean, J. and Keshavan, M. 2017. The neurobiology of depression: An integrated view. *Asian J Psychiatr.* 27:101–111.

Devi, N., Kumar, S., Pandey, S.K. and Singh, V. 2018. 1(3)-Formyl-β-carbolines: Potential aldo-X precursors for the synthesis of β-carboline based molecular architectures. *Asian J. Org. Chem.* 7:6–36.

Dhara, S., Singha, R., Ahmed, A., Mandal, H., Ghosh, M., Nuree, Y. and Ray, J.K. 2014. Synthesis of α, β and γ-carbolines via Pd-mediated Csp2-H/N–H activation *RSC Adv.* 4:45163–45167.

Dhiman, S., Mishra, U.K. and Ramasastry, S.S.V. 2016. One-Pot Trimetallic relay catalysis: A unified approach for the synthesis of β-Carbolines and other [c]-fused pyridines. *Angew. Chem. Int. Ed.* 55(7737):7868–7872.

Ding, S., Shi, Z. and Jiao, N. 2010. Pd(II)-Catalyzed synthesis of carbolines by iminoannulation of internal alkynes via Direct C–H bond cleavage using dioxygen as oxidant. *Org. Lett.* 12:1540–1543.

Domínguez, G. and Pérez-Castells, J. 2011. Chemistry of β-carbolines as synthetic intermediates. *Eur. J. Org. Chem.* 2011:7243–7253.

Eagon, S. and Anderson, M.O. 2014. Microwave-assisted synthesis of Tetrahydro-b-Carbolines and β –Carbolines. *Eur. J. Org. Chem.* 8(2014):1653–1665.

Eagon, S. and Anderson, M.O. 2014. Microwave-assisted synthesis of Tetrahydro-β-carbolines and β-Carbolines. *Eur. J. Org. Chem.* 2014(8):1653–1665.

Ebrahimi-Ghiri, M., Nasehi, M. and Zarrindast, M.-R. 2019. Anxiolytic and antidepressant effects of ACPA and harmaline co-treatment. *Behav. Brain. Res.* 364:296–302.

Farzin, D. and Mansouri, N. 2006. Antidepressant-like effect of harmane and other β-carbolines in the mouse forced swim test. *Eur Neuropsychopharmacol.* 16(5):324–328.

Ferraz, C.A.A., De Oliveira Junior, R.G., Picot, L., Da Silva Almeida, J.R.G. and Nunes, X.P. 2019. Pre-clinical investigations of β-carboline alkaloids as antidepressant agents: A systematic review. *Fitoterapia* 137:104196.

Fisar, Z. 2016. Drugs related to monoamine oxidase activity. *Prog. Neuropsychopharmacol. Biol. Psychiatry.* 69:112–124.

França, P.H., Barbosa, D.P., Silva D.L., Ribeiro, E.A., Santana, A.E., Santos B.V., Barbosa-F, J.M., Quintans, J.S., Barreto, R.S. and Quintans-J, L.J. 2014. Indole alkaloids from marine sources as potential leads against infectious diseases. *BioMed Res. Int.* 2014:375423.

Freis, E.D. 1954. Mental depression in hypertensive patients treated for long periods with large doses of reserpine. *N. Engl. J. Med.* 251(25):1006–1008.

Glennon, R.A., Dukat, M., Grella, B., Hong, S.-S., Costantino, L., Teitler, M., Smith, C., Egan, C., Davis, K. and Mattson, M.V. 2000. Binding of β-carbolines and related agents at serotonin (5-HT2 and 5-HT1A), dopamine (D2) and benzodiazepine receptors. *Drug and Alcohol Dependence*, 60(2):121–132.

Gold, P.W., Machado-vieira, G. and Pavlatou, M.G. 2015. Clinical and biochemical manifestations of depression: Relation to the neurobiology of stress. *Neural Plast.* 2015:7–9.

Goldberg, J.S., Bell, C.E. and Pollard, D.A. 2014. Revisiting the monoamine hypothesis of depression: A new perspective. *Perspect. Medicinal. Chem.* 6:PMC.S11375.

Guiard, B.P. and Giovanni, G.D. 2015. Central serotonin-2A (5-HT2A) receptor dysfunction in depression and epilepsy: The missing link? *Front. Pharmacol.* 6:1–17.

Hamid, H.A., Ramli, A.N.M. and Yusoff, M.M. 2017. Indole alkaloids from plants as potential leads for antidepressant drugs: A mini review. *Front. Pharmacol.* 8: 1–7.

Hamon, M. and Blier, P. 2013. Monoamine neurocircuitry in depression and strategies for new treatments. *Prog.* 45:54–63.

He, Y., Agarwal, P.K., Kiran, I.N.C., Yu, R., Cao, B., Zou, C., Zhou, X., Xu, H., Xu, B., Zhu, L., Lan, Y. and Nicolaou, K.C. 2016. Efficient synthesis of dimeric oxazoles, piperidine and tetrahydroisoquinolines from N-Substituted 2-oxazolones. Chem. *Eur. J* 22(23):7696–7701.

Herraiz, T. and Guillen, H. 2018. Monoamine Oxidase-A Inhibition and associated antioxidant activity in plant extracts with potential antidepressant actions. *BioMed Res. Int.* 2018:1–10.

Herraiz, T., González, D., Ancin-Azpilicueta, C., Aran, V.J. and Guillen, H. 2010. β-Carboline alkaloids in Peganum harmala and inhibition of human monoamine oxidase (MAO). *Food. Chem. Toxicol.* 48(3):839–845.

Heshmati, P., Nasehi, M. and Zarrindast, M.R. 2014. An overview of cognitive aspects of βcarbolines. *J. Paramed. Sci.* 5:99–104.

Hong-J, B., Kyung-H, J., Gi-S, M., Sung-K, M. and Hyang-Y, L. 2020. A facile and efficient method for the synthesis of crystalline tetrahydro β-carbolines via the Pictet-Spengler reaction in water. *Sci. Rep.* 10:1057–1062.

Jaramillo, M.C. and Zhang, D.D. 2013. The emerging role of the Nrf2-Keap1 signaling pathway in cancer. *Genes. Dev.* 27:2179–2191.

Jesulola, E., Micalos, P. and Baguley, I.J. 2018. Understanding the pathophysiology of depression: From monoamines to the neurogenesis hypothesis model – are we there yet? *Behav. Brain Res.* 341:79–90.

Ji, Y.P., Biao, C., Wanhua, W., Cheng, Y., Guoqing, Z. and Jason, J.C. 2021. Design and synthesis of fluorescent 1,3-Diaryl-β-carbolines and 1,3-Diaryl-3,4-dihydro-β-carbolines. *ACS Omega.* 6:12238–12249.

Jiang, X.-L., Shen, H.-W. and Yu, A.-M. 2009. Pinoline may be used as a probe for CYP2D6 activity. *Drug Metab. Dispos.* 37(3):443–446.

Kamlah, A., Lirk, F. and Bracher, F. 2015. A new approach to 1-substituted β-carbolines and isoquinolines utilizing tributyl[(Z)-2-ethoxyvinyl]stannane as a C-3,C-4 building block. *Tetrahedron* 72(6):837–845.

Kang, J., Tae, N., Min, B.S., Choe, J. and Lee, J.H. 2012. Malabaricone C suppresses lipopolysaccharide-induced inflammatory responses via inhibiting ROS mediated Akt/IKK/NF-kB signaling in murine macrophages. *Int. Immunopharmacol.* 14:302–310.

Kim, H., Sablin, S.O. and Ramsay, R.R. 1997. Inhibition of monoamine Oxidase A by β-Carboline derivatives. *Arch. Biochem. Biophys.* 337(1):137–142.

Laskin, D.L., Sunil, V.R., Gardner, C.R. and Laskin, J.D. 2011. Macrophages and tissue injury: Agents of defense or destruction? *Annu. Rev. Pharmaco.l Toxicol.* 51:267–288.

Maines, M.D. and Panahian, N. 2001. The heme oxygenase system and cellular defense mechanisms. Do HO-1 and HO-2 have different functions? *Adv. Exp. Med. Biol.* 502:249–272.

Maity, P., Adhikari, D. and Jana, A.K. 2019. An overview on synthetic entries to tetrahydro-β-carbolines. *Tetrahedron* 75:965–1028.

McIsaac, W.M., Taylor, D., Walker, K.E. and Ho, B.T. 1972. 6-Methoxy-1,2,3,4-tetrahydro-?-carboline?? Serotonin elevator. *J. Neurochem.* 19(4):1203–1206.

Medzhitov, R. 2008. Origin and physiological roles of inflammation. *Nature* 454:428–435.

Mei-L, Y., Ping-C, K., Tsong-L, H., Wen-F, C., Keduo, Q., Chin-Y, L., Kuo-H, L. and Tian-S, W. 2011. Synthesis, in vitro anti-inflammatory and cytotoxic evaluation, and mechanism of action studies of 1-benzoyl-b-carboline and 1-benzoyl-3-carboxy-b-carboline derivatives. *Bioorg. Med. Chem.* 19:1674–1682.

Morales-García, J.A., De La Fuente Revenga, M., Alonso-Gil, S., Rodriguez-Franco, M.I., Feilding, A., Perez-Castillo, A. and Riba, J. 2017. The alkaloids of Banisteriopsis caapi, the plant source of the Amazonian hallucinogen Ayahuasca, stimulate adult neurogenesis in vitro. *Sci. Rep.* 7(1):5309–5321.

Motterlini, R. and Foresti, R. 2014. Heme oxygenase-1 as a target for drug discovery. *Antioxid. Redox. Signal.* 20:1810–1826.

Muhammad, N., Saeed, M., Khan, H. and Haq, I. 2013. Evaluation of n-hexane extract of Viola betonicifolia for its neuropharmacological properties. *J. Nat. Med.* 67(1):1–8.

Ngoc, P.B., Pham, T.B., Nguyen, H.D., Tran, T.T., Chu, H.H., Chau, V.M., Lee, J.H. and Nguyen, T.D. 2016. A new anti-inflammatory β-carboline alkaloid from the hairy-root cultures of Eurycoma longifolia, *Nat. Prod. Res.* 30:1360–1365.

Nissen, F., Richard, V., Alayrac, C. and Witulski, B. 2011. Synthesis of β- and γ-carbolines viaruthenium and rhodium catalysed [2+2+2] cycloadditions of yne-ynamides with methylcyanoformate *Chem. Commun.* 47:6656–6658.

Pahkla, R., Kask, A. and Rago, L. 2000. Differential effects of beta-carbolines and antidepressants on rat exploratory activity in the elevated zero-maze. *Pharmacol.* 65(4):737–742.

Paine, A., Eiz-Vesper, B., Blasczyk, R. and Immenschuh, S. 2010. Signaling to heme oxygenase-1 and its anti-inflammatory therapeutic potential. *Biochem. Pharmacol.* 80:1895–1903.

Piechowska, P., Zawirska-Wojtasiak, R. and Mildner-Szkudlarz, S. 2019. Bioactive β-carbolines in food: A review. *Nutrients* 11:814–823.

Rajesh, R.P. and Murugan A. 2019. Spectroscopic identification of brominated, nonbrominated alkaloids and evaluation of antimicrobial activity of Eudistomin-I, Eudistomin H, from green ascidian Eudistoma viride. *J. Appl. Pharmaceut. Sci.* 9:116–123.

Royer, J., Bonin, M. and Micouin, L. 2004. Chiral heterocycles by iminium ion cyclization. *Chem. Rev.* 104:2311–2352.

Royer, J., Bonin, M. and Micouin, L. 2004. Chiral heterocycles by iminium ion cyclization. *Chem. Rev.* 104:2311–2352.

Saha, S., Reddy, C.V.R. and Patro, B. 2011. Facile two-step synthesis of crispine a and harmicine by cyclopropylimine rearrangement." *Tetrahedron Lett.* 52(31): 4014–4016.

Sathiyalingama, S. and Stefan, R. 2022. Synthesis of α- and β-Carbolines by a Metalation/Negishi CrossCoupling/SNAr Reaction Sequence. Adv. *Synth. Catal.* 364:1769–1774.

Sheng, T., Yu, C., Wang, Y., Li, S., Wu, H., Gu, Q., Meng, G., Herron, A.N. and Gao, X. 2023. Chromatography-free synthesis of β-Carboline 1-Hydrazides and an investigation of the mechanism of their bioactivity: The discovery of β-carbolines as promising antifungal and antibacterial candidates. *J. Med. Chem.* 66(13):9040–9056.

Singh, V. and Batra, S. 2012. 1-Formyl-9H-b-Carboline: A useful scaffold for synthesizing substituted- and fused β –Carbolines. Curr. *Org. Synth.* 9(4): 513–528.

Singh, V., Devi, N., Kumar, S. and Pandey, S.K. 2017. 1(3)-Formyl-Carbolines: Potential Aldo-X precursors for the synthesis of *β* -Carboline-based molecular architectures. *Asian. J. Org. Chem.* 7(1):6–36.

Skalisz, L.L., Beijamini, V., Joca, S.L., Vital, M.A.B.F., Da Cunha, C. and Andreatini, R. 2002. Evaluation of the face validity of reserpine administration as an animal model of depression–Parkinson's disease association. *Prog* 26(5):879–883.

Smirnova, O.B., Golovko, T.V. and Granik, V.G. 2011. Carbolines. Part 2: Comparison of some of the properties of α-, γ-, and δ-carbolines. *Pharm. Chem. J.* 45:389–400.

Spindler, A., Stefan, K. and Wiese, M. 2016. Synthesis and investigation of Tetrahydro-*β* carboline derivatives as inhibitors of the breast cancer resistance protein (ABCG2). *J. Med. Chem.* 59(13):6121–1635.

Tabas, I. and Glass, C.K. 2013. Anti-inflammatory therapy in chronic disease: Challenges and opportunities. *Science* 339:166–172.

Tatsui, G. and Über, D. 1928. Synthese von Carbolinderivaten. *Yakugaku Zasshi* 48:453–459.

Thatikayala, M., Wadhwa, P., Kaur, P., Singh, P.K., Yadav, A., Kaushik, M. and Sahu, S.K. 2022. Beta-carboline as a promising heterocyclic nucleus: Synthetic aspects, pharmacological potential and structure activity relationship. *Eur. J. Med. Chem.* 6:100096–100118.

Tran, T.V. Malainer, C. Schwaiger, S. Atanasov, A.G. Heiss, E.H. Dirsch, V.M. and Stuppner, H. 2014. NF-kB inhibitors from Eurycoma longifolia. *J. Nat. Prod.* 77:483–488.

Verniest, G., England, D., De Kimpe, N. and Padwa, A. 2010. Synthesis of substituted β-carbolines via gold(III)-catalyzed cycloisomerization of N-propargylamides. *Tetrahedron* 66:1496–1502.

Viswanathan, N.K., Chirgwin, M.E., Gibbs, J., Kalaj, B.N., Durham, S., Tran, J., Gomez, M., Lazaro, H., Chen, M., Christopher, R., Emily, M., Derbyshire, R. and Eagon, S. 2023. Synthesis and activity of β-carboline antimalarials targeting the Plasmodium falciparum heat shock 90 protein. *Bioorg. Med. Chem. Lett.* 92:129410–129413.

Wang, Z.X., Xiang, J.C., Cheng, Y., Ma, J.T., Wu, Y.D. and Wu, AX. 2018. Direct biomimetic synthesis of β-carboline alkaloids from two amino acids. *J. Org. Chem.* 83:12247–12254.

Wen, L.R., Dou, Q., Wang, Y.C., Zhang, J.W., Guo, W.S. and Li, M. 2017. Synthesis of 1-Thio-substituted isoquinoline derivatives by tandem cyclization. *J. Org. Chem.* 82(3):1428–1436.

Wu, C.Y. and Sun, C.M. 2002. Parallel synthesis of 1,2,3,4-tetrahydro-βcarbolines usingmicrowave radiation. *Synlett.* 10:1709–1711.

Yamazaki, Y. and Kawano, Y. 2011. Inhibitory effects of herbal alkaloids on the tumor necrosis factor-α and nitric oxide production in lipopolysaccharide-stimulated RAW264 macrophages. *Chem. Pharm. Bull.* 59(3):388–391.

Yang, M.L., Kuo, P.C., Damu, A.G., Chang, R.J., Chiou, W.F. and Wu, T.S. 2006. A versatile route to the synthesis of 1-substituted β-carbolines by a single step Pictet–Spengler cyclization.*Tetrahedron* 62:10900–10906.

Yang, M.L., Kuo, P.C., Hwang, T.L., Chiou, W.F., Qian, K., Lai, C.Y., Lee, K.H. and Wu, T.S. 2011. Synthesis, in vitro anti-inflammatory and cytotoxic evaluation, and mechanism of action studies of 1-benzoyl-β-carboline and 1-benzoyl-3-carboxy-β-carboline derivatives. *Bioorg. Med. Chem.* 19(5):1674–1682.

Yeh, W.P., Chang, W.J., Sun, M.L. and Sun, C.M. 2007. Microwave-assisted traceless synthesis of hydantion-fused β-carboline scafold. *Tetrahedron* 63:11809–11816.

Zeng, H., Zemin, W. and Li, C.J. 2019. Two-in-One strategy for palladium-catalyzed C–H functionalization in water. *Angew. Chem. Int. Ed.* 58:2859–2863.

Zhao, F., Gao, Z., Jiao, W., Chen, L., Chen, L. and Yao, X. 2012. In vitro anti-inflammatory effects of beta-carboline alkaloids, isolated from Picrasma quassioides, through inhibition of the iNOS pathway. *Planta Med.* 78(18):1906–1911.

Zhihui, S., Xiaoyu, Z., Xinyan, Li., Weilin, W., Jiaqiao, D., Bing, C. and Mingqin, Z. 2023. Synthesis of Tetrahydro-β-carbolines by a Manganese-catalysed oxidative pictet–spengler reaction. *Chem. Eur. J.* 29:e202203758.

Zhu, Y.P., Liu, M.C., Cai, Q., Jia, F.C. and Wu, A.X. 2013. A cascade coupling strategy for one-pot total synthesis of –Carboline and Isoquinoline-Containing natural products and derivatives. *Chem. Eur. J.* 19(31):10132–10137. doi:10.1002/chem.201301734

9 β-Carbolines as Anti-HIV Agents
Synthesis and Biological Study

Jaymin Parikh, Keyur Bhatt, and Kuldeep V Joshi

9.1 BRIEF OVERVIEW OF DRUG DISCOVERY AND THE IMPORTANCE OF SCAFFOLDS

The process of drug discovery is intricate and ever evolving, encompassing the discernment, formulation, and advancement of novel therapeutic substances aimed at aliveating medical conditions. It encompasses a range of scientific disciplines, including biology, chemistry, pharmacology, and clinical research (Sneader 2005; Drews 2000). The objective of drug discovery is to look for molecules that can selectively interact with disease-associated targets and modulate their activity, leading to a desired therapeutic effect (Hughes et al. 2011). Identifying and refining molecular frameworks is a pivotal facet of the drug discovery process. A scaffold can be defined as a core structure or framework of a molecule that imparts essential characteristics and properties to the compound. It forms the basis upon which modifications and functional groups can be added to improve the drug-like properties, such as potency, selectivity, solubility, and bioavailability (HARVEY 2008; Grabley and Thiericke 1998). Machine learning and AI have become increasingly valuable in drug discovery, aiding in the analysis of vast biological and chemical data sets, predicting molecular interactions, and fast-tracking the identification and optimization of probable drug scaffolds with higher efficiency and accuracy (Dara et al. 2022; Vijayan et al. 2022; Sliwoski et al. 2014).

Scaffolds play a pivotal role in drug discovery due to their ability to provide structural diversity, target specificity, favorable pharmacokinetic properties, intellectual property protection, and drug-likeness (Barnes, Kumar, and Davis 2016; Wang et al. 2017; Davison and Brimble 2019). They often possess favorable pharmacokinetic properties, influencing a drug's absorption, distribution, metabolism, and excretion. Proprietary scaffolds can establish a competitive advantage through intellectual property protection. Scaffolds conforming to drug-like properties are more likely to possess favorable safety and pharmacological profiles (Li et al. 2020; Zheng, Tice, and Singh 2014). Overall, scaffolds are essential tools in the discovery and development of new drugs. The choice of a suitable scaffold in drug discovery is a crucial

DOI: 10.1201/9781351058032-9

step that influences the success of the subsequent optimization and development process. Medicinal chemists evaluate various scaffolds based on their chemical diversity, target specificity, pharmacokinetic properties, and safety considerations (Zhang and Pike 2021; Li Petri et al. 2021). By leveraging the advantages of different scaffolds, researchers can design and synthesize a diverse array of compounds with the capacity to become efficient and nontoxic therapeutic agents.

β-carbolines have emerged as a powerful scaffold in the field of drug discovery, garnering significant attention due to their plethora of therapeutic activities and structural versatility (Kamboj et al. 2021). These heterocyclic compounds possess a tricyclic core structure containing of a pyridine ring fused with a pyrrole ring, which imparts unique chemical and biological properties. The concept of a powerful scaffold refers to a molecular framework that possesses inherent properties or structural features that make it particularly well-suited for drug discovery (Ayipo et al. 2021; Ayipo, Osunniran, and Mordi 2021). Privileged scaffolds exhibit a high degree of promiscuity, interacting with multiple biological targets and displaying a broad spectrum of pharmacological activities. β-carbolines have been recognized as a powerful scaffold due to their remarkable capability to engage with various molecular targets and modulate biological processes (Selvaraj et al. 2023).

The structural diversity of β-carbolines allows for modifications at multiple positions on the scaffold, enabling the synthesis of diverse derivatives with distinct biological profiles. This flexibility in structural modification has facilitated their exploration in a wide range of therapeutic areas, including neurological disorders, cancer, infectious diseases, and metabolic disorders (Luo and Song 2021; Ayipo, Osunniran, and Mordi 2021). By modifying different regions of the β-carboline scaffold, medicinal chemists can fine-tune the compound's properties and optimize its activity against specific targets or disease pathways (Karpov, Rominger, and Müller 2005; Drung et al. 2014; Kumar et al. 2017).

The biological activities exhibited by β-carbolines stem from their interactions with various cellular processes. These compounds can act as enzyme inhibitors, receptor ligands, or DNA intercalators, among other mechanisms. The presence of nitrogen atoms and aromatic rings in the β-carboline scaffold contributes to its ability to interact with biomolecules and participate in diverse molecular recognition events (Patel et al. 2012; Horton et al. 2017; Deveau et al. 2001). This versatility in mechanisms of action allows β-carbolines to exert their effects through different pathways, making them valuable tools for understanding disease mechanisms and developing novel therapeutic interventions. The powerful nature of β-carbolines is further supported by their ability to engage with a wide range of biological targets. They can interact with enzymes, receptors, ion channels, transporters, and nucleic acids, among other biomolecules. This broad target engagement allows for the exploration of various disease pathways and the potential identification of novel therapeutic interventions.

Furthermore, the β-carboline framework demonstrates advantageous characteristics for drug development, encompassing a high level of oral bioavailability, metabolic stability, and cell permeability. These properties are crucial for the successful development of drug candidates, as they ensure that the compounds can reach their intended targets in the body and exert the desired therapeutic effects. The inherent drug-like characteristics of β-carbolines contribute to their potential as active compounds in drug discovery campaigns. In summary, β-carbolines represent a powerful scaffold in drug discovery due to their structural diversity, target promiscuity, and biological activities. Their unique tricyclic core structure, flexibility in structural modifications, and versatile mechanisms of action make them valuable tools for developing novel therapeutics. By exploring the potential of β-carbolines as a powerful scaffold, researchers can uncover new therapeutic opportunities and contribute to the advancement of modern drug discovery.

9.2 DISCOVERY AND SIGNIFICANCE OF β-CARBOLINES AS A PRIVILEGED SCAFFOLD

Harmine, a tricyclic beta-carboline alkaloid, was first extracted from the seeds of Peganum harmala in 1847. Over the course of history, harmine has found application in ceremonial and therapeutic formulations in various areas, including the Middle East, Central Asia, and South America. It is naturally found in diverse flora, aquatic life forms, bugs, animals, alongside human tissues, and bodily fluids. Harmine demonstrates a diverse therapeutic activity, encompassing antimicrobial, antiplasmodial, antifungal, antioxidative, antitumor, antimutagenic, cytotoxic, and hallucinogenic impacts. One of the notable actions of beta-carboline compounds, including harmine, is their ability to act as inverse agonists at the benzodiazepine position of γ-aminobutyric acid type A (GABA-A) receptors (Herraiz et al. 2010; Aarons, Victor Rossi, and Orzechowski 1977; Khan et al. 2013). In contrast to the anxiolytic benzodiazepines, beta-carbolines produce effects that are opposite to anxiety reduction. Additionally, these compounds are linked with the modulation of monoaminergic pathways through various mechanisms. This includes the inhibition of monoamine oxidase (MAO) A or B enzymes, blocking the reuptake of monoamines, and directly activating monoamine receptors (Yonezawa et al. 2011; Réus et al. 2012; Egusa et al. 2011; Onishi et al. 2012).

During the mid-20th century, β-carbolines gained attention in the field of neuroscience due to their interaction with neurotransmitter systems, particularly the serotonin system (Sarter, Schneider, and Stephens 1988; Park et al. 2003; Bonnet et al. 2004). Researchers found that β-carbolines could act as inhibitors of the enzyme MAO, which is responsible for the breakdown of neurotransmitters like serotonin. MAO inhibitors were later developed as antidepressant drugs, with examples such as isocarboxazid and phenelzine (*Advanced Journal of Chemistry, Section B* 2019; Mashkovskii, Andreeva, and Polezhaeva 1980; Leonard 1994). β-carbolines also attracted interest in the field of neuropharmacology because of their affinity for certain receptors in the brain, such as the benzodiazepine receptor. Compounds like harmane and harmine have been shown to have interactions with these receptors, leading to a development of sedative and anxiolytic Tranylcypromine. This research contributed to a better understanding of the benzodiazepine receptor and its role in the central nervous system. Another significant historical aspect of β-carbolines is their association with hallucinogenic substances. The Banisteriopsis caapi vine, which contains harmine and other β-carbolines (Figure 9.1), is a fundamental component of the traditional ayahuasca brew used by indigenous communities in the Amazon basin

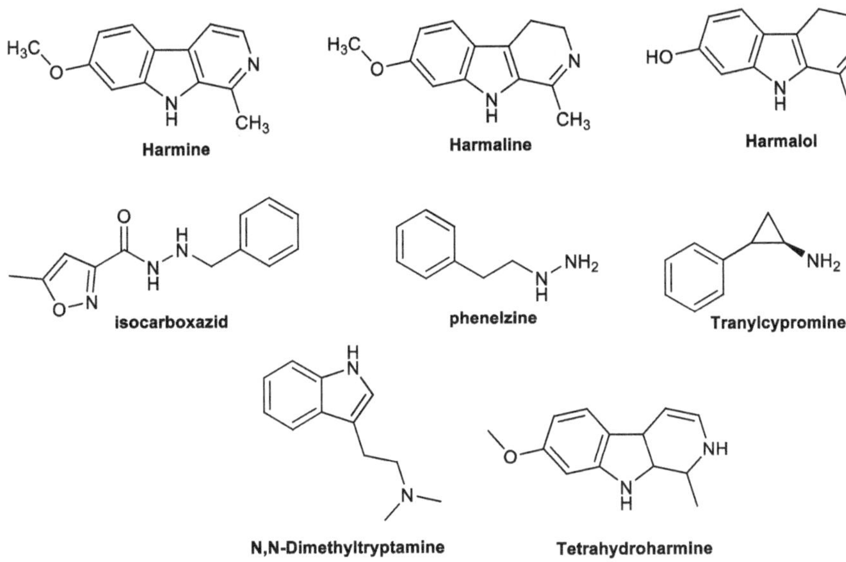

FIGURE 9.1 Medicinally important β-carbolines based compounds.

for spiritual and healing purposes (McKenna, Towers, and Abbott 1984; Berlowitz, Egger, and Cumming 2022). β-carbolines present in ayahuasca act as reversible inhibitors of the enzyme MAO, allowing the psychoactive compound DMT (N,N-Dimethyltryptamine) to exert its effects when consumed orally.

In recent years, β-carbolines have continued to be investigated for their potential therapeutic applications. Researchers have explored their anticancer properties, antioxidant effects, and neuroprotective capabilities. Several research studies have proposed that β-carbolines may exhibit neuroprotective characteristics in relation to conditions such as Alzheimer's disease and Parkinson's disease. However, it is essential to conduct additional research to gain a comprehensive understanding of their mechanisms of action and the potential clinical uses of these compounds. Overall, β-carbolines have a fascinating historical perspective, encompassing traditional medicine, pharmacology, neuroscience, and entheogenic practices. Their biological activities and potential therapeutic benefits continue to be the subject of scientific exploration, highlighting the enduring interest in these compounds.

9.3 SYNTHESIS OVERVIEW OF β-CARBOLINES AS ANTI-HIV AGENTS

9.3.1 Isolation from Natural Sources

Natural sources, particularly marine organisms like sponges, have been found to be rich reservoirs of diverse bioactive compounds, including β-carboline derivatives. Researchers have extensively investigated various marine species for the presence of these alkaloids with potential anti-HIV activity. For instance, manzamines and manzamine-related alkaloids have been isolated from Indo-Pacific sponges and have shown potent anti-HIV, antitumor, antibacterial, and cytotoxic activities (Yousaf et al. 2004; Peng et al. 2003). Similarly, the compound flazin was isolated from the fruiting bodies of Suillus granulatus and displayed moderate anti-HIV-1 activity (Su et al. 2002; Dong et al. 2007; Nakatsuka et al. 1986). The isolation of these compounds from natural sources allows for the identification of new and unique chemical entities with potential therapeutic applications.

9.3.2 Chemical Synthesis

The synthesis of β-carboline derivatives is of interest due to their various biological activities, including anti-HIV potential. Several synthetic routes have been explored to prepare 1-substituted β-carbolines, which are of particular interest for medicinal chemistry research. These routes include the Pictet–Spengler reaction, Bischler–Napieralski reaction, cyclization of 2-acyltryptamine, and dehydrogenation of appropriate 1,2,3,4-tetrahydro-9H-β-carboline derivatives (Peduto et al. 2012; Suzuki et al. 2005; Hagen et al. 1989; Hagen and Cook 1988). However, some of these synthetic methods pose challenges in terms of yields, regioselectivity, and efficiency. The use of microwave-assisted reactions has shown promise in improving the yields of certain β-carboline derivatives.

In summary, the exploration of β-carboline derivatives for anti-HIV studies involves two main approaches: isolation from natural sources, particularly marine organisms, and synthesis through various chemical reactions. Both approaches contribute to the discovery and development of potential anti-HIV agents with diverse chemical structures and biological activities.

9.4 BIOLOGICAL DISCUSSION OF β-CARBOLINES AS ANTI-HIV AGENTS

The following section present an overview of the research conducted thus far on β-carboline-based anti-HIV activities. The structures and corresponding EC_{50} values of these compounds are provided in the figure (Name or coding of structures are given as presented in original research paper). This study aims to provide a concise overview of the progress made in identifying and evaluating β-carboline derivatives as potential agents against HIV.

Junko Ishida et al. pioneered the reports of β-carbolines as anti-HIV agents. They discussed the isolation of 1-methoxycanthinone (1) and 5-methoxycanthinone (2) and other compounds, out of which **1-methoxycanthinone and 5-methoxycanthinone** are the β-carbolines (Figure 9.2) (Xu et al. 2000). The study evaluated the biological activities of these compounds, focusing on their anti-HIV and cytotoxic effects. The results showed that **1-methoxycanthinone**, a β-carboline derivative, exhibited EC50 value of 0.26 μg/mL and a therapeutic index (TI) greater than 39 establishing itself as potent anti-HIV agent. This suggests its potential as an anti-AIDS drug candidate. **5-methoxycanthinone**, a β-carboline derivative, exhibited potent anti-HIV activity against HIV-infected H9 cells, with an impressive TI of over 391. The IC50 value (concentration that inhibited uninfected H9 cell growth by 50%) was greater than

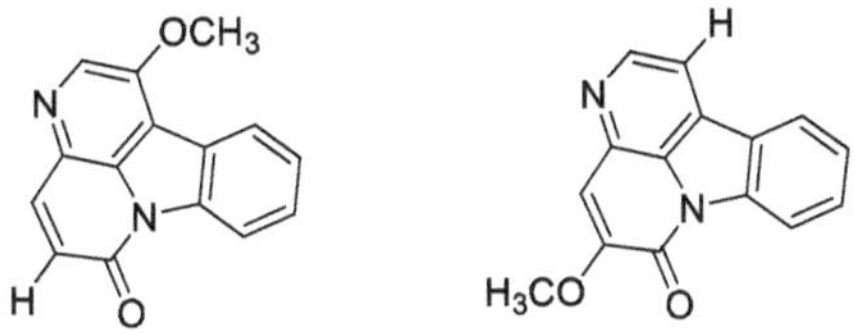

1-methoxycanthinone
EC_{50} (μM)= 0.256

5-methoxycanthinone

FIGURE 9.2 Structure of 1-methoxycanthinone, and 5-methoxycanthinone.

100 μg/mL, while the EC50 value (concentration that inhibited viral replication by 50%) was 0.256 μg/mL. This discovery highlights the potential of β-carbolines as a new class of compounds for anti-AIDS drug development. The study suggests that further research, including structure–activity relationship studies of β-carbolines, is warranted to explore their potential as anti-HIV agents. The findings provide clues about the therapeutic potential of β-carboline derivatives and their potential application in the development of therapeutic compounds for the treatment of HIV/AIDS.

Further, their group discussed the evaluation of the anti-HIV potency of β-carboline derivatives (**derivatives 1 to 28,** Figures 9.3 and 9.4), particularly focusing on harman (**structure a,** Figure 9.3) and its analogues (Ishida et al. 2001). Harman, a β-carboline compound, was identified from Symplocos setchuensis and observed to inhibit HIV replication in H9 lymphocyte cells. Among the 28 derivatives of harman evaluated, compound 19 (N-butylharmine) exhibited anti-HIV activity (EC50=0.037 μM) with a therapeutic index (TI) of 210. This compound demonstrated the most significant potency in the series. Other compounds, such as harmine (Figure 9.3, 3), N-ethylharmine (Figure 9.3, 12), 6-bromoharmine hydrobromide (Figure 9.4, 18), along with 33, showed considerable activity with moderate TI values. Other derivatives were less potent compared to harman. The study revealed several trends in the anti-HIV activity of the β-carboline derivatives. Functionalization with methoxy group at the 7-position of harmine (**3**) resulted in improved activity, while the isopropoxy, hexyloxy, decyloxy, hexadecyloxy, and 7-camphanoyl derivatives were inactive. The addition of a methyl group at the 1-position of harman was found to be important for activity. Hydrogenation of harman to produce derivative **15** (3,4-dihydro β-carboline) resulted in lower anti-HIV activity. Bromination of 9 at C-6 resulted in the most potent compound (**18**), with improved water solubility. The fully conjugated derivative (**27**) maintained some activity. Compound 28 highlights the significance of β-carbolines, as the benzimidazole-containing compound exhibited no anti-HIV activity. Additionally, alkylating the nitrogen of indole (**11, 12**, and **13**) enhances the anti-HIV activity, with derivative **13** (N-butyl) exhibiting greater potency than **12** (N-ethyl), suggesting the importance of the alkyl chain length.

Karumanchi V. Rao et al conducted two studies focusing on the investigation of manzamine-type alkaloids for their bioactivity against HIV-1 (Figure 9.5) (Rao et al. 2006; Rao et al. 2004). In the first study, from Indonesian spone, several new manzamine-type alkaloids were isolated. They evaluated anti-HIV-1 activities of these manzamines and their oxa-derivatives. They observed that manzamine A and 8-hydroxymanzamine

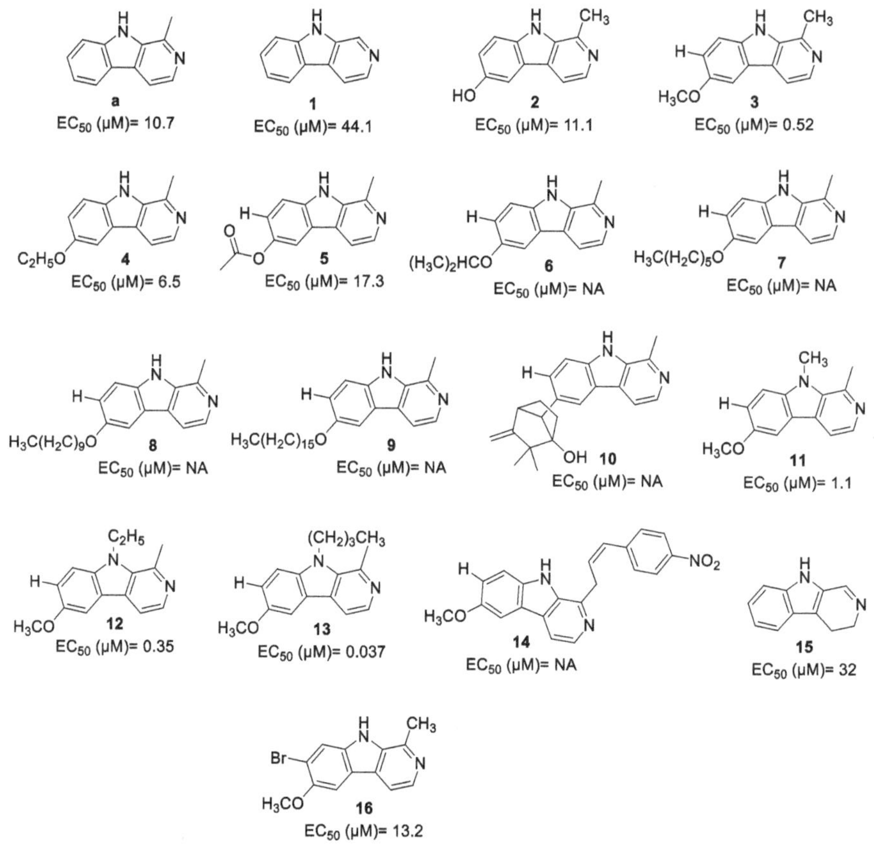

FIGURE 9.3 Structure of harman (5) and its analogues from 7 to 22 evaluated for anti-HIV study.

A displayed greater potency than manzamines E, F, and Y, emphasizing the significance of functionalization of hydroxyl group and introduction of the eight-membered ring on β-carboline moiety for HIV-1 activity. In the second study, a series of manzamine alkaloids were isolated from a sponge species of the genus Acanthostrongylophora. The researchers found that manzamine A, manzamine A N-oxide, and 8-hydroxymanzamine A exhibited higher anti-HIV-1 activity compared to manzamines E and F, indicating the crucial role of the eight-membered ring in their effectiveness. The findings contribute to the understanding of structure–activity relationships of manzamine alkaloids and their potential as antimicrobial and anti-HIV agents.

P.-W. Hsieh et al. reported two carbolines, drymaritin and an isomer of canthin-4-one, isolated from Drymaria diandra (Figure 9.6) (Hsieh et al. 2004). They investigated the anti-HIV activity in H9 lymphocytes and found that drymaritin exhibited significant anti-HIV effects, with an EC_{50} value 0.699 μg/mL and a therapeutic index (TI) of 20.6. The canthin-4-one isomer also showed anti-HIV activity with EC_{50} and TI values same as drymaritin, indicating the potential of this series of compounds.

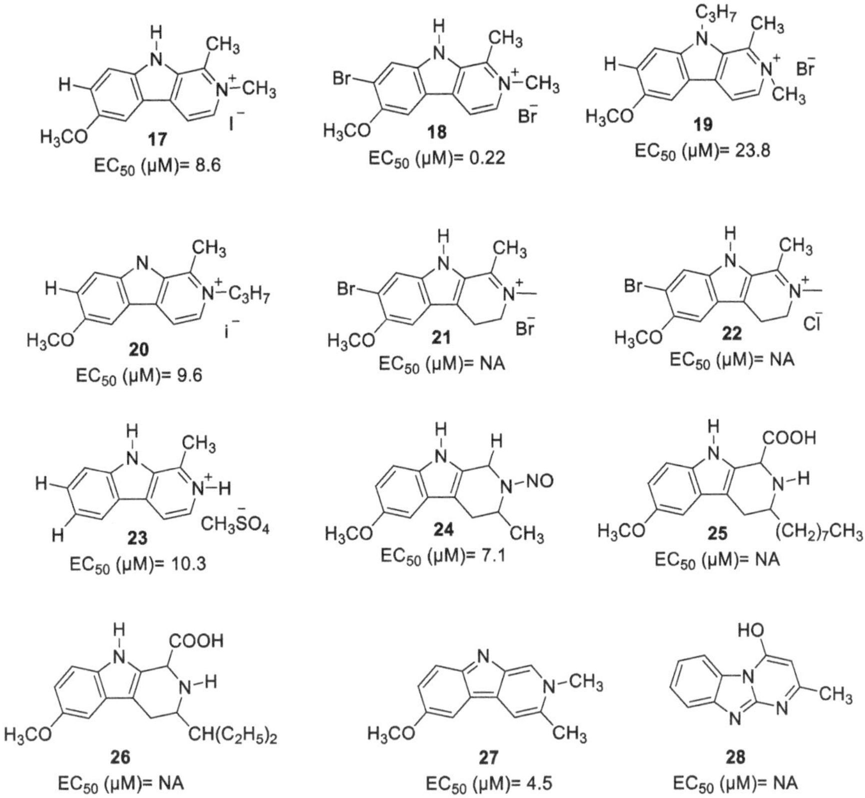

FIGURE 9.4 Structure of harman and its analogues from 23 to 34 evaluated for anti-HIV study.

In a study by Jian-Guo Tang, flazin and its derivatives were subjected to the anti-HIV activity test. Flazin, isolated from the fruiting bodies of Suillus granulatus, exhibited weak anti-HIV activity with an EC_{50} of 2.36 μM (AZT (3'-azido-3'-deoxythymidine) was used as a standard with an EC50 value of 0.009 μM (Tang et al. 2008). To explore the structure–activity relationship of flazin, 46 derivatives were developed and evaluated for their anti-HIV activities (Figures 9.7, 9.8, 9.9, 9.10). Among them, the most effective derivative was found to be flazinamide (**9a**), which had an EC_{50} value of 0.38 μM and TI of 312. Substitutions such as amide modification at C(3) and S-atom in the 1'-position enhanced the anti-HIV activity, while the introduction of a carboxylate or CH_2OH group at C(3) yielded inactive compounds. The study suggests that flazinamide (**9a**) can be viewed as a hopeful anti-HIV agent. The SAR analysis provides insights into the role of functionalization in enhancing the anti-HIV activity of flazin and its derivatives, guiding the design and development of more potent molecules for anti-HIV therapy.

Ashok et al. designed and explored novel β-carboline derivatives as inhibitors of HIV-1 reverse transcriptase (RT) (Ashok, Sharma, et al. 2015). They aimed to address

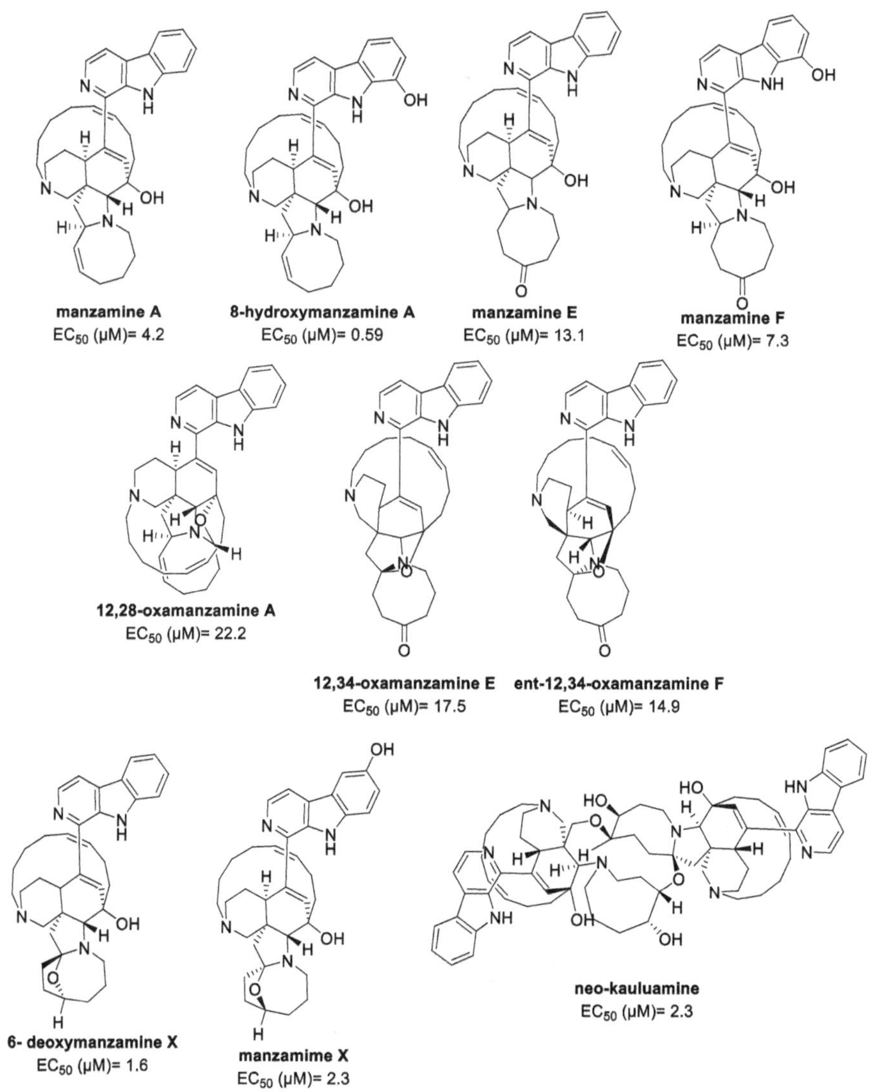

FIGURE 9.5 Structure of Manzamine alkaloids explored for their anti-HIV activity by Rao et al. 2004.

the issue of resistance that arises with the use of existing HIV-1 RT inhibitors. Using Autodock v4.2 they carried out molecular docking studies to assess the binding affinity of the designed analogues against wild and mutant strains of HIV-1 RT. The results of the docking studies indicated that several β-carboline analogues demonstrated significant binding free energy against wild and mutant strains of HIV-1 RT. Compounds 9H3NH4CN, 9H6CNP, and 9H4CN exhibited notable binding free energy values against both strains, comparable to standard drugs. Additionally, these drugs projected

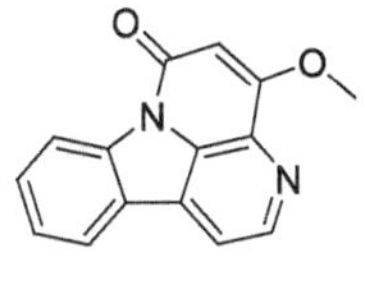

drymaritin
EC_{50} (μM)= 0.699

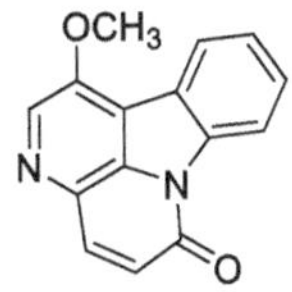

isomer of canthin-4-one
EC_{50} (μM)= 0.699

FIGURE 9.6 Structure of drymaritin and an isomer of canthin-4-one explored for their anti-HIV activity.

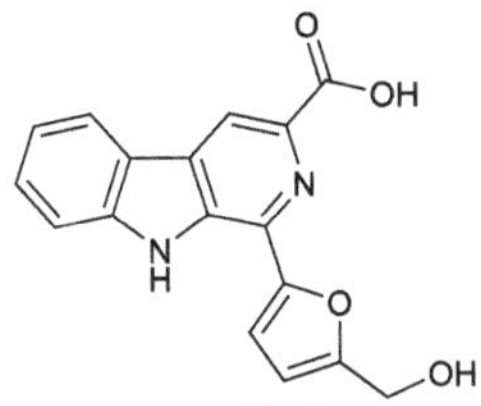

Flazin
EC_{50} (μM) = 2.36

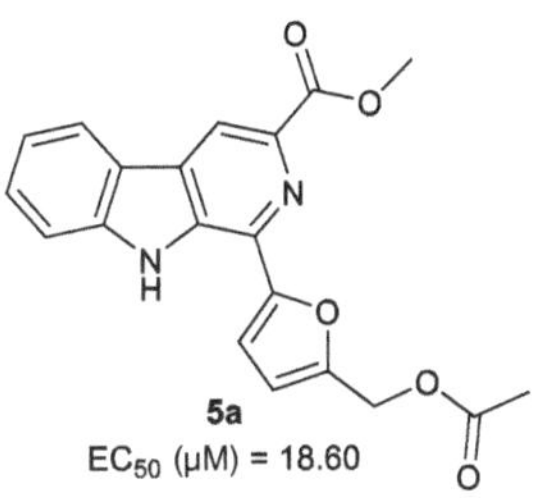

5a
EC_{50} (μM) = 18.60

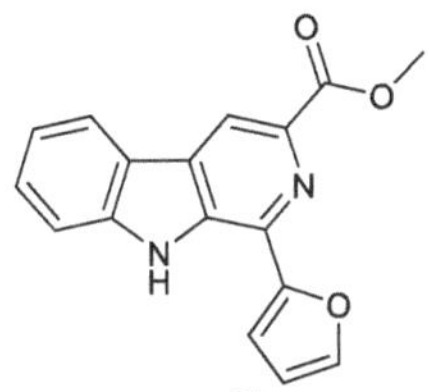

5b
EC_{50} (μM) = 10.10

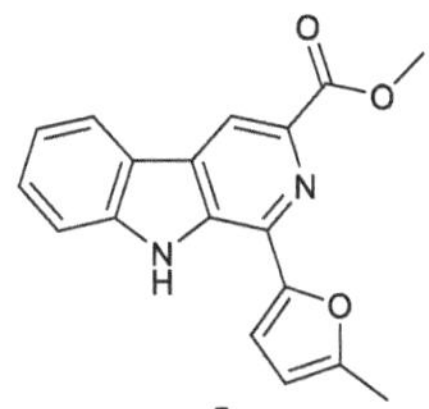

5c
EC_{50} (μM) =16.80

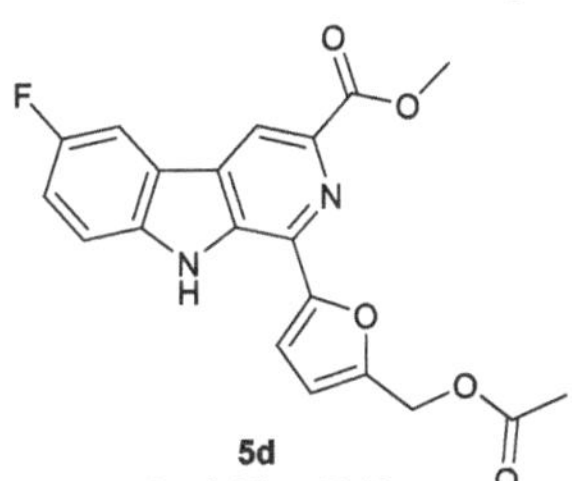

5d
EC_{50} (μM) = 16.10

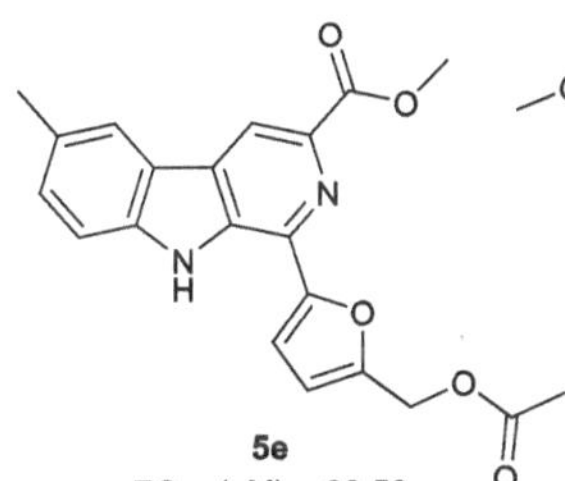

5e
EC_{50} (μM) = 30.50

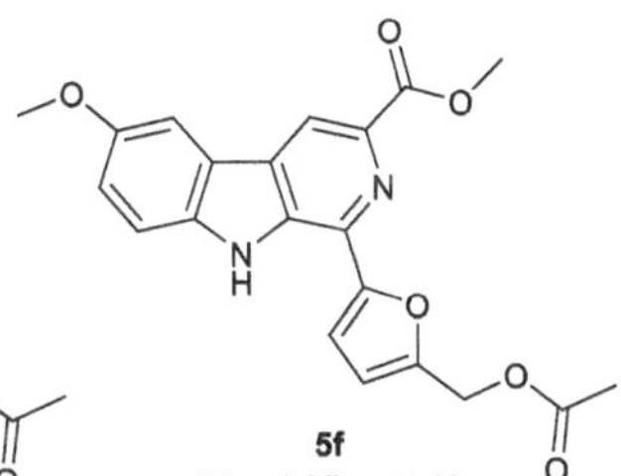

5f
EC_{50} (μM) = 14.50

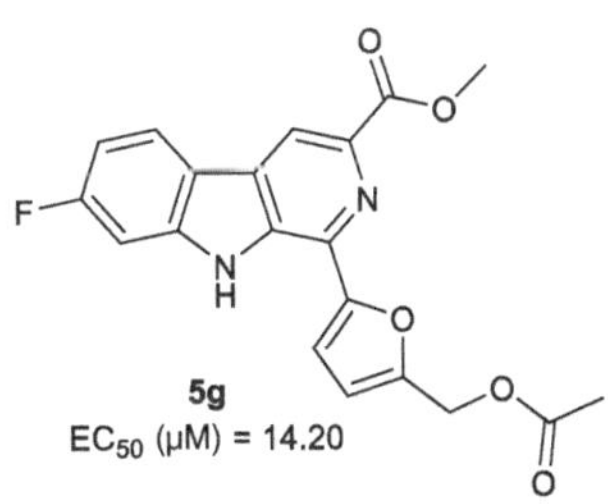

5g
EC_{50} (μM) = 14.20

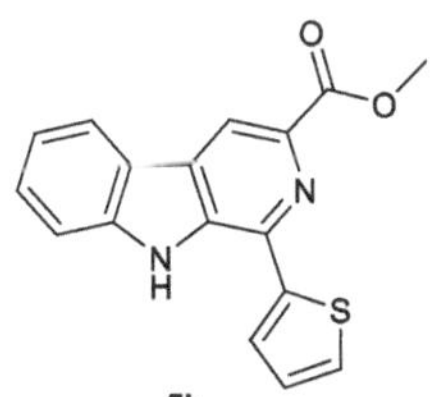

5h
EC_{50} (μM) = 2.31

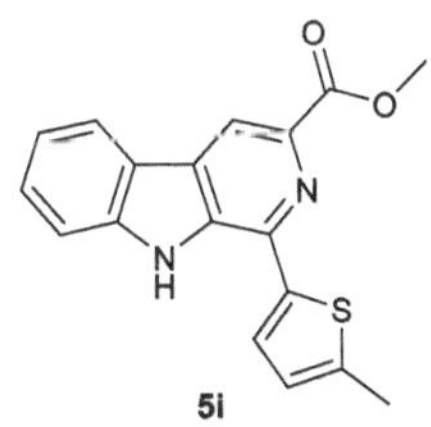

5i
EC_{50} (μM) = 3.58

FIGURE 9.7 Structure and EC_{50} of Flazin and its analogues (5a-5i).

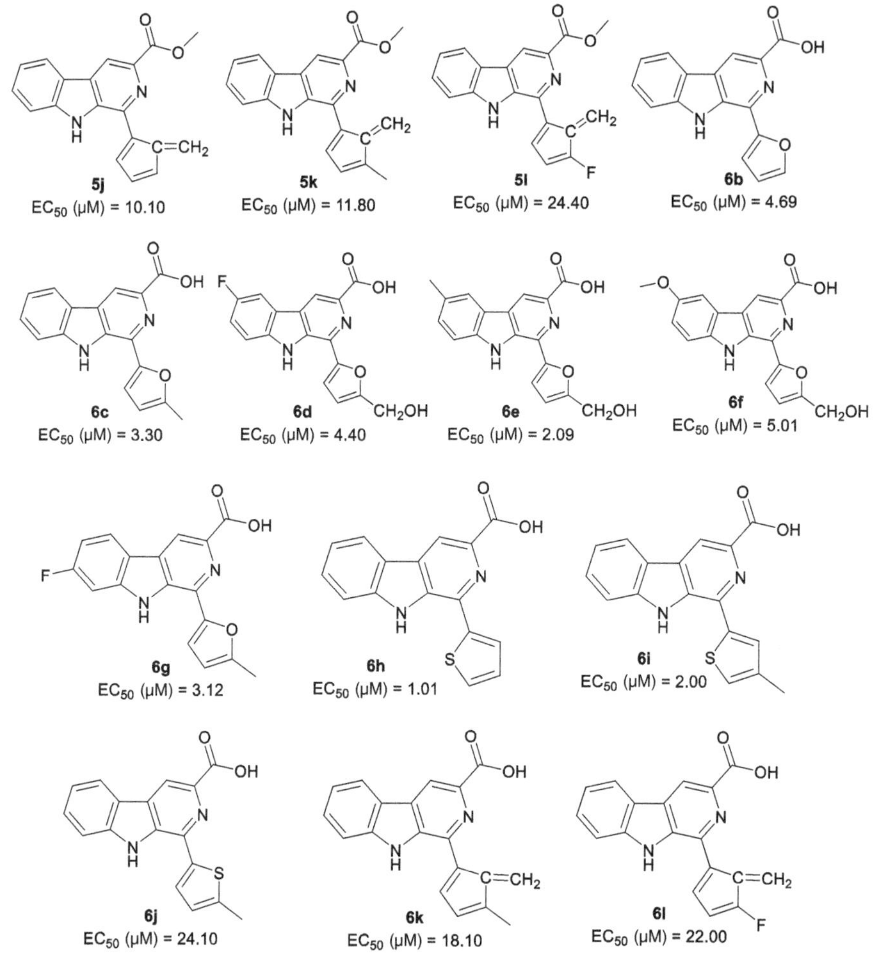

FIGURE 9.8 Structure and EC_{50} of Flazin and its analogues (5j-6l).

inhibitory constant values (Ki) indicate encouraging action against both stains of HIV-1 RT (Figure 9.11, 9.12, 9.13, 9.14, 9.15). The designed derivatives also demonstrated a variety of molecular interactions with target protein, including electrostatic and hydrogen bonding interaction. The hydrogen bonding interactions with hydrophilic amino acids of the HIV-1 RT enzyme were discovered to be caused by hydrophilic body substitution and hydrogen bond donor group of the phenyl ring. Additionally, the target protein's non-nucleoside inhibitor binding pocket (NNIBP) aromatic amino acids were contacted by hydrophobic wings of β-carbolines derivatives through π-π and π-cation interaction. Like previous non-nucleoside reverse transcriptase inhibitors (NNRTIs), the designed β-carboline derivatives takes up the butterfly like shape, which allowed for interaction between amino acid residues of HIV-1 RT. The proposed analogues' potential as next-generation NNRTIs with promising anti-HIV-1

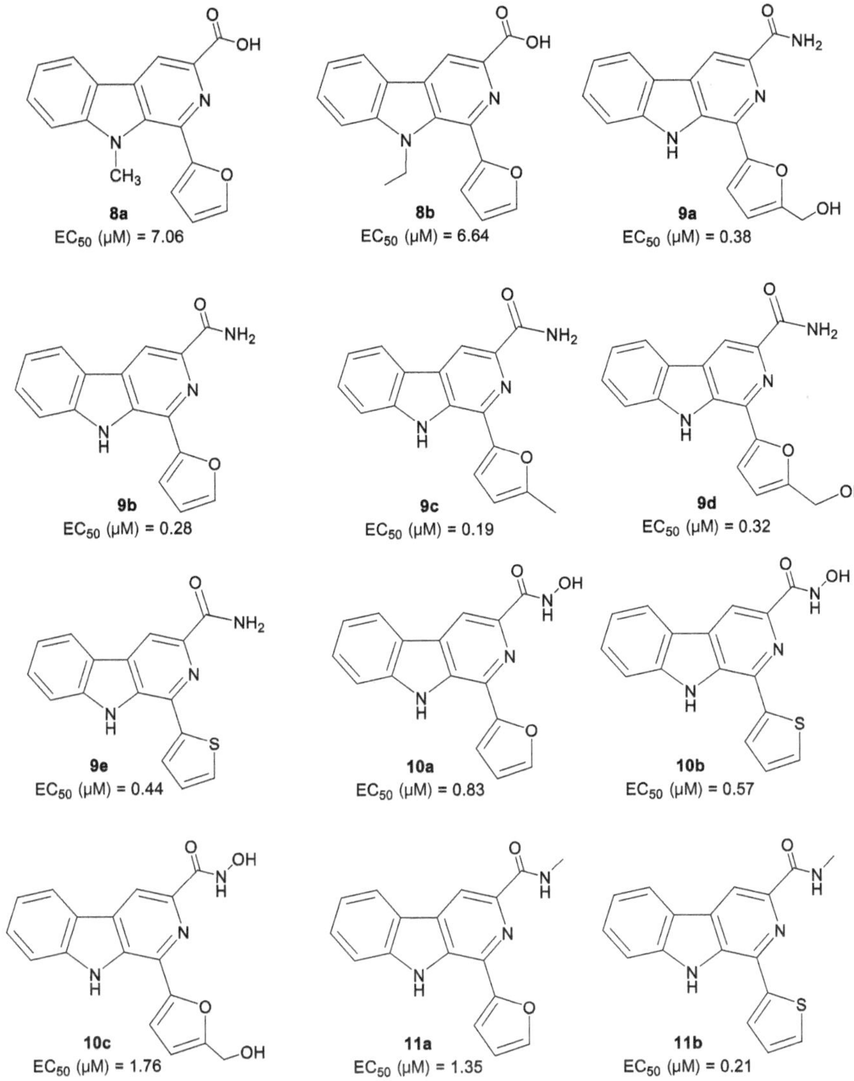

FIGURE 9.9 Structure and EC_{50} of Flazin and its analogues (8a-11b).

effectiveness and enhanced pharmacokinetic profiles was shown by assessments of the Lipinski drug similarity and ADMET (absorption, distribution, metabolism, excretion, and toxicity) features. To corroborate their docking findings their active methyl groups then carried out the experimental studies as discussed in the following.

Further, P. Ashok et al carried out a study focusing on the synthesis of new β-carboline derivatives (29-44) (Figure 9.16, 9.17) and their evaluation as inhibitors of HIV-1 and 2 strains (Ashok, Chander, et al. 2015). From those synthesized derivatives, **30**, **34**, **39,** and **40** exhibited selective inhibition of the HIV-2 strain with

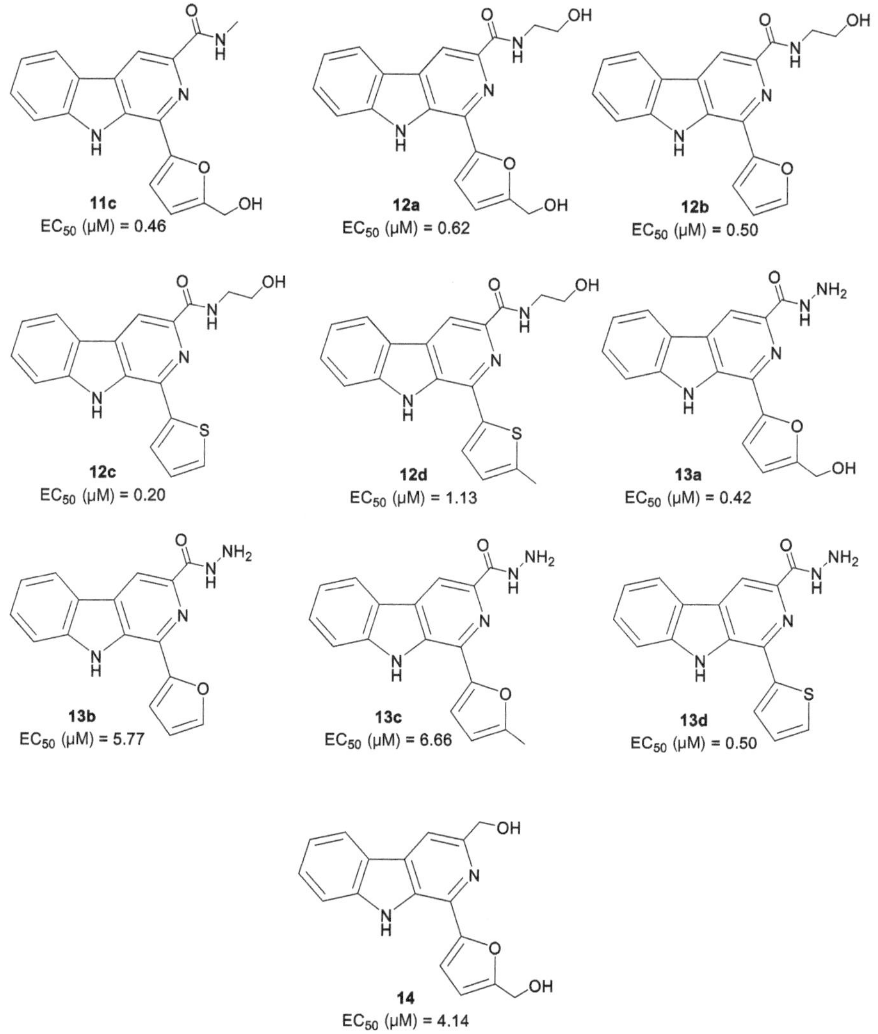

FIGURE 9.10 Structure and EC_{50} of Flazin and it's analogues (11c-14).

comparable activity to nucleoside reverse transcriptase inhibitors Nevirapine (EC_{50} HIV 1–0.027 μM and HIV-2 >4), Lamivudine (EC_{50} HIV 1–0.89 μM and HIV 2–3.56 μM), Zidovudine (EC_{50} HIV 1–0.0019 μM and HIV 2–0.0016 μM), and Dideoxyinosine (EC_{50} HIV 1– 2.56 μM and HIV 2–2.27 μM). These compounds showed EC_{50} values ranging from 2.6 to 5.4 μM against HIV-2, with selectivity indices greater than or equal to 72. Interestingly, these compounds did not demonstrate in vitro HIV-2 reverse transcriptase inhibition, suggesting that their specific anti-HIV-2 activity may involve a different mechanism. The structure–activity relationship analysis revealed that electron-donating groups on the phenyl ring increased the anti-HIV potency, while electron-withdrawing groups decreased the potency. Substitution with fluorine

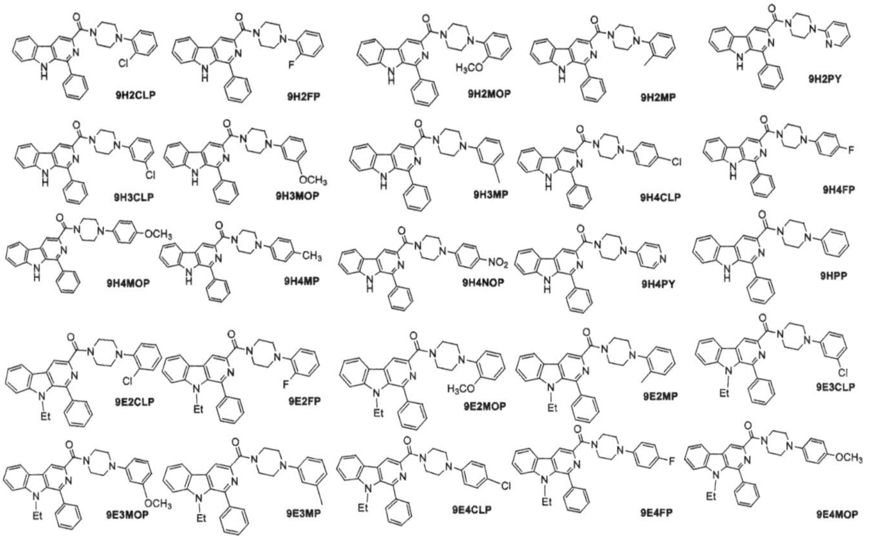

FIGURE 9.11 β-carboline analogues as inhibitors of HIV-1 reverse transcriptase (RT) (Series 1).

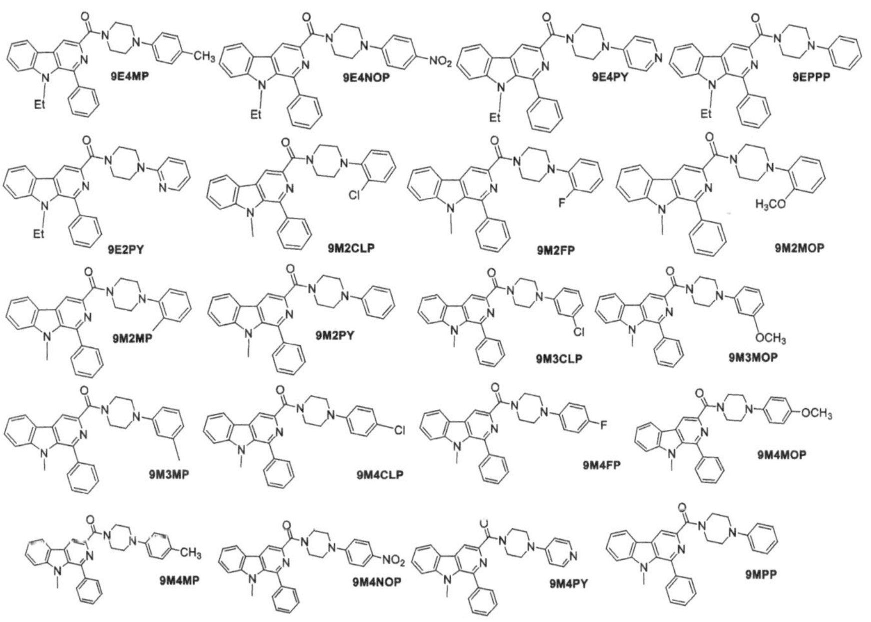

FIGURE 9.12 β-carboline analogues as inhibitors of HIV-1 reverse transcriptase (RT) (Series 1).

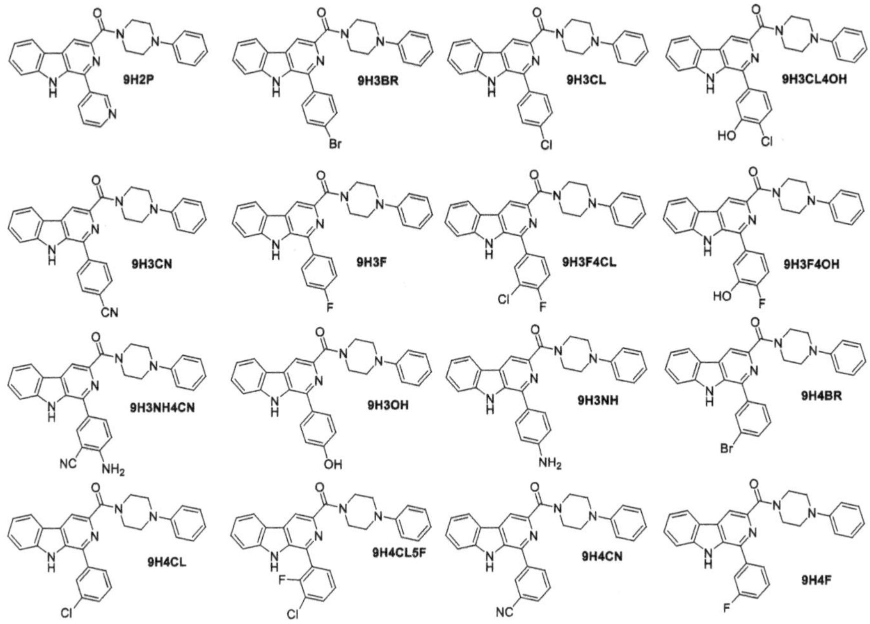

FIGURE 9.13 β-carboline analogues as inhibitors of HIV-1 reverse transcriptase (RT) (Series 2).

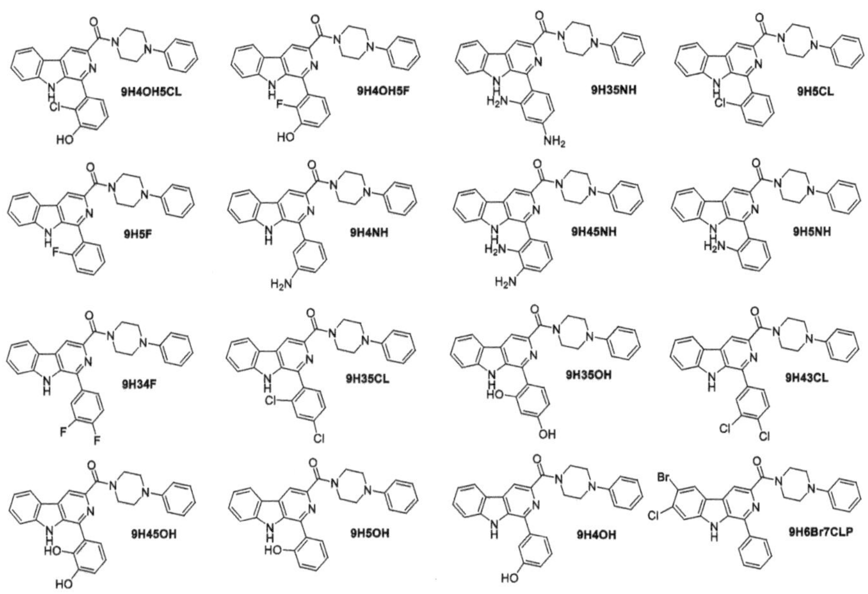

FIGURE 9.14 β-carboline analogues as inhibitors of HIV-1 reverse transcriptase (RT) (Series 2).

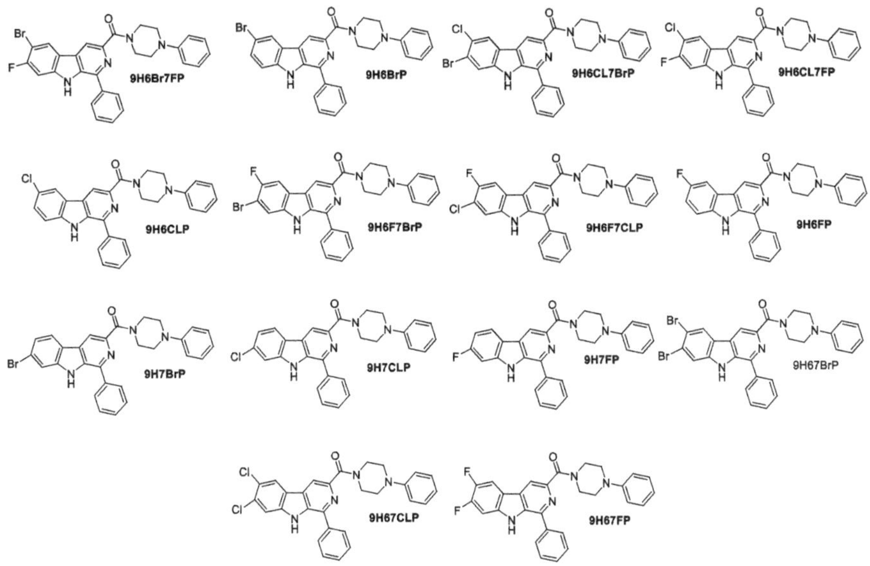

FIGURE 9.15 β-carboline analogues as inhibitors of HIV-1 reverse transcriptase (RT) (Series 2).

groups at specific positions resulted in enhanced potency, particularly with compound **39**. Substitution of the benzene ring with benzyl and heteroaryl rings increased cytotoxicity without significantly affecting the anti-HIV activity. None of the substances, however, had anti-HIV-1 activity below their lethal concentration, revealing a distinct anti-HIV-2 activity spectrum for these - β-carboline derivatives.

Keyur et al. developed a series of β-carboline derivatives using an aromatization and chemoselective alkylation method (Brahmbhatt et al. 2010). The developed compounds were tested for their anti-HIV activity against the HIV-1 NL4.3 virus in human CD4+ T cell line (CEM-GFP) (Figures 9.18 and 9.19). Compound **58**, which is a 1-formyl-β-carboline-3-carboxylic acid methyl ester, showed inhibition of HIV with an IC_{50} of 2.9 μM. Cytotoxicity of the synthesized analogues was assessed using a cell viability assay, and noncytotoxic concentrations were selected for further evaluation of anti-HIV activity. Among the tested compounds, **45**, **48**, **54**, and **57** showed inhibition of HIV-1 NL4.3. From this result, alkylated derivatives of 8c, 13, and 10 were synthesized to explore their activity. However, N-alkylation of 8c and 13 showed loss of activity, indicating an alternative mechanism or site of interaction in contrast to harmine derivatives. Notably, compound **58**, the 1-formyl-β-carboline-3-carboxylic acid methyl ester, demonstrated significant anti-HIV activity with an IC_{50} of 2.9 μM. This compound is particularly interesting as there are no previous reports of a 1-formyl derivative of β-carboline showing anti-HIV activity in the literature. In conclusion, this study presented the synthesis and biological evaluation of β-carboline derivatives for their anti-HIV activity. Among the tested compounds, compound **58** showed promising activity against HIV-1.

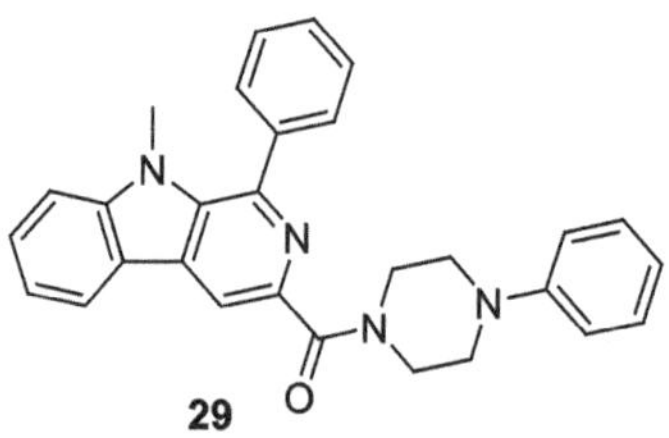

29

EC_{50} (μM) (HIV-1) = **>224**

EC_{50} (μM) (HIV-2) = **29.4 ± 5.9**

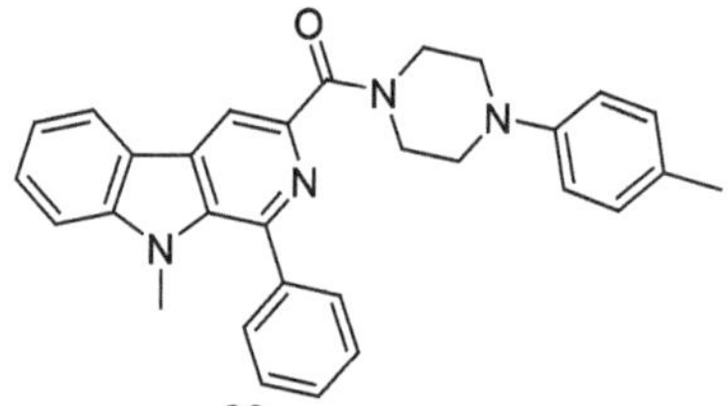

30

EC_{50} (μM) (HIV-1) =**>29.6**

EC_{50} (μM) (HIV-2) = **3.3 ± 0.02**

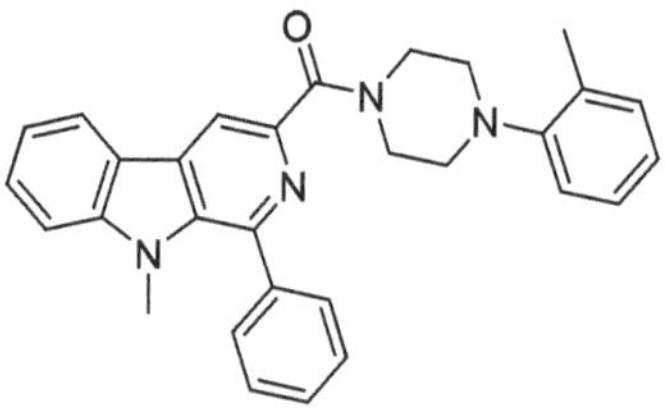

31

EC_{50} (μM) (HIV-1) = **>16.9**

EC_{50} (μM) (HIV-2) = **>16.9**

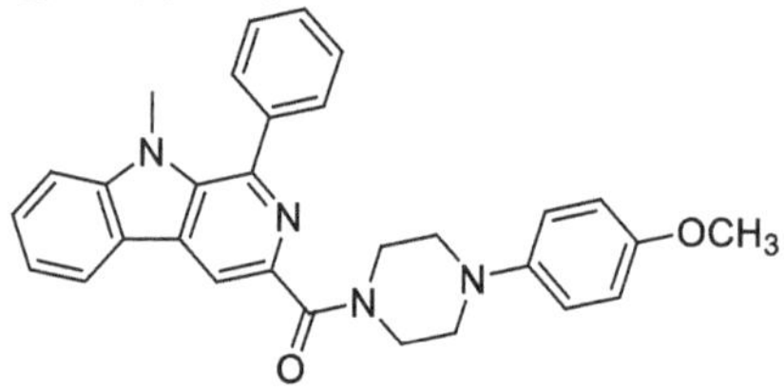

32

EC_{50} (μM) (HIV-1) = **>210.0**

EC_{50} (μM) (HIV-2) = **>210.0**

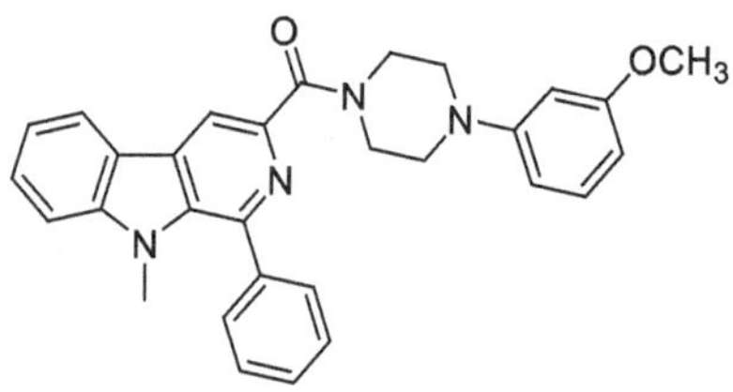

33

EC_{50} (μM) (HIV-1) = **>133.9**

EC_{50} (μM) (HIV-2) =**>133.9**

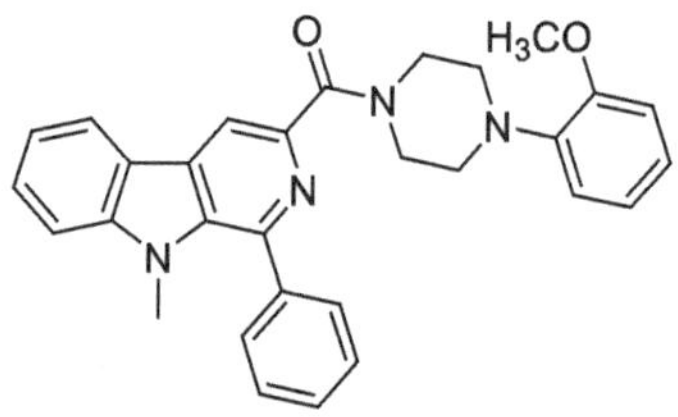

34

EC_{50} (μM) (HIV-1) = **>25.3**

EC_{50} (μM) (HIV-2) = **3.2 ± 0.22**

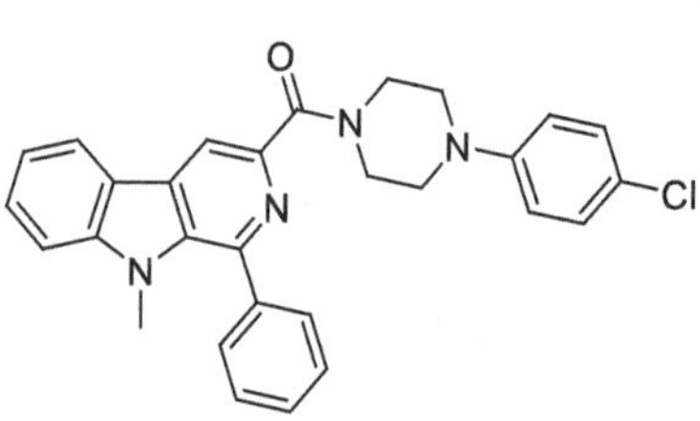

35

EC_{50} (μM) (HIV-1) = **>208.3**

EC_{50} (μM) (HIV-2) = **>208.3**

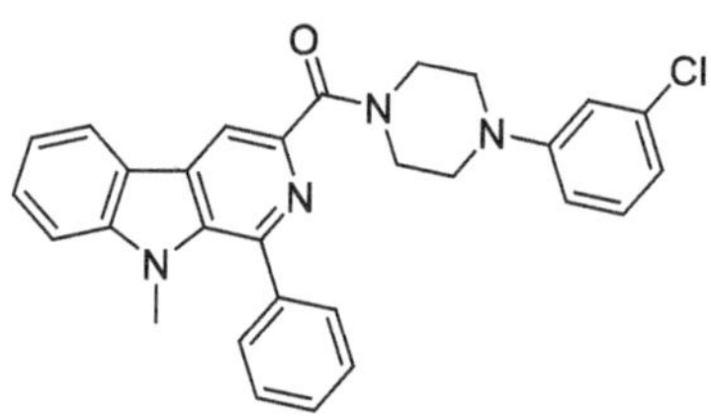

36

EC_{50} (μM) (HIV-1) = **>110.6**

EC_{50} (μM) (HIV-2) = **>110.6**

FIGURE 9.16 Synthesized β-carboline derivatives by P. Ashok et al (29–36).

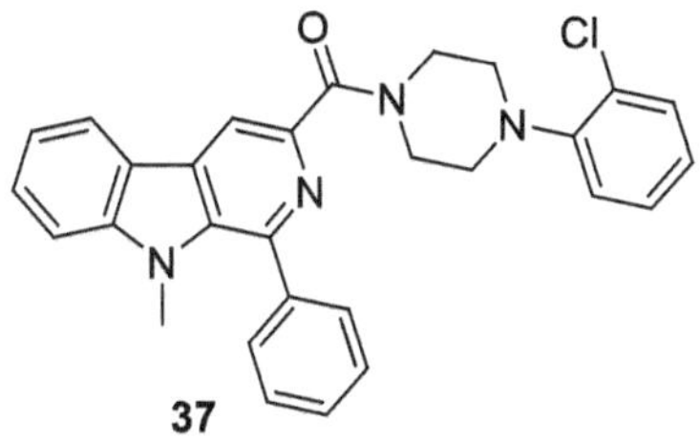

37

EC_{50} (μM) (HIV-1) = **>23.6**
EC_{50} (μM) (HIV-2) = **>23.6**

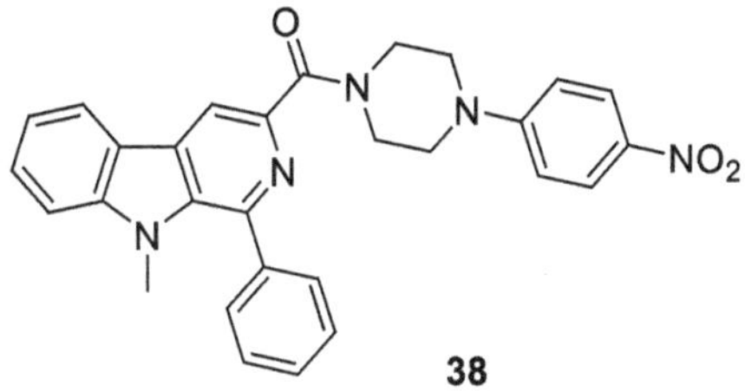

38

EC_{50} (μM) (HIV-1) = **>203.6**
EC_{50} (μM) (HIV-2) = **>203.6**

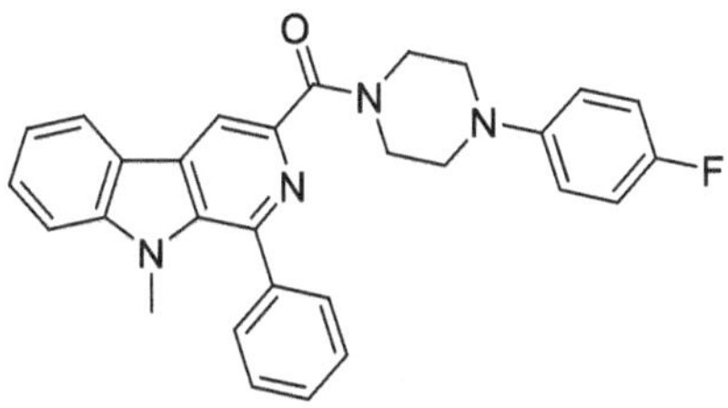

39

EC_{50} (μM) (HIV-1) = **>189**
EC_{50} (μM) (HIV-2) = **2.6 ± 0.02**

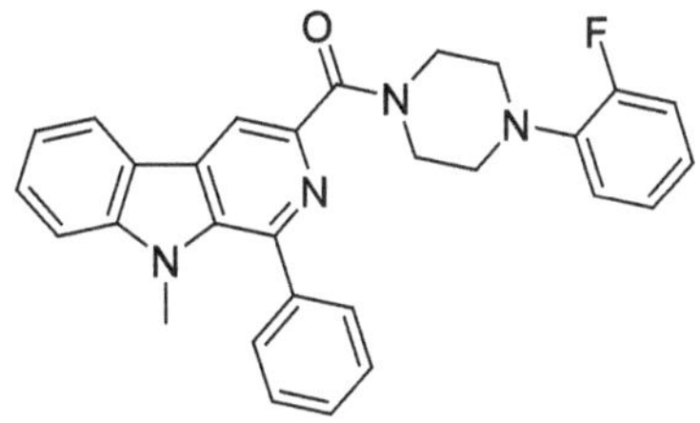

40

EC_{50} (μM) (HIV-1) = **>50.6**
EC_{50} (μM) (HIV-2) = **5.4 ± 0.23**

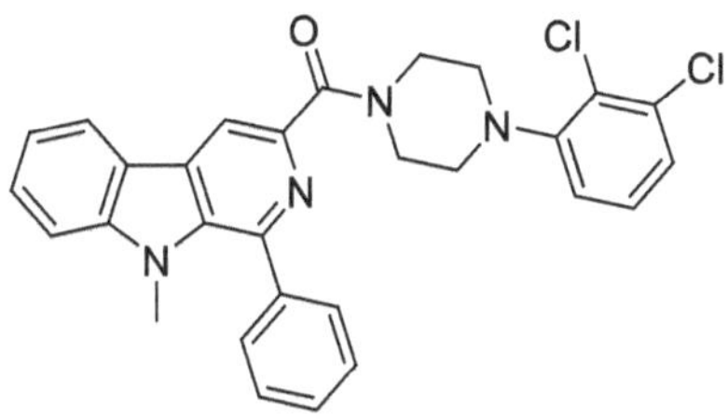

41

EC_{50} (μM) (HIV-1) =**>63.9**
EC_{50} (μM) (HIV-2) = **>63.9**

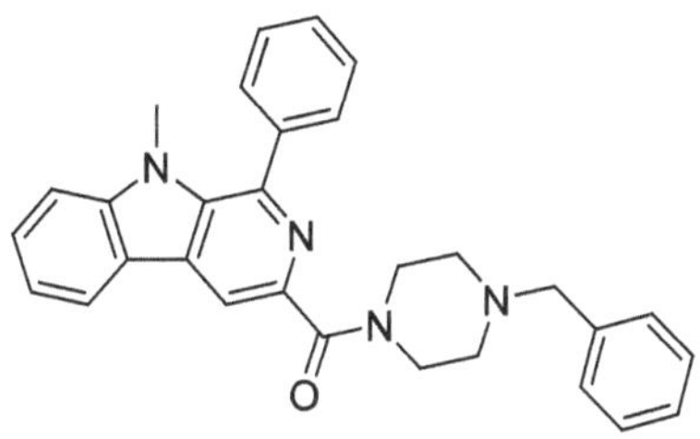

42

EC_{50} (μM) (HIV-1) = **>21.7**
EC_{50} (μM) (HIV-2) = **>21.7**

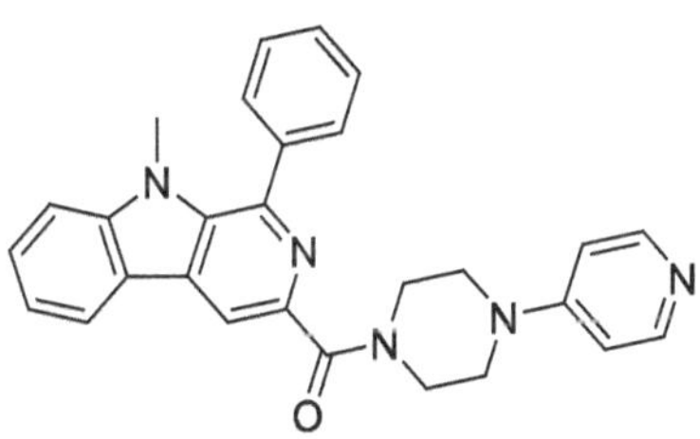

43

EC_{50} (μM) (HIV-1) = **>4.02**
EC_{50} (μM) (HIV-2) = **>4.02**

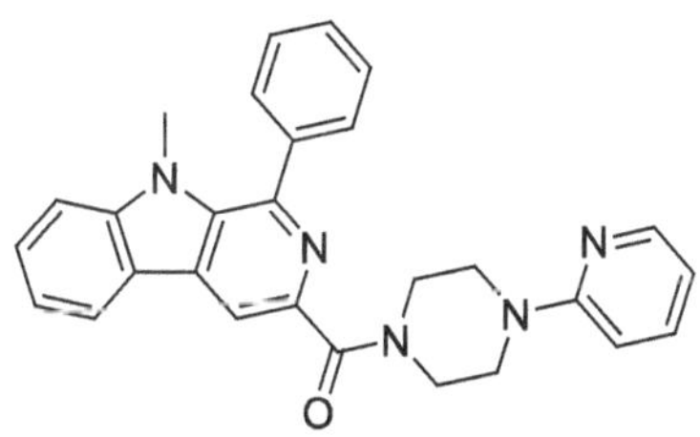

44

EC_{50} (μM) (HIV-1) = **>19.2**
EC_{50} (μM) (HIV-2) = **>19.2**

FIGURE 9.17 Synthesized β-carboline derivatives by P. Ashok et al (37–44).

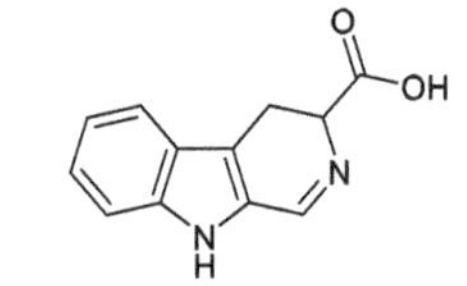

45
% Inhibition p24 ELISA = 16.48

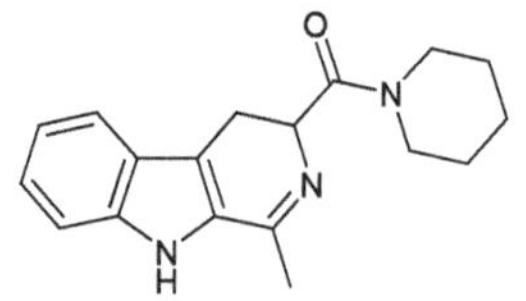

46
% Inhibition p24 ELISA = NA

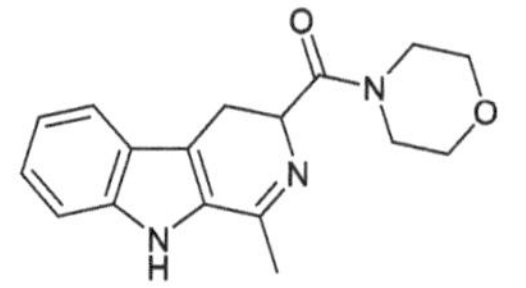

47
% Inhibition p24 ELISA = NA

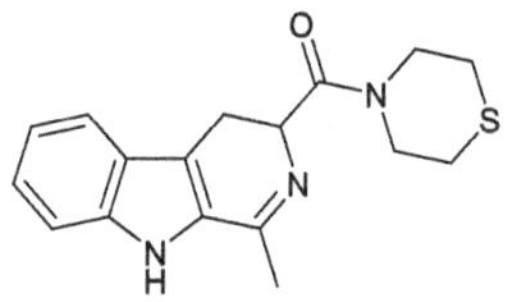

48
% Inhibition p24 ELISA = 16.9

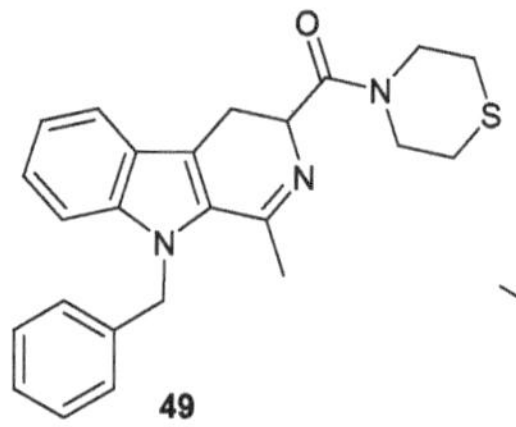

49
% Inhibition p24 ELISA = NA

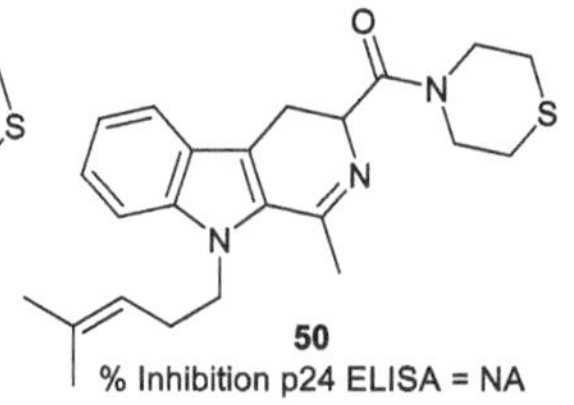

50
% Inhibition p24 ELISA = NA

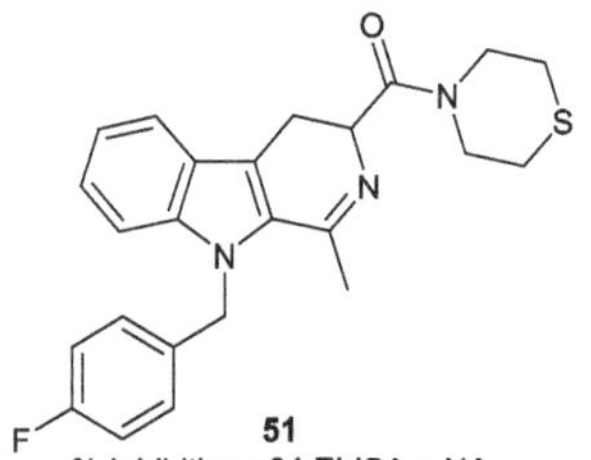

51
% Inhibition p24 ELISA = NA

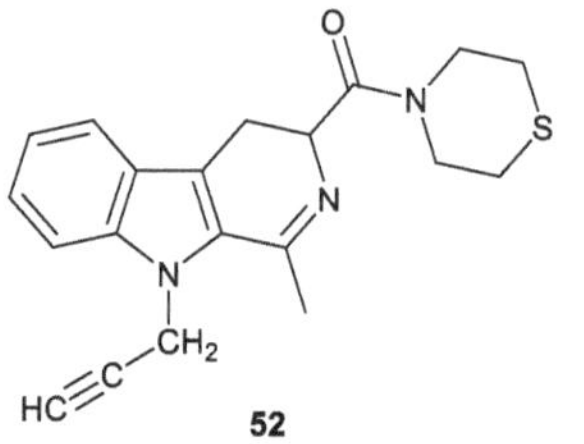

52
% Inhibition p24 ELISA = NA

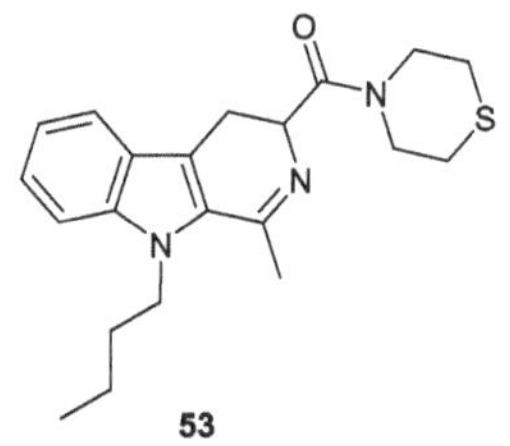

53
% Inhibition p24 ELISA = NA

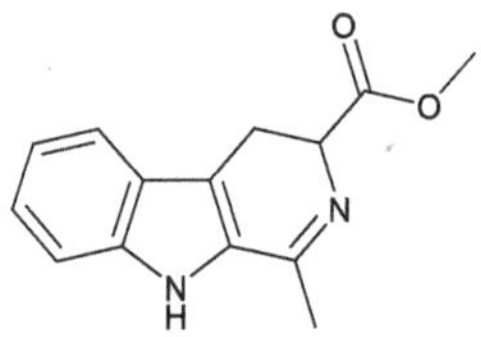

54
% Inhibition p24 ELISA = 17.26

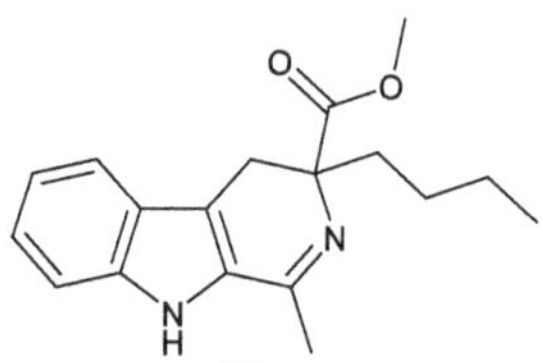

55
% Inhibition p24 ELISA = 12.17

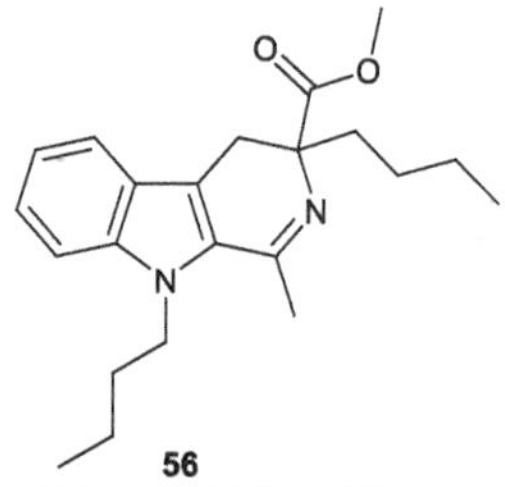

56
% Inhibition p24 ELISA = NA

FIGURE 9.18 β-carboline derivatives developed by Keyur et al. with % inhibition of p24 ELISA.

11.5 ROLE OF FUNCTIONALIZATION

Functionalization of the β-carboline scaffold plays a pivotal role in determining the anti-HIV activity of these compounds, as elucidated by the above studies. The findings highlight several key trends and insights into the impact of functional groups on the potency of β-carbolines as anti-HIV agents.

Firstly, the introduction of a methoxy group at the 7-position of harmine and other derivatives has been shown to enhance their anti-HIV activity. Compounds with 7-methoxy substitutions, such as 7-hydroxy, 7-ethoxy, and 7-acetoxy derivatives, display similar or improved activity compared to the parent compound (harman). On the

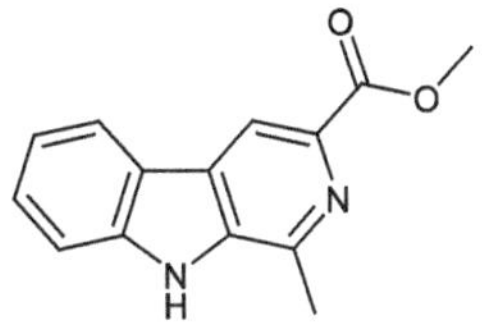

57
% Inhibition p24 ELISA = 21.3

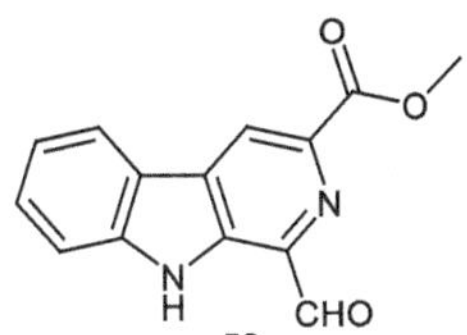

58
% Inhibition p24 ELISA = 78.17

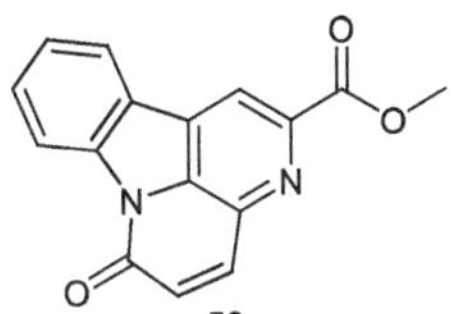

59
% Inhibition p24 ELISA = NA

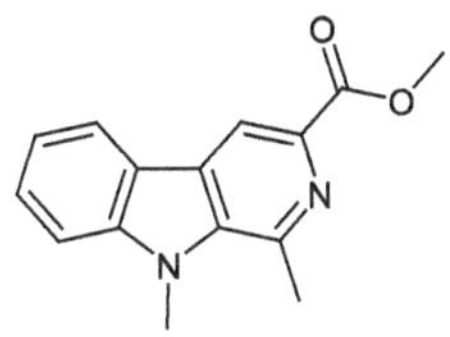

60
% Inhibition p24 ELISA = NA

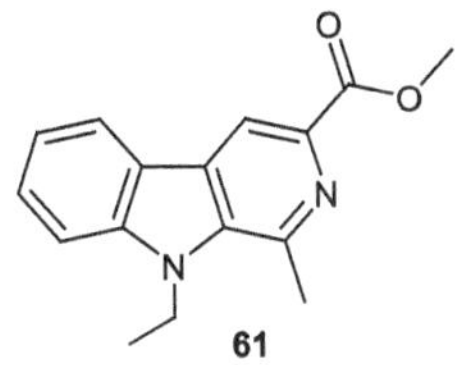

61
% Inhibition p24 ELISA = NA

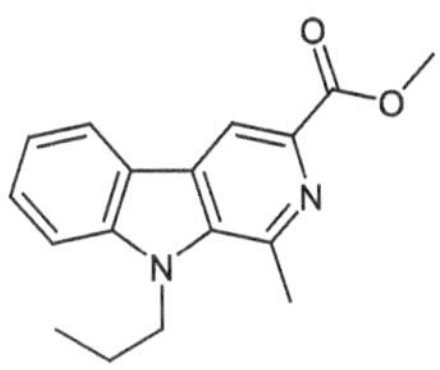

62
% Inhibition p24 ELISA = NA

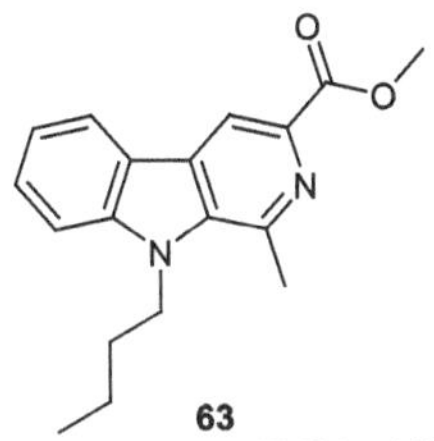

63
% Inhibition p24 ELISA = NA

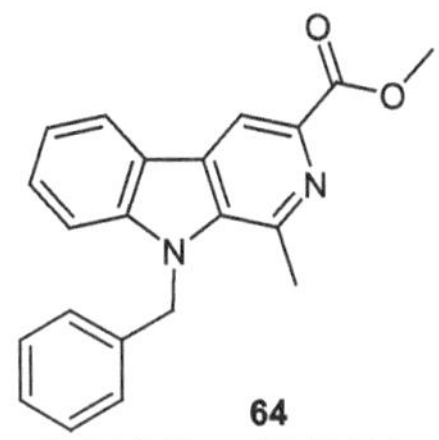

64
% Inhibition p24 ELISA = NA

FIGURE 9.19 β-carboline derivatives developed by Keyur et al. with % inhibition of p24 ELISA.

other hand, substitutions such as isopropoxy, hexyloxy, decyloxy, hexadecyloxy, and 7-camphanoyl at the 7-position render the compounds inactive. This indicates that the position and nature of the substituent significantly influence the anti-HIV activity of β-carbolines.

Secondly, the presence of a methyl group at the 1-position of β-carbolines, exemplified by harman (5), is critical for their anti-HIV potency. Removal of the methyl group in compound 7 results in a substantial reduction (approximately four-fold) in anti-HIV activity. This underscores the importance of the 1-methyl group in the mechanism of action and highlights its role in enhancing the anti-HIV activity of β-carbolines.

Furthermore, the unsaturated nature of the β-carboline scaffold contributes to its anti-HIV effectiveness. Hydrogenation of the β-carboline ring system, leading to the formation of 3,4-dihydro β-carboline derivatives, results in decreased anti-HIV activity compared to their unsaturated counterparts. This suggests that the unsaturated structure plays a crucial role in the potency of β-carbolines against HIV.

In addition, bromination at the C-6 position of harmine (9) has been observed to enhance the anti-HIV activity. Compound 6-bromoharmine hydrobromide (24), a water-soluble derivative, exhibits higher potency than the corresponding free base (22). The introduction of a bromine atom at the C-6 position likely improves the anti-HIV activity, potentially through enhanced solubility and interactions with the viral target.

Moreover, alkylating the indole nitrogen of harman (5) with longer alkyl chains, such as N-ethyl (18) and N-butyl (19), increases the anti-HIV activity. Notably, compound 19 (N-butyl) exhibits higher potency than 18 (N-ethyl), suggesting that the length of the alkyl chain at this position influences the anti-HIV activity of β-carbolines.

Overall, the trends in functionalization of the β-carboline scaffold provide valuable insights into the structure–activity relationship of these compounds as anti-HIV agents. Methoxylation at the 7-position and the presence of a methyl group at the 1-position generally enhance the anti-HIV activity, while hydrogenation and certain substitutions may decrease or abolish their potency. Bromination at the C-6 position and alkylations at the indole nitrogen can improve the anti-HIV activity, with the length of the alkyl chain playing a role in determining their potency.

Understanding these trends and mechanisms of action enables the design and optimization of β-carboline derivatives with enhanced anti-HIV properties. By strategically modifying the functional groups and positions, it may be possible to develop more potent and selective β-carboline-based anti-HIV agents. Further research is warranted to fully elucidate the precise mechanisms of action, address challenges such as drug resistance and potential toxicity, and conduct preclinical and clinical trials to evaluate their efficacy and safety. The multifaceted activities and structural versatility of β-carbolines position them as a promising class of compounds in the pursuit of effective treatments against HIV/AIDS.

9.6 SUMMARY AND FUTURE OUTLOOK

The studies on the anti-HIV activity of β-carbolines highlight their potential as a novel class of compounds for the development of anti-HIV drugs. Various β-carboline derivatives are evaluated for their anti-HIV effects, and several trends and mechanisms of action were identified. The findings from these studies open up exciting prospects for the development of β-carboline derivatives as effective anti-HIV drugs. Nonetheless, additional investigation is required to gain a comprehensive understanding of how they work and enhance their effectiveness. There are still future directions and considerations that should be addressed including:

1. Structure–activity relationship studies: Further investigations should be conducted to explore the relationships between the structure of β-carboline derivatives and their anti-HIV activity. This will provide valuable insights for designing and synthesizing more potent analogues.
2. Mechanistic studies: Detailed studies are required to elucidate the precise mechanism of action of β-carbolines against HIV, particularly their interaction with viral targets and their effects on viral replication.
3. Drug resistance: The potential for the development of drug resistance should be addressed. Ongoing studies should focus on assessing the susceptibility of HIV strains to β-carbolines and monitoring the emergence of resistant strains.

4. Preclinical and clinical trials: Promising β-carboline derivatives should undergo preclinical testing to assess their efficacy and safety in animal models. Subsequently, clinical trials can be initiated to evaluate their effectiveness in humans.

Overall, the future outlook for β-carbolines as anti-HIV drugs is promising. The identified trends and mechanisms of action provide a foundation for the design and development of more potent and selective compounds. With further research and optimization, β-carbolines have the potential to contribute to the development of new therapeutic options for the treatment of HIV/AIDS.

9.7 CONCLUSION

In conclusion, this chapter underscores the significant potential of β-carbolines as a promising class of compounds for the development of anti-HIV agents. Their diverse biological activities, structural versatility, and interactions with viral targets position them as valuable candidates in the fight against HIV/AIDS. The studies discussed in the chapter reveal trends in their structure–activity relationships, highlighting the importance of specific substitutions and positions in modulating their anti-HIV activity. Methoxylation at the 7-position and the presence of a methyl group at the 1-position generally enhance their potency, while hydrogenation and certain substitutions decrease their anti-HIV activity. β-Carbolines interact with viral enzymes involved in HIV replication, inhibit viral entry into host cells, and exhibit immunomodulatory effects. However, detailed investigation is needed to fully comprehend the mechanisms of action, optimize their potency, and address challenges such as drug resistance and potential toxicity. Preclinical and clinical trials are essential to evaluate their efficacy, safety, and potential as anti-HIV therapeutics. Overall, β-carbolines hold promise as a novel class of compounds in the ongoing quest for effective treatments against HIV/AIDS.

REFERENCES

Aarons, Darryl H., G. Victor Rossi, and Raymond F. Orzechowski. 1977. 'Cardiovascular Actions of Three Harmala Alkaloids: Harmine, Harmaline, and Harmalol'. *Journal of Pharmaceutical Sciences* 66 (9): 1244–1248. doi:10.1002/jps.2600660910

Advanced Journal of Chemistry, Section B. 2019. 'Comparative Examination of Moclobemide, Tranylcypromine, Phenelzine and Isocarboxazid for Monoamine Oxidase–A Inhibition' 1 (1): 23–28. doi:10.33945/SAMI/AJCB.2019.1.5

Ashok, Penta, Subhash Chander, Jan Balzarini, Christophe Pannecouque, and Sankaranarayanan Murugesan. 2015. 'Design, Synthesis of New β-Carboline Derivatives and Their Selective Anti-HIV-2 Activity'. *Bioorganic & Medicinal Chemistry Letters* 25 (6): 1232–1235. doi:10.1016/j.bmcl.2015.01.058

Ashok, Penta, Hardik Sharma, Hiren Lathiya, Subhash Chander, and Sankaranarayanan Murugesan. 2015. 'In-Silico Design and Study of Novel Piperazinyl β-Carbolines as

Inhibitor of HIV-1 Reverse Transcriptase'. *Medicinal Chemistry Research* 24 (2): 513–522. doi:10.1007/s00044-014-1141-7

Ayipo, Yusuf Oloruntoyin, Mohd Nizam Mordi, Muzaimi Mustapha, and Thenmoly Damodaran. 2021. 'Neuropharmacological Potentials of β-Carboline Alkaloids for Neuropsychiatric Disorders'. *European Journal of Pharmacology* 893 (February): 173837. doi:10.1016/j.ejphar.2020.173837

Ayipo, Yusuf Oloruntoyin, Wahab Adesina Osunniran, and Mohd Nizam Mordi. 2021. 'Metal Complexes of β-Carboline: Advances in Anticancer Therapeutics'. *Coordination Chemistry Reviews* 432 (April): 213746. doi:10.1016/j.ccr.2020.213746

Barnes, Emma C., Rohitesh Kumar, and Rohan A. Davis. 2016. 'The Use of Isolated Natural Products as Scaffolds for the Generation of Chemically Diverse Screening Libraries for Drug Discovery'. *Natural Product Reports* 33 (3): 372–381. doi:10.1039/C5NP00121H

Berlowitz, Ilana, Klemens Egger, and Paul Cumming. 2022. 'Monoamine Oxidase Inhibition by Plant-Derived β-Carbolines; Implications for the Psychopharmacology of Tobacco and Ayahuasca'. *Frontiers in Pharmacology* 13 (May). doi:10.3389/fphar.2022.886408

Bonnet, Ragnhild, Sladan Pavlovic, Lehmann Jochen, and Hans Peter Rommelspacher. 2004. 'The Strong Inhibition of Triosephosphate Isomerase by the Natural β-Carbolines May Explain Their Neurotoxic Actions'. *Neuroscience* 127 (2): 443–453. doi:10.1016/j.neuroscience.2004.05.002

Brahmbhatt, Keyur G., Nafees Ahmed, Sudeep Sabde, Debashis Mitra, Inder Pal Singh, and Kamlesh K. Bhutani. 2010. 'Synthesis and Evaluation of β-Carboline Derivatives as Inhibitors of Human Immunodeficiency Virus'. *Bioorganic & Medicinal Chemistry Letters* 20 (15): 4416–4419. doi:10.1016/j.bmcl.2010.06.052

Dara, Suresh, Swetha Dhamercherla, Surender Singh Jadav, CH Madhu Babu, and Mohamed Jawed Ahsan. 2022. 'Machine Learning in Drug Discovery: A Review'. *Artificial Intelligence Review* 55 (3): 1947–1999. doi:10.1007/s10462-021-10058-4

Davison, Emma K, and Margaret A Brimble. 2019. 'Natural Product Derived Privileged Scaffolds in Drug Discovery'. *Current Opinion in Chemical Biology* 52 (October): 1–8. doi:10.1016/j.cbpa.2018.12.007

Deveau, Amy Morin, Marc A. Labroli, Christine M. Dieckhaus, Mary T. Barthen, Kirsten S. Smith, and Timothy L. Macdonald. 2001. 'The Synthesis of Amino-Acid Functionalized β-Carbolines as Topoisomerase II Inhibitors'. *Bioorganic & Medicinal Chemistry Letters* 11 (10): 1251–1255. doi:10.1016/S0960-894X(01)00136-6

Dong, Zejun J, Jian-Fei Wang, Rongxuan Wang, Liming Yang, Yong-Tang Zheng, and Jian-Kui Liu. 2007. 'Chemical Constituents of the Fruiting Bodies from the Basidiomycete Suillus Granulatus'. *Chin Tradit Herb Drugs* 38: 17–19.

Drews, Jurgen. 2000. 'Drug Discovery: A Historical Perspective'. *Science* 287 (5460). American Association for the Advancement of Science: 1960–1964.

Drung, Binia, Christoph Scholz, Valéria A. Barbosa, Azadeh Nazari, Maria H. Sarragiotto, and Boris Schmidt. 2014. 'Computational & Experimental Evaluation of the Structure/Activity Relationship of β-Carbolines as DYRK1A Inhibitors'. *Bioorganic & Medicinal Chemistry Letters* 24 (20): 4854–4860. doi:10.1016/j.bmcl.2014.08.054

Egusa, Hiroshi, Masanori Doi, Makio Saeki, Sho Fukuyasu, Yoshihiro Akashi, Yoshifumi Yokota, Hirofumi Yatani, and Yoshinori Kamisaki. 2011. 'The Small Molecule Harmine Regulates NFATc1 and Id2 Expression in Osteoclast Progenitor Cells'. *Bone* 49 (2): 264–274. doi:10.1016/j.bone.2011.04.003

Grabley, Susanne, and Ralf Thiericke. 1998. *Drug Discovery from Nature*. Springer Science & Business Media.

Hagen, Timothy J., and James M. Cook. 1988. 'Synthesis of 1-Methoxy Canthine-6-One'. *Tetrahedron Letters* 29 (20): 2421–2424. doi:10.1016/S0040-4039(00)87897-1

Hagen, Timothy J, Krishnaswamy Narayanan, Jeffrey Names, and James M. Cook. 1989. 'DDQ Oxidations in the Indole Area. Synthesis of 4-Alkoxy-. Beta.-Carbolines Including the Natural Products Crenatine and 1-Methoxycanthin-6-One'. *The Journal of Organic Chemistry* 54 (9). ACS Publications: 2170–2178.

Harvey, A. 2008. 'Natural Products in Drug Discovery'. *Drug Discovery Today* 13 (19–20): 894–901. doi:10.1016/j.drudis.2008.07.004

Herraiz, Tomas, Diana González, Carmen Ancín-Azpilicueta, Vicente J. Arán, and Hugo Guillén. 2010. 'β-Carboline Alkaloids in Peganum Harmala and Inhibition of Human Monoamine Oxidase (MAO)'. *Food and Chemical Toxicology* 48 (3): 839–845. doi:10.1016/j.fct.2009.12.019

Horton, William, Abha Sood, Swarada Peerannawar, Nandor Kugyela, Aditya Kulkarni, Rekha Tulsan, Chris D. Tran, et al. 2017. 'Synthesis and Application of β-Carbolines as Novel Multi-Functional Anti-Alzheimer's Disease Agents'. *Bioorganic & Medicinal Chemistry Letters* 27 (2): 232–236. doi:10.1016/j.bmcl.2016.11.067

Hsieh, Pei-Wen, Fang-Rong Chang, Kuo-Hsiung Lee, Tsong-Long Hwang, Sue-Ming Chang, and Yang-Chang Wu. 2004. 'A New Anti-HIV Alkaloid, Drymaritin, and a New C-Glycoside Flavonoid, Diandraflavone, from *Drymaria d Iandra*'. *Journal of Natural Products* 67 (7): 1175–1177. doi:10.1021/np0400196

Hughes, James P., Steve Rees, Sarkis Barret Kalindjian, and Karen L. Philpott. 2011. 'Principles of Early Drug Discovery'. *British Journal of Pharmacology* 162 (6): 1239–1249. doi:10.1111/j.1476-5381.2010.01127.x

Ishida, Junko, Hui-Kang Wang, Masayoshi Oyama, Mark L. Cosentino, Chang-Qi Hu, and Kuo-Hsiung Lee. 2001. 'Anti-AIDS Agents. 46. Anti-HIV Activity of Harman, an Anti-HIV Principle from *Symplocos Setchuensis*, and Its Derivatives'. *Journal of Natural Products* 64 (7): 958–960. doi:10.1021/np0101189

Kamboj, Aarzoo, Binita Sihag, Deshkanwar Singh Brar, Arshpreet Kaur, and Deepak B. Salunke. 2021. 'Structure Activity Relationship in β-Carboline Derived Anti-Malarial Agents'. *European Journal of Medicinal Chemistry* 221 (October): 113536. doi:10.1016/j.ejmech.2021.113536

Karpov, Alexei S., Frank Rominger, and Thomas J. J. Müller. 2005. 'A Diversity Oriented Four-Component Approach to Tetrahydro-β-Carbolines Initiated by Sonogashira Coupling'. *Organic & Biomolecular Chemistry* 3 (24): 4382. doi:10.1039/b511861a

Khan, Farhan A., Aneela Maalik, Zafar Iqbal, and Imran Malik. 2013. 'Recent Pharmacological Developments in β-Carboline Alkaloid "Harmaline"'. *European Journal of Pharmacology* 721 (1–3): 391–394. doi:10.1016/j.ejphar.2013.05.003

Kumar, Sumit, Amandeep Singh, Kewal Kumar, and Vipan Kumar. 2017. 'Recent Insights into Synthetic β -Carbolines with Anti-Cancer Activities'. *European Journal of Medicinal Chemistry* 142 (December): 48–73. doi:10.1016/j.ejmech.2017.05.059

Leonard, Brian E. 1994. 'Biochemical Strategies for the Development of Antidepressants'. *CNS Drugs* 1 (4): 285–304. doi:10.2165/00023210-199401040-00006

Li Petri, Giovanna, Maria Valeria Raimondi, Virginia Spanò, Ralph Holl, Paola Barraja, and Alessandra Montalbano. 2021. 'Pyrrolidine in Drug Discovery: A Versatile Scaffold for Novel Biologically Active Compounds'. *Topics in Current Chemistry* 379 (5): 34. doi:10.1007/s41061-021-00347-5

Li, Yibo, Jianxing Hu, Yanxing Wang, Jielong Zhou, Liangren Zhang, and Zhenming Liu. 2020. 'DeepScaffold: A Comprehensive Tool for Scaffold-Based De Novo Drug Discovery Using Deep Learning'. *Journal of Chemical Information and Modeling* 60 (1): 77–91. doi:10.1021/acs.jcim.9b00727

Luo, Bo, and Xinqiang Song. 2021. 'A Comprehensive Overview of β-Carbolines and Its Derivatives as Anticancer Agents'. *European Journal of Medicinal Chemistry* 224 (November): 113688. doi:10.1016/j.ejmech.2021.113688

Mashkovskii, Mairla Deona, Natalia I. Andreeva, and Alex Ivanova Polezhaeva. 1980. 'Monoamine Oxidase Inhibitors as Drugs'. *Pharmaceutical Chemistry Journal* 14 (2): 81–86. doi:10.1007/BF00765902

McKenna, Dennis J., G.H.N. Towers, and F. Abbott. 1984. 'Monoamine Oxidase Inhibitors in South American Hallucinogenic Plants: Tryptamine and β-Carboline Constituents of Ayahuasca'. *Journal of Ethnopharmacology* 10 (2): 195–223. doi:10.1016/0378-8741(84)90003-5

Nakatsuka, Shin-ichi, Bai-nian Feng, Toshio Goto, and Kiyoshi Kihara. 1986. 'Structures of Flazin and YS, Highly Fluorescent Compounds Isolated from Japanese Soy Sauce'. *Tetrahedron Letters* 27 (29). Elsevier: 3399–3402.

Onishi, Yoshiaki, Katsutaka Oishi, Yasuhiro Kawano, and Yoshimitsu Yamazaki. 2012. 'The Harmala Alkaloid Harmine Is a Modulator of Circadian *Bmal1* Transcription'. *Bioscience Reports* 32 (1): 45–52. doi:10.1042/BSR20110002

Park, Tai Hwan, Oh Sang Kwon, Se Young Park, Eun Sook Han, and Chung Soo Lee. 2003. 'N-Methylated β-Carbolines Protect PC12 Cells from Cytotoxic Effect of MPP+ by Attenuation of Mitochondrial Membrane Permeability Change'. *Neuroscience Research* 46 (3): 349–358. doi:10.1016/S0168-0102(03)00097-X

Patel, K, M Gadewar, R Tripathi, SK Prasad, and Dinesh Kumar Patel. 2012. 'A Review on Medicinal Importance, Pharmacological Activity and Bioanalytical Aspects of Beta-Carboline Alkaloid "Harmine"'. *Asian Pacific Journal of Tropical Biomedicine* 2 (8): 660–664. doi:10.1016/S2221-1691(12)60116-6

Peduto, Antonella, Rosanna Filosa, Paolo De Caprariis, Vincenzo De Feo, Shiyou Li, Ping Wang, and Laura De Martino. 2012. '1-Methoxy-Canthin-6-One and Related β-Carbolines: From Natural Compound to Synthesis and Biological Activities'. *Studies in Natural Products Chemistry* 38. Elsevier: 81–104.

Peng, Jiangnan, Jin-Feng Hu, Abul B Kazi, Ze Li, Mitchell Avery, Olivier Peraud, Russell T Hill, Scott G Franzblau, Fangqiu Zhang, and Raymond F. Schinazi. 2003. 'Manadomanzamines A and B: A Novel Alkaloid Ring System with Potent Activity against Mycobacteria and HIV-1'. *Journal of the American Chemical Society* 125 (44). ACS Publications: 13382–13386.

Rao, Karumanchi V., Marwa S. Donia, Jiangnan Peng, Esther Garcia-Palomero, Diana Alonso, Ana Martinez, Miguel Medina, et al. 2006. 'Manzamine B and E and Ircinal A Related Alkaloids from an Indonesian *Acanthostrongylophora* Sponge and Their Activity against Infectious, Tropical Parasitic, and Alzheimer's Diseases'. *Journal of Natural Products* 69 (7): 1034–1040. doi:10.1021/np0601399

Rao, Karumanchi V., Noer Kasanah, Subagus Wahyuono, Babu L. Tekwani, Raymond F. Schinazi, and Mark T. Hamann. 2004. 'Three New Manzamine Alkaloids from a Common Indonesian Sponge and Their Activity against Infectious and Tropical Parasitic Diseases'. *Journal of Natural Products* 67 (8): 1314–1318. doi:10.1021/np0400095

Réus, Gislaine Z., Roberto B. Stringari, Cinara L. Gonçalves, Giselli Scaini, Milena Carvalho-Silva, Gabriela C. Jeremias, Isabela C. Jeremias, et al. 2012. 'Administration of Harmine and Imipramine Alters Creatine Kinase and Mitochondrial Respiratory Chain Activities in the Rat Brain'. *Depression Research and Treatment* 2012: 1–7. doi:10.1155/2012/987397.

Sarter, Martin, Herbert H. Schneider, and David N. Stephens. 1988. 'Treatment Strategies for Senile Dementia: Antagonist β-Carbolines'. *Trends in Neurosciences* 11 (1): 13–17. doi:10.1016/0166-2236(88)90042-2

Selvaraj, Manikandan, Karthik Sadasivam, Muralidharan Jothimani, and Karthikeyan Muthusamy. 2023. 'A Comprehensive Computational Perspective in Drug Discovery for Alzheimer's Disease'. *Combinatorial Chemistry & High Throughput Screening* 26 (12): 2113–2123. doi:10.2174/1386207325666220606142910

Sliwoski, Gregory, Sandeepkumar Kothiwale, Jens Meiler, and Edward W. Lowe. 2014. 'Computational Methods in Drug Discovery'. *Pharmacological Reviews* 66 (1): 334–395. doi:10.1124/pr.112.007336

Sneader, Walter. 2005. *Drug Discovery: A History*. John Wiley & Sons.

Su, Bao-Ning, Leng Chee Chang, Eun Jung Park, Muriel Cuendet, Bernard D. Santarsiero, Andrew D. Mesecar, Rajendra G. Mehta, Harry H. Fong, John M. Pezzuto, and A. Douglas Kinghorn. 2002. 'Bioactive Constituents of the Seeds of Brucea Javanica'. *Planta Medica* 68 (8): 730–733. doi:10.1055/s-2002-33798

Suzuki, Hideharu, Makoto Adachi, Yuki Ebihara, Hiroshi Gyoutoku, Hinako Furuya, Yasuoki Murakami, and Hiroaki Okuno. 2005. 'A Total Synthesis of 1-Methoxycanthin-6-One: An Efficient One-Pot Synthesis of the Canthin-6-One Skeleton from β-Carboline-1-Carbaldehyde'. *Synthesis* 2005 (01). © Georg Thieme Verlag Stuttgart· New York: 28–32.

Tang, Jian-Guo, Yun-Hua Wang, Rui-Rui Wang, Ze-Jun Dong, Liu-Meng Yang, Yong-Tang Zheng, and Ji-Kai Liu. 2008. 'Synthesis of Analogues of Flazin, in Particular, Flazinamide, as Promising Anti-HIV Agents'. *Chemistry & Biodiversity* 5 (3): 447–460. doi:10.1002/cbdv.200890044

Vijayan, R.S.K., Jan Kihlberg, Jason B. Cross, and Vasanthanathan Poongavanam. 2022. 'Enhancing Preclinical Drug Discovery with Artificial Intelligence'. *Drug Discovery Today* 27 (4): 967–984. doi:10.1016/j.drudis.2021.11.023

Wang, Lingle, Yuqing Deng, Yujie Wu, Byungchan Kim, David N. LeBard, Dan Wandschneider, Mike Beachy, Richard A. Friesner, and Robert Abel. 2017. 'Accurate Modeling of Scaffold Hopping Transformations in Drug Discovery'. *Journal of Chemical Theory and Computation* 13 (1): 42–54. doi:10.1021/acs.jctc.6b00991

Xu, Zhihong, Fang-Rong Chang, Hui-Kang Wang, Yoshiki Kashiwada, Andrew T. McPhail, Kenneth F. Bastow, Yoko Tachibana, Mark Cosentino, and Kuo-Hsiung Lee. 2000. 'Anti-HIV Agents 45 and Antitumor Agents 205. Two New Sesquiterpenes, Leitneridanins A and B, and the Cytotoxic and Anti-HIV Principles from *Leitneria f Loridana*'. *Journal of Natural Products* 63 (12): 1712–1715. doi:10.1021/np000260u

Yonezawa, Takayuki, Shin-ichi Hasegawa, Midori Asai, Tadashi Ninomiya, Toshinori Sasaki, Byung-Yoon Cha, Toshiaki Teruya, et al. 2011. 'Harmine, a β-Carboline Alkaloid, Inhibits Osteoclast Differentiation and Bone Resorption in Vitro and in Vivo'. *European Journal of Pharmacology* 650 (2–3): 511–518. doi:10.1016/j.ejphar.2010.10.048.

Yousaf, Muhammad, Nicholas L Hammond, Jiangnan Peng, Subagus Wahyuono, Kylie A McIntosh, William N Charman, Alejandro M. S Mayer, and Mark T Hamann. 2004. 'New Manzamine Alkaloids from an Indo-Pacific Sponge. Pharmacokinetics, Oral Availability, and the Significant Activity of Several Manzamines against HIV-I, AIDS Opportunistic Infections, and Inflammatory Diseases'. *Journal of Medicinal Chemistry* 47 (14). ACS Publications: 3512–3517.

Zhang, Yun, and Andrew Pike. 2021. 'Pyridones in Drug Discovery: Recent Advances'. *Bioorganic & Medicinal Chemistry Letters* 38 (April): 127849. doi:10.1016/j.bmcl.2021.127849

Zheng, Yajun, Colin M. Tice, and Suresh B. Singh. 2014. 'The Use of Spirocyclic Scaffolds in Drug Discovery'. *Bioorganic & Medicinal Chemistry Letters* 24 (16): 3673–3682. doi:10.1016/j.bmcl.2014.06.081

Index

Note: Page numbers in *italic* refers to figures/schemes.

M

N

O

P

R

S

T

tryptamine and tetrahydro-β-C biphenyl derivatives 99

Y

FSC
www.fsc.org
MIX
Papier | Fördert gute Waldnutzung
FSC® C083411

Kolibri 360 GmbH
Ferdinand-Jühlke-Straße 7
99095 Erfurt, Deutschland
produktsicherheit@kolibri360.de